Yunzhi Zou

Multi-Variable Calculus

A First Step

DE GRUYTER

Mathematics Subject Classification 2010
Primary: 26B12, 26B20, 26B15; Secondary: 26B05, 26B10

Author
Prof. Yunzhi Zou
Department of Mathematics
Sichuan University
610065 Chengdu
People's Republic of China
zouyz@scu.edu.cn

ISBN 978-3-11-067414-9
e-ISBN (PDF) 978-3-11-067437-8
e-ISBN (EPUB) 978-3-11-067443-9

Library of Congress Control Number: 2019953764

Bibliographic information published by the Deutsche Nationalbibliothek
The Deutsche Nationalbibliothek lists this publication in the Deutsche Nationalbibliografie; detailed bibliographic data are available on the Internet at http://dnb.dnb.de.

© 2020 Walter de Gruyter GmbH, Berlin/Boston
Cover image: shulz/iStock/Getty Images Plus
Typesetting: VTeX UAB, Lithuania
Printing and binding: CPI books GmbH, Leck

www.degruyter.com

Contents

Introduction — IX

1		Vectors and the geometry of space — 1
1.1		Vectors — 1
1.1.1		Concepts of vectors — 1
1.1.2		Linear operations involving vectors — 2
1.1.3		Coordinate systems in three-dimensional space — 3
1.1.4		Representing vectors using coordinates — 5
1.1.5		Lengths, direction angles — 7
1.2		Dot product, cross product, and triple product — 9
1.2.1		The dot product — 9
1.2.2		Projections — 12
1.2.3		The cross product — 13
1.2.4		Scalar triple product — 17
1.3		Equations of lines and planes — 18
1.3.1		Lines — 18
1.3.2		Planes — 23
1.4		Curves and vector-valued functions — 30
1.5		Calculus of vector-valued functions — 32
1.5.1		Limits, derivatives, and tangent vectors — 32
1.5.2		Antiderivatives and definite integrals — 35
1.5.3		Length of curves, curvatures, TNB frame — 37
1.6		Surfaces in space — 42
1.6.1		Graph of an equation $F(x, y, z) = 0$ — 42
1.6.2		Cylinder — 44
1.6.3		Quadric surfaces — 46
1.6.4		Surface of revolution — 46
1.7		Parameterized surfaces — 49
1.8		Intersecting surfaces and projection curves — 50
1.9		Regions bounded by surfaces — 56
1.10		Review — 57
1.11		Exercises — 59
1.11.1		Vectors — 59
1.11.2		Lines and planes in space — 60
1.11.3		Curves and surfaces in space — 61
2		Functions of multiple variables — 65
2.1		Functions of multiple variables — 65
2.1.1		Definitions — 65
2.1.2		Graphs and level curves — 67

2.1.3	Functions of more than two variables —— 69	
2.1.4	Limits —— 70	
2.1.5	Continuity —— 75	
2.2	Partial derivatives —— 76	
2.2.1	Definition —— 76	
2.2.2	Interpretations of partial derivatives —— 80	
2.2.3	Partial derivatives of higher order —— 82	
2.3	Total differential —— 83	
2.3.1	Linearization and differentiability —— 83	
2.3.2	The total differential —— 89	
2.3.3	The linear/differential approximation —— 90	
2.4	The chain rule —— 92	
2.4.1	The chain rule with one independent variable —— 92	
2.4.2	The chain rule with more than one independent variable —— 94	
2.4.3	Partial derivatives for abstract functions —— 97	
2.5	The Taylor expansion —— 98	
2.6	Implicit differentiation —— 101	
2.6.1	Functions implicitly defined by a single equation —— 101	
2.6.2	Functions defined implicitly by systems of equations —— 103	
2.7	Tangent lines and tangent planes —— 106	
2.7.1	Tangent lines and normal planes to a curve —— 106	
2.7.2	Tangent planes and normal lines to a surface —— 109	
2.8	Directional derivatives and gradient vectors —— 113	
2.9	Maximum and minimum values —— 122	
2.9.1	Extrema of functions of two variables —— 122	
2.9.2	Lagrange multipliers —— 130	
2.10	Review —— 136	
2.11	Exercises —— 138	
2.11.1	Functions of two variables —— 138	
2.11.2	Partial derivatives and differentiability —— 139	
2.11.3	Chain rules and implicit differentiation —— 140	
2.11.4	Tangent lines/planes, directional derivatives —— 141	
2.11.5	Maximum/minimum problems —— 142	
3	**Multiple integrals —— 145**	
3.1	Definition and properties —— 145	
3.2	Double integrals in rectangular coordinates —— 150	
3.3	Double integral in polar coordinates —— 157	
3.4	Change of variables formula for double integrals —— 161	
3.5	Triple integrals —— 165	
3.5.1	Triple integrals in rectangular coordinates —— 165	
3.5.2	Cylindrical and spherical coordinates —— 175	

3.6	Change of variables in triple integrals —— 179	
3.7	Other applications of multiple integrals —— 181	
3.7.1	Surface area —— 181	
3.7.2	Center of mass, moment of inertia —— 187	
3.8	Review —— 188	
3.9	Exercises —— 191	
3.9.1	Double integrals —— 191	
3.9.2	Triple integrals —— 192	
3.9.3	Other applications of multiple integrals —— 193	
4	**Line and surface integrals —— 195**	
4.1	Line integral with respect to arc length —— 195	
4.1.1	Definition and properties —— 196	
4.1.2	Evaluating a line integral, $\int_C f(x,y)ds$, in $\mathbb{R}^2$ —— 197	
4.1.3	Line integrals $\int_C f(x,y,z)ds$ in $\mathbb{R}^3$ —— 199	
4.2	Line integral of a vector field —— 201	
4.2.1	Vector fields —— 201	
4.2.2	The line Integral of a vector field along a curve C —— 202	
4.3	The fundamental theorem of line integrals —— 208	
4.4	Green's theorem: circulation-curl form —— 216	
4.4.1	Positive oriented simple curve and simply connected region —— 216	
4.4.2	Circulation around a closed curve —— 217	
4.4.3	Circulation density —— 217	
4.4.4	Green's theorem: circulation-curl form —— 219	
4.4.5	Applications of Green's theorem in circulation-curl form —— 222	
4.5	Green's theorem: flux-divergence form —— 231	
4.5.1	Flux —— 231	
4.5.2	Flux density – divergence —— 232	
4.5.3	The divergence-flux form of Green's theorem —— 233	
4.6	Source-free vector fields —— 235	
4.7	Surface integral with respect to surface area —— 237	
4.8	Surface integrals of vector fields —— 241	
4.8.1	Orientable surfaces —— 241	
4.8.2	Flux integral $\iint_S (\mathbf{F} \cdot \mathbf{N})dS$ —— 242	
4.9	Divergence theorem —— 248	
4.9.1	Divergence of a three-dimensional vector field —— 248	
4.9.2	Divergence theorem —— 250	
4.10	Stokes theorem —— 256	
4.10.1	The curl of a three-dimensional vector field —— 256	
4.10.2	Stokes theorem —— 258	
4.11	Review —— 265	

4.12	Exercises —— 268
4.12.1	Line integrals —— 268
4.12.2	Surface integrals —— 269

5	**Introduction to ordinary differential equations** —— 273
5.1	Introduction —— 273
5.2	First-order ODEs —— 275
5.2.1	General and particular solutions and direction fields —— 275
5.2.2	Separable differential equations —— 277
5.2.3	Substitution methods —— 279
5.2.4	Exact differential equations —— 281
5.2.5	First-order linear differential equations —— 283
5.3	Second-order ODEs —— 287
5.3.1	Reducible second-order equations —— 287
5.3.2	Second-order linear differential equations —— 291
5.3.3	Variation of parameters —— 307
5.4	Other ways of solving differential equations —— 308
5.4.1	Power series method —— 309
5.4.2	Numerical approximation: Euler's method —— 310
5.5	Review —— 313
5.6	Exercises —— 315
5.6.1	Introduction to differential equations —— 315
5.6.2	First-order differential equations —— 315
5.6.3	Second-order differential equations —— 316

Further reading —— 319

Index —— 321

Introduction

Calculus has been widely applied to an incredible number of disciplines since its inception in the seventeenth century. In particular, the marvelous Maxwell equations revealed the laws that govern electric and magnetic fields, which led to the forecasting of the existence of the electromagnetic waves. The industrial revolution witnessed the many applications of calculus. The power of calculus never diminishes, even in today's scientific world. For this reason, there is no doubt that calculus is one of the most important courses for undergraduate students at any university in the world.

On the other hand, during the past century, especially since the 2000s, many Chinese and other non-English speaking people have gone to English speaking countries to further their studies, and more are on their way. Also, as global cooperations and communications become important for people to tackle big problems, there are needs for people to know and understand each other better. Fortunately, Sichuan University has a long history of global connections. Its summer immersion program is well known for its size and popularity. Each year, it sends and hosts thousands of students from different parts of the world. We believe that there are other similar situations where students come and go to different places or countries without disrupting their studies. For those students, a suitable textbook is helpful.

However, there are many challenges in developing such a suitable book. First of all, for most freshmen whose English is not their first language, the textbook should employ English as plain as possible. Second, the textbook should take into account what students have learned in high school and what they need in a calculus course. Third, there must be a smooth transition from the local standards to those globally accepted. Furthermore, such a book must have some new insights to inject new energy into the many already existing texts. This includes, but is not limited to, addressing discovery over rote learning; being as concise as possible while covering the essential content required by most local and global universities; and being printed in color as most texts in English are. The book *Single Variable Calculus: A First Step*, which was the first such calculus text in China, has provided a response to these challenges since it was published by the World Publishing Company in 2015 and by De Gruyter in 2018. The present book, *Multivariable calculus: a first step*, makes sure that these efforts continue.

With more than 10 years in teaching calculus courses to students at the Wu Yuzhang honors college at Sichuan University, I have had the chance to work with local students using books and resource materials in English. We adopted or referred students to many calculus books, for example *Thomas' Calculus*, 10th edition, by Finney, Weir, and Giordano; *Calculus*, 5th edition, by Stewart; *Calculus*, by Larson, Edwards, and Hosteltle; *Calculus: Ideas and Applications*, by Himonnas and Howard; *Calculus: Early Transcendentals*, 2nd edition, by Briggs, Cochran, and Gillett; and other books in

Chinese, including the calculus books published by the Mathematics departments at Sichuan University and Tongji University. They are all good textbooks, and I acknowledge that I was inspired much by them, not only by their structure that builds the contents, but also by the nice problems that enhance understanding. Most of the exercises in this text were developed over the past decade. My many teaching assistants contributed a lot by helping create or collecting resources in the past years. Among them are Zengbao Wu, Liang Li, Mengxin Li, Bo Qian, Xi Zhu, and Yang Yang. Most of the exercises were inspired by the books mentioned before. Some are original, while others we think original may be similar to those found in existing books or other resources. Those are usually classic problems and can be found in many places.

My thanks also go to students in the years of 2015, 2016, 2017, and 2018 at Wu Yuzhang College who helped proofread the manuscript or notes and provided useful feedback. I also appreciate my wonderful colleagues Wengui Hu, Li Ren, and Hao Wang, who worked with me teaching calculus using the early versions of this book. In particular, Wengui contributed many good problems. I received valuable suggestions from my dear friend Dr. Harold Reiter at UNCC and Dr. Wenyuan Liao at the University of Calgary, who always lends me a hand in solving problems arising in using LaTeX or MATLAB. Professor Xiaozhan Xu provided me with many excellent PPTs and animations for teaching the course.

My special thanks go to my dear friend Dr. Jonathan Kane, whose talents in mathematics and English improved the manuscript a lot, with thorough and professional edits. Also the ideas of adding contents such as moment of inertia, the torus problem, and solving simple PDEs were due to Jon.

I enjoyed very much working with the wonderful people mentioned above. I am sure that without them this book would not have achieved this level. The following list is in alphabetical order.

1. Dr. Wengui Hu, Associate Professor of Mathematics
   Sichuan University, China
2. Dr. Jonathan Kane, Emeritus Professor of Mathematics
   University of Wisconsin, Whitewalter, USA
3. Dr. Wenyuan Liao, Associate Professor of Mathematics
   University of Calgary, Canada
4. Dr. Harold Reiter, Professor of Mathematics,
   University of North Carolina at Charlotte, USA
5. Dr. Li Ren, Associate Professor of Mathematics
   Sichuan University, China
6. Dr. Hao Wang, Associate Professor of Mathematics
   Sichuan University, China
7. Mr. Xiaozhan Xu, Professor of Mathematics
   Sichuan University, China

8. Mr. Zengbao Wu, Mr. Liang Li, Ms. Mengxin Li, Mr. Bo Qian, Mr. Xi Zhu, Mr. Yang Yang, and Mr. Yi Guo
   Graduate students working as teacher assistants

I also would like to thank my Mathematics department and the academic affairs office at Sichuan University. I always have their encouragement and generous support, which make me happy to devote time and energy in writing this book and make the publication of the work possible.

We have been working hard on this version; however, there might still be typos and even mistakes. The responsibility for those errors in this book lie entirely with me. I will be happy to receive comments and feedback anytime whenever they arise. I can be reached via zouyz@scu.edu.cn.

Sincerely,
Yunzhi Zou
Professor of Mathematics
Sichuan University
Chengdu, P.R. China
zouyz@scu.edu.cn
610065

1 Vectors and the geometry of space

In this chapter we introduce vectors and coordinate systems for three-dimensional space. They are very helpful in our study of multivariable calculus. In particular, vectors provide simple descriptions and insight concerning curves and planes. We also introduce some surfaces in space. The graph of a function of two variables is a surface in space which gives additional insight into the properties of the function.

1.1 Vectors

1.1.1 Concepts of vectors

The term *vector* is used to indicate a quantity that has both a *magnitude* and a *direction*, for instance, displacement, acceleration, velocity, and force. Scientists often represent a vector geometrically by an arrow (a directed line segment). The arrow of the directed line segment points in the direction of the vector, while the length of the arrow represents the magnitude of the vector. We denote vectors by letters that have an arrow overbar, such as $\vec{a}, \vec{b}, \vec{i}, \vec{k}, \vec{v}$. For example, suppose an object moves along a straight line from point A to point B. The vector $\vec{s}$ representing this displacement geometrically has *initial point* A (the tail) and *terminal point* B (the head), and we indicate this by writing $\vec{s} = \overrightarrow{AB}$ (as shown in Figure 1.1(a)). We also denote vectors by printing the letters in boldface, such as **a**, **b**, **i**, **k**, **v**. In this book, we use both notations. We denote the *magnitude* (also called the *length*) of a vector $\vec{a}$ (or **a**) by $|\vec{a}|$ (or $|\mathbf{a}|$). If $|\vec{a}| = 1$, then we say that $\vec{a}$ is a *unit vector*.

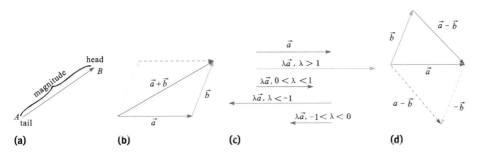

Figure 1.1: Vectors, addition, scalar multiplication, and subtraction.

We say that two vectors $\vec{a}$ and $\vec{b}$ are *equivalent* (or *equal*) if they have the same length and the same direction, and we write $\vec{a} = \vec{b}$. Note that two vectors with the same length and direction are considered equal even when the vectors are in two different locations. The *zero vector*, denoted by $\vec{0}$ or **0**, has length 0, and, consequently, it is the only vector with no specific direction. If nonzero vectors $\vec{a}$ and $\vec{b}$ have the same direction or if $\vec{a}$ has exactly the opposite direction to that of $\vec{b}$, then we say that they are parallel, and we write $\vec{a} \parallel \vec{b}$.

1.1.2 Linear operations involving vectors

We assume that vectors considered here can be represented by directed line segments or arrows in two-dimensional space, $\mathbb{R}^2$, or three-dimensional space, $\mathbb{R}^3$. However, vectors can be defined much more generally without reference to the directed line segments.

Definition 1.1.1 (Vector addition). If $\vec{a}$ and $\vec{b}$ are vectors positioned so the initial point of $\vec{b}$ is at the terminal point of $\vec{a}$, then the sum $\vec{a} + \vec{b}$ is the vector from the initial point of $\vec{a}$ to the terminal point of $\vec{b}$.

This definition of vector addition is illustrated in Figure 1.1(b), and you can see why this definition is sometimes called the *triangle law* or *parallelogram law*.

Note. If the initial point of $\vec{b}$ is not at the terminal point of $\vec{a}$, then a copy of $\vec{b}$ (same length and direction) can be made with its initial point at the terminal point of $\vec{a}$, and the sum can be created using $\vec{a}$ and this copy of $\vec{b}$.

Vector addition satisfies the following laws for any three vectors $\vec{a}, \vec{b}, \vec{c}$:
(1) Commutative law: $\vec{a} + \vec{b} = \vec{b} + \vec{a}$.
(2) Associative law: $(\vec{a} + \vec{b}) + \vec{c} = \vec{a} + (\vec{b} + \vec{c})$.

Definition 1.1.2 (Scalar multiplication, negative of a vector). If λ is a scalar (a number) and $\vec{a}$ is a vector, then the **scalar multiple** $\lambda\vec{a}$ is also a vector. If $\lambda > 0$, then $\lambda\vec{a}$ has the same direction as the vector $\vec{a}$ and has length λ times the length of $\vec{a}$. If $\lambda < 0$, then $\lambda\vec{a}$ has the reverse direction to the direction of $\vec{a}$ and has length that is $|\lambda|$ times the length of $\vec{a}$. If $\lambda = 0$ or $\vec{a} = \vec{0}$ (zero vector), then $\lambda\vec{a} = \vec{0}$. In particular, the vector $-\vec{a}$ is called the **negative** of $\vec{a}$, and it means the scalar multiple $(-1)\vec{a}$ has the same length as $\vec{a}$ but points in the opposite direction.

Scalar multiplication satisfies the following laws for any two vectors $\vec{a}, \vec{b}$ and any two scalars λ, μ:
(3) Associative law: $\lambda(\mu\vec{a}) = (\lambda\mu)\vec{a} = \mu(\lambda\vec{a})$.
(4) Distributive laws: $(\lambda + \mu)\vec{a} = \lambda\vec{a} + \mu\vec{a}$ and $\lambda(\vec{a} + \vec{b}) = \lambda\vec{a} + \lambda\vec{b}$.

By the distributive law (4) $\vec{b} + (-\vec{b}) = 1\vec{b} + (-1)\vec{b} = (1 - 1)\vec{b} = \vec{0}$, so $\vec{b}$ and $-\vec{b}$ act as negatives of each other. Also, we can see that two nonzero vectors are parallel to each other if they are scalar multiples of one another. The zero vector is considered to be parallel to all other vectors. It is easy to establish the following theorem.

Theorem 1.1.1. *Suppose $\vec{a}$ and $\vec{b}$ are two nonzero vectors. Then $\vec{a} \parallel \vec{b}$ if and only if there exists a number $\lambda \neq 0$ such that $\vec{a} = \lambda \vec{b}$.*

The *difference* or *subtraction* $\vec{a} - \vec{b}$ of two vectors is defined in terms of a vector sum as

$$\vec{a} - \vec{b} = \vec{a} + (-\vec{b}).$$

Hence, we can construct $\vec{a} - \vec{b}$ geometrically by first drawing the negative $-\vec{b}$ of $\vec{b}$, and then adding $-\vec{b}$ to $\vec{a}$ using the parallelogram law as in Figure 1.1(d). This shows that the vector $\vec{a} - \vec{b}$ is the vector from the head of $\vec{b}$ to the head of $\vec{a}$. The operation of subtracting two vectors does not satisfy the commutative law (1) or the associative law (2), but it does satisfy the distributive law (4), $\lambda(\vec{a} - \vec{b}) = \lambda\vec{a} - \lambda\vec{b}$.

1.1.3 Coordinate systems in three-dimensional space

To locate a point in a plane in a two-dimensional Cartesian coordinate system with perpendicular x- and y-axes, two numbers or *coordinates* are necessary, and this is why a plane is called two-dimensional. That is, the point can be represented as an ordered pair (a, b) of real numbers where the x-coordinate, a, is the directed distance from the y-axis to the point, and the y-coordinate, b, is the directed distance from the x-axis to the point.

To locate a point in three-dimensional space, three *coordinates* are required. We start with a fixed point, O, called the origin. We then draw three number lines that all pass through O and are perpendicular to each other. Usually, we put two number lines: one horizontal and one vertical. We call the three number lines the coordinate axes and label them as the x-axis, the y-axis, and the z-axis in a way that satisfies the *right-hand rule*. This rule helps determine the direction of the z-axis. If you curl your right-hand fingers naturally in a 90° rotation from the positive x-axis to the positive y-axis, then the direction that your thumb points is the positive direction of the z-axis, as shown in Figure 1.2(a). The three axes determine three coordinate planes called the xy-plane, the xz-plane, and the yz-plane, as shown in Figure 1.2(b). Therefore, the

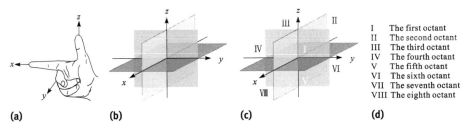

Figure 1.2: Three-dimensional coordinate system, axes, coordinate planes, and octants.

space is divided into eight octants. We label them the first octant, the second octant, the third octant, the fourth octant, the fifth octant, the six octant, the seventh octant, and the eighth octant in a way that is shown in Figure 1.2(c).

To locate a point P in space, we project the point onto the three coordinate planes. If the directed distance from the yz-plane to the point P is a, the directed distance from the xz-plane to the point P is b, and the directed distance from the xy-plane to the point P is c, then we say that the point P has x-coordinate a, y-coordinate b, and z-coordinate c, and we use the ordered triple (a, b, c) to represent these coordinates. This can be seen by drawing a rectangular box where O and P are two end points of the main diagonal, as shown in Figure 1.3(a). This coordinate system is called the three-dimensional Cartesian coordinate system. For example, to locate the point with coordinates $(1, 2, -1)$, we start from the origin and go along the x-axis for 1 unit; then turn left and go parallel to the y-axis for 2 units; then go downward for 1 unit arriving at $(1, 2, -1)$, which is in the fifth octant as shown in Figure 1.3(b).

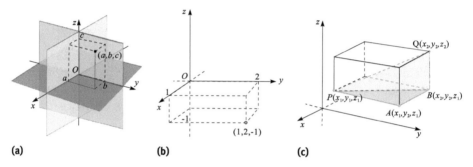

Figure 1.3: Three-dimensional coordinate system, coordinates, points, distance between two points.

Note that there is a one-to-one correspondence between points in the space and the set of all ordered triples (a, b, c). Sometimes, we use $\mathbb{R}^3$ to denote the Cartesian product $\mathbb{R} \times \mathbb{R} \times \mathbb{R} = \{(x, y, z) | x, y, z \in \mathbb{R}\}$.

Distance between two points in space

In a two-dimensional plane, by using the Pythagorean theorem, we have the following formula for the distance between two points (x_1, y_1) and (x_2, y_2) in the plane:

$$\text{distance between two points } d = \sqrt{(x_1 - x_2)^2 + (y_1 - y_2)^2}.$$

In three-dimensional space, for any two points $P(x_1, y_1, z_1)$ and $Q(x_2, y_2, z_2)$, we have a rectangular box with P and Q as the two endpoints of a main diagonal, as shown in Figure 1.3(c). Then we apply the Pythagorean theorem twice to get

$$\text{distance between } P \text{ and } Q = \sqrt{(x_1 - x_2)^2 + (y_1 - y_2)^2 + (z_1 - z_2)^2}. \tag{1.1}$$

1.1.4 Representing vectors using coordinates

It is extremely useful to represent vectors using coordinates. First, we have three *standard basis vectors* called $\vec{i}, \vec{j}$, and $\vec{k}$, which are three unit vectors in the positive directions of the x-, y-, and z-axes, respectively. If those vectors have their tails at the origin O, then their heads will be the points $(1,0,0)$, $(0,1,0)$, $(0,0,1)$, respectively, as shown in Figure 1.4(a).

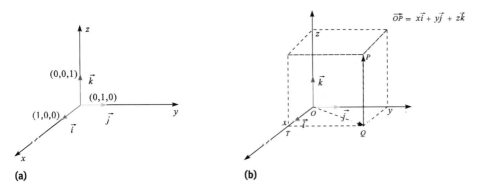

Figure 1.4: Three-dimensional coordinate system, basis vectors, position vectors.

Definition 1.1.3. A vector $\overrightarrow{OP}$ with initial point O, the origin, and terminal point $P(x, y, z)$ is called the *position vector* of the point $P(x, y, z)$.

By the definition of vector addition, we must have $\overrightarrow{OP} = x\vec{i} + y\vec{j} + z\vec{k}$. This follows from the box determined by the vector $\overrightarrow{OP}$ (see Figure 1.4(b)), because the parallelogram rule for addition gives

$$\overrightarrow{OP} = \overrightarrow{OQ} + \overrightarrow{QP}$$
$$= \overrightarrow{OT} + \overrightarrow{TQ} + \overrightarrow{QP},$$

where $\overrightarrow{OT}$ is along the x-axis with length x and is $x\vec{i}$, $\overrightarrow{TQ}$ is parallel to the y-axis with length y and is $y\vec{j}$, and $\overrightarrow{QP}$ is parallel to the z-axis with length z and is $z\vec{k}$. The numbers x, y, and z are referred to as the *components* of the vector $\overrightarrow{OP}$.

If we add two vectors expressed in the $\vec{i}, \vec{j}, \vec{k}$ format, then the commutative and associative laws of vector addition show that adding two vectors can be done by adding their components, i.e.,

$$(x_1\vec{i} + y_1\vec{j} + z_1\vec{k}) + (x_2\vec{i} + y_2\vec{j} + z_2\vec{k}) = (x_1 + x_2)\vec{i} + (y_1 + y_2)\vec{j} + (z_1 + z_2)\vec{k}. \quad (1.2)$$

By the distributive law one can see that multiplying a vector by a scalar λ is the same as multiplying each component by λ, i. e.,

$$\lambda(x\vec{i} + y\vec{j} + z\vec{k}) = \lambda x\vec{i} + \lambda y\vec{j} + \lambda z\vec{k}. \tag{1.3}$$

Example 1.1.1. If $\vec{a} = 5\vec{i} + 2\vec{j} - 3\vec{k}$ and $\vec{b} = 4\vec{i} - 9\vec{k}$, express the vector $2\vec{a} + 3\vec{b}$ in terms of $\vec{i}, \vec{j}$, and $\vec{k}$.

Solution. Using properties of vectors, we have

$$\begin{aligned} 2\vec{a} + 3\vec{b} &= 2(5\vec{i} + 2\vec{j} - 3\vec{k}) + 3(4\vec{i} - 9\vec{k}) \\ &= 10\vec{i} + 4\vec{j} - 6\vec{k} + 12\vec{i} - 27\vec{k} \\ &= 22\vec{i} + 4\vec{j} - 33\vec{k}. \end{aligned}$$

Now we use the notation $\langle x, y, z \rangle$ to denote a position vector with its head at the point (x, y, z), and this is the coordinate representation of this position vector. Since any vector in space can be translated so that its initial point is the origin, any vector in space can be represented in the form $\langle x, y, z \rangle$. We now give definitions for vector operations using its coordinates representation as follows.

Definition 1.1.4. If $\vec{a} = \langle x_1, y_1, z_1 \rangle$ and $\vec{b} = \langle x_2, y_2, z_2 \rangle$ are two position vectors and λ is a real number, then

$$\begin{aligned} \vec{a} + \vec{b} &= \langle x_1 + x_2, y_1 + y_2, z_1 + z_2 \rangle, \\ \vec{a} - \vec{b} &= \langle x_1 - x_2, y_1 - y_2, z_1 - z_2 \rangle, \\ \lambda \vec{a} &= \langle \lambda x_1, \lambda y_1, \lambda z_1 \rangle. \end{aligned}$$

Note that those operations also work for two-dimensional vectors; the only difference is that there is no z-component (or the z-component is always 0). Also, from the definition, we know that

$$\vec{a} = \vec{b} \iff x_1 = x_2,\ y_1 = y_2,\ \text{and}\ z_1 = z_2, \tag{1.4}$$

that is, their corresponding components are identical.

Example 1.1.2. Consider any vector $\overrightarrow{PQ}$, where the initial point is $P(x_1, y_1, z_1)$ and the terminal point is $Q(x_2, y_2, z_2)$. Then find coordinates of the midpoint of the line segment $\overline{PQ}$.

Solution. Since

$$\overrightarrow{PQ} = \overrightarrow{OQ} - \overrightarrow{OP} = \langle x_2 - x_1, y_2 - y_1, z_2 - z_1 \rangle,$$

if $M(x,y,z)$ is the midpoint of the line segment PQ, then $2\overrightarrow{PM} = \overrightarrow{PQ}$, so we have

$$\langle 2(x - x_1), 2(y - y_1), 2(z - z_1)\rangle = \langle x_2 - x_1, y_2 - y_1, z_2 - z_1\rangle.$$

This means

$$2\langle x, y, z\rangle = \langle x_2 - x_1, y_2 - y_1, z_2 - z_1\rangle + 2\langle x_1, y_1, z_1\rangle$$
$$= \langle x_2 + x_1, y_2 + y_1, z_2 + z_1\rangle.$$

Hence, we can deduce that the formula for the midpoint M is

$$M\left(\frac{x_1 + x_2}{2}, \frac{y_1 + y_2}{2}, \frac{z_1 + z_2}{2}\right). \tag{1.5}$$

1.1.5 Lengths, direction angles

Length and distance formula

The length of a vector is the length of the line segment whose endpoints are the head and tail of the vector. By using the Pythagorean theorem, we have the following theorem.

Theorem 1.1.2. *If a vector is represented by $\vec{a} = \langle x, y, z\rangle$, then*

$$|\vec{a}| = \sqrt{x^2 + y^2 + z^2}.$$

If $|\vec{a}| = 1$, then $\vec{a}$ is a unit vector. If $\vec{a}$ is not the zero vector, $\frac{1}{|\vec{a}|}\vec{a}$ is the unit vector in the direction of $\vec{a}$.

Example 1.1.3. Find the unit vector in the direction of

$$(1)\vec{a} = \langle 1, 2, -1\rangle \quad \text{and} \quad (2)\mathbf{b} = 4\mathbf{i} - \mathbf{j} - 8\mathbf{k}.$$

Solution.
1. The length of $\vec{a}$ is $|\vec{a}| = \sqrt{1^2 + 2^2 + (-1)^2} = \sqrt{6}$. So the unit vector $\vec{e}$ in the direction of $\vec{a}$ is

$$\vec{e} = \frac{1}{|\vec{a}|}\vec{a} = \frac{1}{\sqrt{6}}\langle 1, 2, -1\rangle = \left\langle \frac{1}{\sqrt{6}}, \frac{2}{\sqrt{6}}, \frac{1}{\sqrt{6}}\right\rangle.$$

2. The given vector has length

$$|4\mathbf{i} - \mathbf{j} - 8\mathbf{k}| = \sqrt{4^2 + (-1)^2 + (-8)^2} = \sqrt{81} = 9.$$

So the unit vector with the same direction is

$$\frac{1}{9}(4\mathbf{i} - \mathbf{j} - 8\mathbf{k}) = \frac{4}{9}\mathbf{i} - \frac{1}{9}\mathbf{j} - \frac{8}{9}\mathbf{k}.$$

We have seen the distance formula before. Now we can derive it from the length of a vector as well. The distance between the two points $P(x_1, y_1, z_1)$ and $Q(x_2, y_2, z_2)$ is, therefore, the length of the vector $\vec{PQ}$, so it is

$$|\vec{PQ}| = |\vec{OQ} - \vec{OP}| = \sqrt{(x_1 - x_2)^2 + (y_1 - y_2)^2 + (z_1 - z_2)^2}. \tag{1.6}$$

Example 1.1.4. Find a point P on the y-axis such that $|PA| = |PB|$, where $A(-4, 1, 7)$ and $B(3, 5, 2)$ are two points.

Solution. We assume the point P has the coordinates $(0, y, 0)$. From the distance formula, we have

$$\sqrt{(-4-0)^2 + (1-y)^2 + (7-0)^2} = \sqrt{(3-0)^2 + (5-y)^2 + (2-0)^2}.$$

Solving for y, we obtain $y = \frac{-7}{2}$. Therefore, the point P has coordinates $(0, -\frac{7}{2}, 0)$.

Direction angles and direction cosines
Let $\vec{a} = \vec{OA}$ and $\vec{b} = \vec{OB}$ be two vectors in a plane or space as in Figure 1.5(a) and (b).

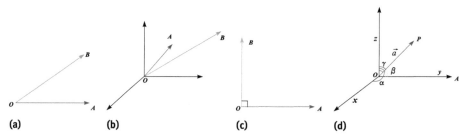

Figure 1.5: Angle between two vectors, perpendicular vectors, and direction angles.

Definition 1.1.5 (Angle between two vectors, direction angle, and direction cosines). If $\vec{a}$ and $\vec{b}$ are two vectors with a common tail, then:
1. The **angle between the vectors** $\vec{a}$ and $\vec{b}$ is the angle θ between 0 and π formed using the two vectors as sides.
2. The two vectors $\vec{a}$ and $\vec{b}$ are called **perpendicular (orthogonal)** if and only if the angle between them is $\frac{\pi}{2}$.
3. The angle between a vector $\vec{a}$ and the x-axis is the angle between $\vec{a}$ and the unit base vector $\vec{i}$.
4. The angle between a vector $\vec{a}$ and the y-axis is the angle between $\vec{a}$ and the unit base vector $\vec{j}$.
5. The angle between a vector $\vec{a}$ and the z-axis is the angle between $\vec{a}$ and the unit base vector $\vec{k}$.
6. The **direction angles** α, β, and γ of a vector $\vec{a}$ are the angles between $\vec{a}$ and the x-, y-, and z-axes, respectively; $\cos \alpha$, $\cos \beta$, and $\cos \gamma$ are called **direction cosines** of $\vec{a}$.

From Figure 1.5(d), if the vector $\vec{a} = \langle x, y, z \rangle$ has direction angles α, β, γ, then we have

$$\cos \alpha = \frac{x}{|\vec{a}|}, \quad \cos \beta = \frac{y}{|\vec{a}|}, \quad \text{and} \quad \cos \gamma = \frac{z}{|\vec{a}|}. \tag{1.7}$$

Since

$$\cos^2 \alpha + \cos^2 \beta + \cos^2 \gamma = \frac{x^2}{|\vec{a}|^2} + \frac{y^2}{|\vec{a}|^2} + \frac{z^2}{|\vec{a}|^2}$$
$$= \frac{x^2 + y^2 + z^2}{|\vec{a}|^2}$$
$$= 1,$$

it follows that

$$\langle \cos \alpha, \cos \beta, \cos \gamma \rangle = \frac{\langle x, y, z \rangle}{|\vec{a}|} \tag{1.8}$$

is the unit vector in the direction of $\vec{a}$.

Example 1.1.5. If $A(2, 2, \sqrt{2})$ and $B(1, 3, 0)$ are two points, find the length, direction cosines, and direction angles of the vector $\vec{AB}$.

Solution. Because $\vec{AB} = \langle 1 - 2, 3 - 2, 0 - \sqrt{2} \rangle = \langle -1, 1, -\sqrt{2} \rangle$, we have

$$|\vec{AB}| = \sqrt{(-1)^2 + 1^2 + (-\sqrt{2})^2} = 2.$$

The unit vector in the direction of $\vec{AB}$ is

$$\frac{1}{2} \langle -1, 1, -\sqrt{2} \rangle = \left\langle \frac{-1}{2}, \frac{1}{2}, -\frac{\sqrt{2}}{2} \right\rangle.$$

Hence,

$$\cos \alpha = -\frac{1}{2}, \quad \cos \beta = \frac{1}{2}, \quad \text{and} \quad \cos \gamma = -\frac{\sqrt{2}}{2}$$

are the three direction cosines, and

$$\alpha = \frac{2\pi}{3}, \quad \beta = \frac{\pi}{3}, \quad \text{and} \quad \gamma = \frac{3\pi}{4}$$

are the three direction angles with the positive x-, y- and z-axes, respectively.

1.2 Dot product, cross product, and triple product

1.2.1 The dot product

So far we have introduced the two operations on vectors: addition and multiplication by a scalar. Now the following questions arise: How about multiplication? Can

Figure 1.6: Work done by a force, dot product.

we multiply two vectors to obtain a useful quantity? In fact, there are two commonly used useful products of vectors called the dot product and the cross product.

As shown in Figure 1.6, you may already know from physics that the work done, W, by a force **F** applied during a displacement along the vector **s** is

$$W = |\mathbf{F}||\mathbf{s}|\cos\theta,$$

where θ is the angle between the two vectors **F** and **s**. It is, therefore, useful to define a product of two vectors in this way.

Definition 1.2.1 (Dot/scalar/inner product). The dot product **a** · **b** of the two vectors **a** and **b** is defined by

$$\mathbf{a} \cdot \mathbf{b} = |\mathbf{a}||\mathbf{b}|\cos\theta,$$

where θ is the angle between vectors **a** and **b**.

Example 1.2.1. If the two vectors **a** and **b** have length 3 and 4, and the angle between them is $\pi/3$, find **a** · **b**.

Solution. Using the definition, we have

$$\mathbf{a} \cdot \mathbf{b} = |\mathbf{a}||\mathbf{b}|\cos(\pi/3) = 3 \cdot 4 \cdot \frac{1}{2} = 6.$$

Well, this definition looks good as it has a physical basis. However, mathematically, it is not easy to find the dot product directly as we first need to know the angle between the vectors. Using the coordinate representation of a vector, it turns out that there is a remarkable way to compute the dot product, as we will see in the following theorem.

Theorem 1.2.1. *If* $\mathbf{a} = \langle a_1, a_2, a_3 \rangle$ *and* $\mathbf{b} = \langle b_1, b_2, b_3 \rangle$, *then*

$$\mathbf{a} \cdot \mathbf{b} = a_1 b_1 + a_2 b_2 + a_3 b_3.$$

Proof. Suppose the angle between **a** and **b** is θ. Note that the three vectors, **a**, **b**, and **c** = **b** − **a** form the three sides of a triangle. By the cosine law, we have

$$|\mathbf{c}|^2 = |\mathbf{a}|^2 + |\mathbf{b}|^2 - 2|\mathbf{a}||\mathbf{b}|\cos\theta.$$

Since

$$|\mathbf{c}|^2 = |\mathbf{b} - \mathbf{a}|^2 = (b_1 - a_1)^2 + (b_2 - a_2)^2 + (b_3 - a_3)^2$$
$$= b_1^2 - 2b_1 a_1 + a_1^2 + b_2^2 - 2b_2 a_2 + a_2^2 + b_3^2 - 2b_3 a_3 + a_3^2,$$
$$|\mathbf{a}|^2 = a_1^2 + a_2^2 + a_3^2, \quad \text{and}$$
$$|\mathbf{b}|^2 = b_1^2 + b_2^2 + b_3^2,$$

substituting these values into the cosine law equation and canceling out all the squares gives

$$-2b_1 a_1 - 2b_2 a_2 - 2b_3 a_3 - -2|\mathbf{a}||\mathbf{b}|\cos\theta.$$

Therefore, we have

$$\mathbf{a} \cdot \mathbf{b} = a_1 b_1 + a_2 b_2 + a_3 b_3. \qquad \square$$

In view of this theorem, we give the following alternative definition of the dot product.

Definition 1.2.2 (Alternative definition of the dot product). If $\mathbf{a} = \langle a_1, a_2, a_3 \rangle$, $\mathbf{b} = \langle b_1, b_2, b_3 \rangle$, and θ is the angle between the two vectors, then the dot product is defined by

$$\mathbf{a} \cdot \mathbf{b} = |\mathbf{a}||\mathbf{b}|\cos\theta = a_1 b_1 + a_2 b_2 + a_3 b_3.$$

Finding the dot product of **a** and **b** is incredibly easy by using coordinates. We just multiply corresponding components and add. Using this definition, we can deduce the following properties of the dot product.

Theorem 1.2.2 (Properties of the dot product). *If **a**, **b**, and **c** are any three vectors and λ is any scalar, then the dot product satisfies:*
1. $\mathbf{a} \cdot \mathbf{a} = |\mathbf{a}|^2$,
2. $\mathbf{a} \cdot \mathbf{b} = \mathbf{b} \cdot \mathbf{a}$,
3. *if **a** and **b** are two nonzero vectors, then $\mathbf{a} \cdot \mathbf{b} = 0$ means that **a** and **b** are perpendicular to each other,*
4. $(\mathbf{a} + \mathbf{b}) \cdot \mathbf{c} = \mathbf{a} \cdot \mathbf{c} + \mathbf{b} \cdot \mathbf{c}$,
5. $(\lambda \mathbf{a}) \cdot \mathbf{b} = \lambda(\mathbf{a} \cdot \mathbf{b}) = \mathbf{a} \cdot (\lambda \mathbf{b})$,
6. $\mathbf{0} \cdot \mathbf{a} = 0$.

These properties are similar to the rules for real numbers and can be easily proved by using either of the two definitions of the dot product. However, some properties of real number multiplication do not apply to the dot product. For example, if two real numbers satisfy $ab = 0$, then either $a = 0$ or $b = 0$ or both. This is not true for the dot product. If **a** and **b** are two nonzero vectors, then $\mathbf{a} \cdot \mathbf{b} = 0$ indicates the two vectors are perpendicular to each other, and it is not necessary that either $\mathbf{a} = \mathbf{0}$ or $\mathbf{b} = \mathbf{0}$.

By using the dot product, we can find the angle between two vectors, as shown in the following example.

Example 1.2.2. Find the angle between the two vectors $\mathbf{i} + 2\mathbf{j} - \mathbf{k}$ and $2\mathbf{j} - \mathbf{k}$.

Solution. By the definition of the dot product, $\mathbf{u} \cdot \mathbf{v} = |\mathbf{u}| \cdot |\mathbf{v}| \cos \theta$, so $\cos \theta = \frac{\mathbf{u} \cdot \mathbf{v}}{|\mathbf{u}| \cdot |\mathbf{v}|}$. Thus,

$$\cos \theta = \frac{(\mathbf{i} + 2\mathbf{j} - \mathbf{k}) \cdot (2\mathbf{j} - \mathbf{k})}{|\mathbf{i} + 2\mathbf{j} - \mathbf{k}||2\mathbf{j} - \mathbf{k}|} = \frac{1 \cdot 0 + 2 \cdot 2 + (-1) \cdot (-1)}{\sqrt{1^2 + 2^2 + (-1)^2} \sqrt{2^2 + (-1)^2}} \approx 0.913.$$

So the angle $\theta \approx \cos^{-1}(0.913) \approx 0.42$ radians (about $24°$).

1.2.2 Projections

Suppose that $\mathbf{a} = \overrightarrow{OA}$ and $\mathbf{b} = \overrightarrow{OB}$ are two vectors with the same tail O. If S is the foot of the perpendicular from B to the line containing $\overrightarrow{OA}$, then the vector $\overrightarrow{OS}$ is called the *vector projection* of the vector **b** onto the vector **a**, written as $\text{Proj}_\mathbf{a} \mathbf{b}$. If $\mathbf{e} = \frac{\mathbf{a}}{|\mathbf{a}|}$ is the unit vector in the direction of $\overrightarrow{OA}$, then the vector projection is $\lambda \mathbf{e}$, where $\lambda = |\mathbf{b}| \cos \theta$ is the size (positive or negative) of the projection vector and θ is the angle between the two vectors, as shown in Figure 1.7. Hence, the projection of vector **b** onto vector **a** is

$$\text{Proj}_\mathbf{a} \mathbf{b} = \frac{|\mathbf{b}| \cos \theta}{|\mathbf{a}|} \mathbf{a}.$$

The scalar projection of vector **b** onto vector **a** is defined as

$$\text{ProjScal}_\mathbf{a} \mathbf{b} = |\mathbf{b}| \cos \theta.$$

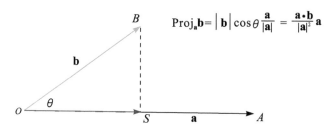

Figure 1.7: Vector projections.

By using the dot product $\mathbf{a} \cdot \mathbf{b} = |\mathbf{a}||\mathbf{b}| \cos \theta$, we have

$$\text{Proj}_\mathbf{a} \mathbf{b} = \frac{\mathbf{a} \cdot \mathbf{b}}{|\mathbf{a}|^2} \mathbf{a} \quad \text{and} \quad \text{ProjScal}_\mathbf{a} \mathbf{b} = \frac{\mathbf{a} \cdot \mathbf{b}}{|\mathbf{a}|}.$$

Example 1.2.3. Show that any vector $\mathbf{r} = \langle x, y, z \rangle$ can be written as

$$\mathbf{r} = \text{Proj}_\mathbf{i} \mathbf{r} + \text{Proj}_\mathbf{j} \mathbf{r} + \text{Proj}_\mathbf{k} \mathbf{r}.$$

Solution. For $\text{Proj}_\mathbf{i} \mathbf{r}$, since $|\mathbf{i}| = 1$, we have

$$\text{Proj}_\mathbf{i} \mathbf{r} = \frac{\mathbf{r} \cdot \mathbf{i}}{|\mathbf{i}|^2} \mathbf{i} = (\mathbf{r} \cdot \mathbf{i})\mathbf{i} = \langle x, y, z \rangle \cdot \langle 1, 0, 0 \rangle \mathbf{i} = x\mathbf{i}.$$

Similarly, $\text{Proj}_\mathbf{j} \mathbf{r} = y\mathbf{j}$ and $\text{Proj}_\mathbf{k} \mathbf{r} = z\mathbf{k}$. Therefore,

$$\mathbf{r} = \text{Proj}_\mathbf{i} \mathbf{r} + \text{Proj}_\mathbf{j} \mathbf{r} + \text{Proj}_\mathbf{k} \mathbf{r}.$$

1.2.3 The cross product

In mechanics, the *moment* of a force $\vec{F}$ acting on a rod $\overrightarrow{OP}$ is the vector with magnitude $|\vec{F}||\overrightarrow{OP}| \sin \theta$, where θ is the angle between the vectors $\vec{F}$ and $\overrightarrow{OP}$. The direction of the moment vector is perpendicular to $\vec{F}$ and $\overrightarrow{OP}$ (see Figure 1.8(a)) and satisfies the right-hand rule: if you curl your right fingers naturally from vector $\vec{F}$ to vector $\overrightarrow{OP}$, then your thumbs points in the direction of the moment vector, as shown in Figure 1.8(b) and (c). Therefore, it makes sense to define a product of two vectors $\vec{a}$ and $\vec{b}$ as follows.

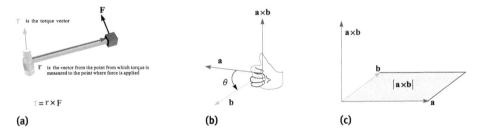

Figure 1.8: Cross product, moment/torque.

Definition 1.2.3 (Cross/vector/outer product). The cross product denoted by $\mathbf{a} \times \mathbf{b}$ of vector $\mathbf{a}$ and vector $\mathbf{b}$ in $\mathbb{R}^3$ is a new vector which is perpendicular to both vector $\mathbf{a}$ and vector $\mathbf{b}$. The length of $\mathbf{a} \times \mathbf{b}$ is

$$|\mathbf{a} \times \mathbf{b}| = |\mathbf{a}||\mathbf{b}| \sin \theta$$

and $\mathbf{a}, \mathbf{b}, \mathbf{a} \times \mathbf{b}$, in that order, satisfy the right-hand rule.

According to the above definition and using Figure 1.4(a), we can see that

$$\mathbf{i} \times \mathbf{i} = \mathbf{0}, \quad \mathbf{i} \times \mathbf{j} = \mathbf{k}, \quad \mathbf{i} \times \mathbf{k} = -\mathbf{j}, \quad \mathbf{j} \times \mathbf{j} = \mathbf{0}, \quad \mathbf{j} \times \mathbf{i} = -\mathbf{k},$$
$$\mathbf{j} \times \mathbf{k} = \mathbf{i}, \quad \mathbf{k} \times \mathbf{i} = \mathbf{j}, \quad \mathbf{k} \times \mathbf{j} = -\mathbf{i}, \quad \text{and} \quad \mathbf{k} \times \mathbf{k} = \mathbf{0}.$$

But in general, how can we compute the cross product? If we try to compute

$$\mathbf{a} \times \mathbf{b} = (a_1\mathbf{i} + a_2\mathbf{j} + a_3\mathbf{k}) \times (b_1\mathbf{i} + b_2\mathbf{j} + b_3\mathbf{k})$$

by using the normal rules for numbers, such as the commutative, associative, and distributive rules, we may find an interesting vector

$$\mathbf{c} = \langle a_2 b_3 - a_3 b_2, a_3 b_1 - b_3 a_1, a_1 b_2 - a_2 b_1 \rangle.$$

This vector, in fact, satisfies conditions that we have set for a cross product, as we will see in the following theorem.

Theorem 1.2.3. *If* $\mathbf{a} = \langle a_1, a_2, a_3 \rangle$, $\mathbf{b} = \langle b_1, b_2, b_3 \rangle$, *and*

$$\mathbf{c} = \langle a_2 b_3 - a_3 b_2, a_3 b_1 - b_3 a_1, a_1 b_2 - a_2 b_1 \rangle,$$

then:
1. *$\mathbf{c}$ is perpendicular to both $\mathbf{a}$ and $\mathbf{b}$.*
2. *$|\mathbf{c}| = |\mathbf{a}||\mathbf{b}| \sin \theta$, where θ is the angle between $\mathbf{a}$ and $\mathbf{b}$.*

Proof. We compute the dot product to show they are perpendicular. We have

$$\mathbf{a} \cdot \mathbf{c} = \langle a_1, a_2, a_3 \rangle \cdot \langle a_2 b_3 - a_3 b_2, a_3 b_1 - b_3 a_1, a_1 b_2 - a_2 b_1 \rangle$$
$$= a_1 a_2 b_3 - a_1 a_3 b_2 + a_2 a_3 b_1 - a_2 b_3 a_1 + a_3 a_1 b_2 - a_3 a_2 b_1$$
$$= 0.$$

Similarly $\mathbf{b} \cdot \mathbf{c} = 0$. Therefore, we claim that $\mathbf{c}$ is perpendicular to both $\mathbf{a}$ and $\mathbf{b}$.

Furthermore,

$$|\mathbf{c}|^2 = (a_2 b_3 - a_3 b_2)^2 + (a_3 b_1 - b_3 a_1)^2 + (a_1 b_2 - a_2 b_1)^2$$
$$= a_2^2 b_3^2 + a_3^2 b_2^2 - 2 a_2 b_3 a_3 b_2 +$$
$$\quad a_3^2 b_1^2 + b_3^2 a_1^2 - 2 a_3 b_1 b_3 a_1 + a_1^2 b_2^2 + a_2^2 b_1^2 - 2 a_1 b_2 a_2 b_1$$
$$= (a_1^2 + a_2^2 + a_3^2)(b_1^2 + b_2^2 + b_3^2) - (a_1 b_1 + a_2 b_2 + a_3 b_3)^2$$
$$= |\mathbf{a}|^2 |\mathbf{b}|^2 - (\mathbf{a} \cdot \mathbf{b})^2$$
$$= |\mathbf{a}|^2 |\mathbf{b}|^2 (1 - \cos^2 \theta) = |\mathbf{a}|^2 |\mathbf{b}|^2 \sin^2 \theta.$$

So $|\mathbf{c}| = |\mathbf{a}||\mathbf{b}| \cdot \sin \theta.$ □

Now the only issue that remains is whether **a**, **b**, and **c**, in that order, satisfy the right-hand rule. This can be seen in a simple case where **a** and **b** are in the first quadrant of the xy-plane with tails at the origin. Then the sign of the term $\frac{a_2}{a_1} - \frac{b_2}{b_1}$ determines the relative positions of **a** and **b**, and the sign of the z-component of **c**, $a_1 b_2 - a_2 b_1$, determines whether **c** points upward or downward. This is exactly the right-hand rule: when you curl your right fingers from **a** to **b**, then your thumb points in the direction of **c**.

In light of the above discussion, we now give an alternative definition of the cross product.

Definition 1.2.4 (Alternative definition of the cross product). Let $\mathbf{a} = \langle a_1, a_2, a_3 \rangle$ and $\mathbf{b} = \langle b_1, b_2, b_3 \rangle$. Then the cross product (also vector product) $\mathbf{a} \times \mathbf{b}$ is defined by

$$\mathbf{a} \times \mathbf{b} = \langle a_2 b_3 - a_3 b_2, a_3 b_1 - b_3 a_1, a_1 b_2 - a_2 b_1 \rangle.$$

Using the knowledge of determinants, we have

$$\mathbf{a} \times \mathbf{b} = \langle a_2 b_3 - a_3 b_2, a_3 b_1 - b_3 a_1, a_1 b_2 - a_2 b_1 \rangle$$

$$= \left\langle \begin{vmatrix} a_2 & a_3 \\ b_2 & b_3 \end{vmatrix}, -\begin{vmatrix} a_1 & a_3 \\ b_1 & b_3 \end{vmatrix}, \begin{vmatrix} a_1 & a_2 \\ b_1 & b_2 \end{vmatrix} \right\rangle = \begin{vmatrix} \mathbf{i} & \mathbf{j} & \mathbf{k} \\ a_1 & a_2 & a_3 \\ b_1 & b_2 & b_3 \end{vmatrix}, \quad (1.9)$$

where $\begin{vmatrix} a & b \\ c & d \end{vmatrix} = ad - bc$. This is much better for remembering the cross product.

Using the definition of the vector product, we have the following theorem.

Theorem 1.2.4 (Properties of the cross product for three-dimensional vectors). *For any three vectors **a**, **b**, and **c** in $\mathbb{R}^3$ and a scalar λ, we have:*

1. $\mathbf{a} \times \mathbf{a} = \mathbf{0}$,
2. *if **a** and **b** are nonzero vectors, then $\mathbf{a} \times \mathbf{b} = \mathbf{0}$ if and only if $\mathbf{a} \parallel \mathbf{b}$,*
3. $\mathbf{b} \times \mathbf{a} = -(\mathbf{a} \times \mathbf{b})$,
4. $\mathbf{a} \times (\mathbf{b} + \mathbf{c}) = \mathbf{a} \times \mathbf{b} + \mathbf{a} \times \mathbf{c}$,
5. $(\mathbf{a} + \mathbf{b}) \times \mathbf{c} = \mathbf{a} \times \mathbf{c} + \mathbf{b} \times \mathbf{c}$,
6. $(\lambda \mathbf{a}) \times \mathbf{b} = \lambda(\mathbf{a} \times \mathbf{b}) = \mathbf{a} \times (\lambda \mathbf{b})$,
7. $\mathbf{a} \cdot (\mathbf{b} \times \mathbf{c}) = (\mathbf{a} \times \mathbf{b}) \cdot \mathbf{c}$,
8. $\mathbf{a} \times (\mathbf{b} \times \mathbf{c}) = (\mathbf{a} \cdot \mathbf{c}) \mathbf{b} - (\mathbf{a} \cdot \mathbf{b}) \mathbf{c}$.

Using one of the definitions of the cross product, we can prove these properties by writing the vectors in their components form. Note that the cross product fails to obey most of the laws satisfied by real number multiplication, such as the commutative and associative laws. Check for yourself that $\mathbf{a} \times (\mathbf{b} \times \mathbf{c}) \neq (\mathbf{a} \times \mathbf{b}) \times \mathbf{c}$ for most vectors **a**, **b**, and **c**.

Example 1.2.4. Find a vector that is perpendicular to the plane containing the three points $P(1, 0, 6)$, $Q(2, 5, -1)$, and $R(-1, 3, 7)$.

Solution. The cross product of the two vectors $\overrightarrow{PQ}$ and $\overrightarrow{PR}$ is such a vector. This is because the cross product is perpendicular to both $\overrightarrow{PQ}$ and $\overrightarrow{PR}$ and is, thus, perpendicular to the plane through the three points P, Q, and R. Since

$$\overrightarrow{PQ} = (2-1)\vec{i} + (5-0)\vec{j} + (-1-6)\vec{k} = \vec{i} + 5\vec{j} - 7\vec{k},$$
$$\overrightarrow{PR} = (-1-1)\vec{i} + (3-0)\vec{j} + (7-6)\vec{k} = -2\vec{i} + 3\vec{j} + \vec{k},$$

we evaluate the cross product of these two vectors using the determinant approach, i.e.,

$$\overrightarrow{PQ} \times \overrightarrow{PR} = \begin{vmatrix} \vec{i} & \vec{j} & \vec{k} \\ 1 & 5 & -7 \\ -2 & 3 & 1 \end{vmatrix} = (5+21)\vec{i} - (1-14)\vec{j} + (3+10)\vec{k}$$
$$= 26\vec{i} + 13\vec{j} + 13\vec{k}.$$

So the vector $\langle 26, 13, 13 \rangle$ is perpendicular to the plane passing through the three points P, Q, and R. In fact, any nonzero scalar multiple of this vector, such as $\langle 2, 1, 1 \rangle$, is also perpendicular to the plane. Figure 1.9 illustrates the vector perpendicular to the plane.

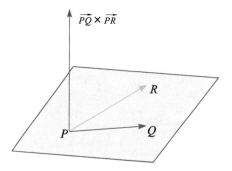

Figure 1.9: Cross product, Example 1.2.4.

Note that the length of the vector $|\mathbf{a} \times \mathbf{b}| = |\mathbf{a}||\mathbf{b}| \sin \theta$ is equal to the area of the parallelogram determined by $\mathbf{a}$ and $\mathbf{b}$, assuming they have the same initial point, as shown in Figure 1.8(d). Therefore, we have the following theorem.

Theorem 1.2.5. *Given two nonzero vectors $\mathbf{a}$ and $\mathbf{b}$ with a common tail, we have*

$$\text{area of a parallelogram with adjacent sides } \mathbf{a} \text{ and } \mathbf{b} = |\mathbf{a} \times \mathbf{b}|,$$
$$\text{area of a triangle with adjacent sides } \mathbf{a} \text{ and } \mathbf{b} = \frac{1}{2}|\mathbf{a} \times \mathbf{b}|.$$

Example 1.2.5. Find the area of the triangle with vertices $P(1, 0, 6)$, $Q(2, 5, -1)$, and $R(-1, 3, 7)$.

Solution. In the previous example, we already computed that $\vec{PQ} \times \vec{PR} = \langle 26, 13, 13 \rangle$. The area of the parallelogram with adjacent sides PQ and PR is the magnitude of the cross product, i. e.,

$$|\vec{PQ} \times \vec{PR}| = \sqrt{(26)^2 + (13)^2 + (13)^2} = 13\sqrt{6}.$$

Thus, the area of the triangle PQR is $\frac{13\sqrt{6}}{2}$.

1.2.4 Scalar triple product

Suppose three nonplanar vectors **a**, **b**, and **c**, have a common tail. What is the volume of the parallelepiped determined by these three vectors as shown in Figure 1.10?

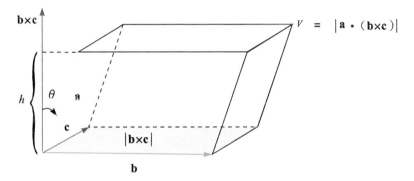

Figure 1.10: Triple product, volume of a parallelepiped.

Consider the base parallelogram; its area is $A = |\mathbf{b} \times \mathbf{c}|$. Let θ be the angle between **a** and $\mathbf{b} \times \mathbf{c}$. Noting that $\mathbf{b} \times \mathbf{c}$ is perpendicular to **b** and **c** and the height h of the parallelepiped is

$$h = |\mathbf{a}||\cos\theta|$$

(we should use $|\cos\theta|$ instead of $\cos\theta$ to ensure that we obtain a positive result when $\theta > \frac{\pi}{2}$), we conclude that the volume V of the parallelepiped is given as follows:

$$V = Ah = |\mathbf{b} \times \mathbf{c}||\mathbf{a}||\cos\theta| = |\mathbf{a} \cdot (\mathbf{b} \times \mathbf{c})|.$$

Thus, we have proved that the volume of the parallelepiped determined by the three vectors **a**, **b**, and **c** with a common tail is given as follows:

$$V = |\mathbf{a} \cdot (\mathbf{b} \times \mathbf{c})|. \tag{1.10}$$

A product like $\mathbf{a} \cdot (\mathbf{b} \times \mathbf{c})$ is called a *scalar triple product* of the three vectors $\vec{a}$, $\vec{b}$, and $\vec{c}$. Note that we can write this scalar triple product as a 3×3 determinant as follows:

$$\mathbf{a} \cdot (\mathbf{b} \times \mathbf{c}) = a_1 \begin{vmatrix} b_2 & b_3 \\ c_2 & c_3 \end{vmatrix} - a_2 \begin{vmatrix} b_1 & b_3 \\ c_1 & c_3 \end{vmatrix} + a_3 \begin{vmatrix} b_1 & b_2 \\ c_1 & c_2 \end{vmatrix} = \begin{vmatrix} a_1 & a_2 & a_3 \\ b_1 & b_2 & b_3 \\ c_1 & c_2 & c_3 \end{vmatrix}.$$

If the above scalar triple product is 0, then it means that the volume of the parallelepiped determined by the three vectors $\mathbf{a}$, $\mathbf{b}$, and $\mathbf{c}$ is 0. Then, we can conclude that the three vectors must be *coplanar* (that is, they lie in the same plane). In terms of linear algebra, they are linearly dependent.

Example 1.2.6. Use the scalar triple product to determine whether the vectors $\mathbf{a} = \langle 2, 0, -7 \rangle$, $\mathbf{b} = \langle 1, -1, -3 \rangle$, and $\mathbf{c} = \langle 1, 1, -1 \rangle$ are coplanar.

Solution. Since

$$\mathbf{a} \cdot (\mathbf{b} \times \mathbf{c}) = \begin{vmatrix} 2 & 0 & -7 \\ 1 & -1 & -3 \\ 1 & 1 & -1 \end{vmatrix} = 2 \begin{vmatrix} -1 & -3 \\ 1 & -1 \end{vmatrix} - 0 \begin{vmatrix} 1 & -3 \\ 1 & -1 \end{vmatrix} - 7 \begin{vmatrix} 1 & -1 \\ 1 & 1 \end{vmatrix}$$
$$= 8 - 0 - 7 \times 2 = -6$$

is not 0, the vectors $\mathbf{a}$, $\mathbf{b}$, and $\mathbf{c}$ are not coplanar.

1.3 Equations of lines and planes

1.3.1 Lines

A line in the two-dimensional xy-plane is determined by a point on the line and the direction of the line (its slope, or angle of inclination, or a vector parallel to the line). The equation of the line can be written by using the usual slope-intercept form $y = mx + b$.

A line L in $\mathbb{R}^3$ is also determined once we know a point $P(x_0, y_0, z_0)$ on L and the direction of L. However, we do not have the concept of "slope of a line" as we do in $\mathbb{R}^2$. In three-dimensional space, the direction of a line L can be conveniently described by a vector $\mathbf{v} = \langle m, n, p \rangle$ parallel to L. If $P(x, y, z)$ is an arbitrary point on L, then the vector $\overrightarrow{P_0P}$ is parallel to $\mathbf{v}$ exactly when the point P is on the line, as shown in Figure 1.11, so for some real number t we have

$$\overrightarrow{P_0P} = t\mathbf{v},$$
$$\langle x - x_0, y - y_0, z - z_0 \rangle = \langle tm, tn, tp \rangle.$$

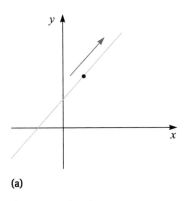

(a)   (b)

Figure 1.11: Lines in space.

Equating the components, we have

$$x - x_0 = tm, \quad y - y_0 = tn, \quad \text{and} \quad z - z_0 = tp,$$

or

$$x = x_0 + tm, \quad y = y_0 + tn \quad \text{and} \quad z = z_0 + tp, \tag{1.11}$$

or

$$\begin{cases} x = x_0 + tm, \\ y = y_0 + tn, \\ z = z_0 + tp. \end{cases} \tag{1.12}$$

Equations (1.11) and (1.12) are called *parametric equations* of the line passing through the point (x_0, y_0, z_0) with the direction vector $\mathbf{v} = \langle m, n, p \rangle$. Note that equation (1.11) can be rewritten as

$$\frac{x - x_0}{m} = \frac{y - y_0}{n} = \frac{z - z_0}{p}, \tag{1.13}$$

which is called *symmetric equations* of the line. If one of m, n, and p is 0, say, $m = 0$, then we can still use the notation symbolically, i.e.,

$$\frac{x - x_0}{0} = \frac{y - y_0}{n} = \frac{z - z_0}{p},$$

but this should be interpreted as $x = x_0$ and $\frac{y - y_0}{n} = \frac{z - z_0}{p}$.

Note. In general, if a vector $\mathbf{v} = \langle m, n, p \rangle$ is used to describe the direction of a line L, then the numbers m, n, and p are called *direction numbers* of L. Since there are many vectors parallel to L, any of them could be used to describe the direction of L. We can also see that any three numbers proportional to m, n, and p are also direction numbers for L. The three direction numbers determine the three direction angles; they are "angles of inclination" with respect to the three coordinate axes. If $\mathbf{v} = \langle m, n, p \rangle$ is a unit vector, then the three direction numbers are actually its three direction cosines.

Also, equation (1.11) can be written in a vector form,

$$\begin{pmatrix} x \\ y \\ z \end{pmatrix} = \begin{pmatrix} x_0 \\ y_0 \\ z_0 \end{pmatrix} + t \begin{pmatrix} m \\ n \\ p \end{pmatrix}, \qquad (1.14)$$

or

$$\mathbf{r} = \langle x_0, y_0, z_0 \rangle + t \langle m, n, p \rangle, \qquad (1.15)$$

or

$$\mathbf{r} = \mathbf{r}_0 + t\mathbf{v}. \qquad (1.16)$$

Equations (1.14)–(1.16) are all called *vector equations* for the line L passing through the point (x_0, y_0, z_0) with direction $\mathbf{v}$.

Example 1.3.1. Find parametric equations, a vector equation, and symmetric equations of the line L which passes through the points $A(1, 2, -1)$ and $B(0, 1, 3)$.

Solution. The vector $\overrightarrow{AB} = \langle 0-1, 1-2, 3-(-1) \rangle = \langle -1, -1, 4 \rangle$ is a direction vector of the line L. Hence, a vector equation of L is

$$\mathbf{r} = \langle 1, 2, -1 \rangle + t \langle -1, -1, 4 \rangle$$

or

$$\begin{pmatrix} x \\ y \\ z \end{pmatrix} = \begin{pmatrix} 1 \\ 2 \\ -1 \end{pmatrix} + t \begin{pmatrix} -1 \\ -1 \\ 4 \end{pmatrix}.$$

This gives the parametric equations of line L

$$x = 1 - t, \quad y = 2 - t, \quad z = -1 + 4t.$$

Symmetric equations of L are obtained by eliminating the parameter t, i.e.,

$$\frac{x-1}{-1} = \frac{y-2}{-1} = \frac{z+1}{4}.$$

The graph of the line is shown in Figure 1.12(a).

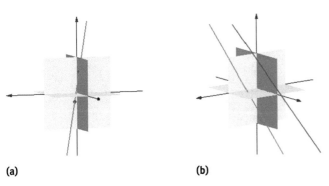

(a) (b)

Figure 1.12: Lines in space, Examples 1.3.1 and 1.3.2.

Example 1.3.2. Show that the lines L_1 and L_2 with parametric equations

$$x = 1 + 2t, \quad y = 2 - t, \quad z = -3 + 4t,$$
$$x = 2 + s, \quad y = 4 - s, \quad z = 4 + 2s$$

are *skew lines*. That is, L_1 and L_2 do not intersect in a point and are not parallel to each other and, therefore, do not lie in the same plane.

Solution. The lines are not parallel because the corresponding direction vectors $\mathbf{v}_1 = \langle 2, -1, 4 \rangle$ and $\mathbf{v}_2 = \langle 1, -1, 2 \rangle$ are not parallel because there is no scalar λ such that $\langle 1, -1, 2 \rangle = \lambda \langle 2, -1, 4 \rangle$. In other words, their components are not proportional. We attempt to solve the system of equations in t and s to find any intersection points. We have

$$1 + 2t = 2 + s,$$
$$2 - t = 4 - s,$$
$$-3 + 4t = 4 + 2s.$$

Solving the first two equations for t and s gives $t = 3$ and $s = 5$, but these values do not satisfy the third equation. Therefore, there are no values of t and s that satisfy all three equations, so the system of equations is inconsistent. Thus, L_1 and L_2 do not intersect and are skew lines. The graphs of the two lines are shown in Figure 1.12(b).

The angle between two lines is the angle between their direction vectors. Therefore, we can use the dot product to find the angle, as shown in the following example.

Example 1.3.3. Find the acute angle between two lines

$$L_1: \frac{x-1}{1} = \frac{y}{-4} = \frac{z+3}{1} \quad \text{and} \quad L_2: x = 2t, y = -2 - 2t, z = -t.$$

Solution. A direction vector of L_1 is $\mathbf{v}_1 = \langle 1, -4, 1 \rangle$ and of L_2 is $\mathbf{v}_2 = \langle 2, -2, -1 \rangle$. If θ is the angle between the two lines, we have

$$\cos\theta = \frac{\mathbf{v}_1 \cdot \mathbf{v}_2}{|\mathbf{v}_1||\mathbf{v}_2|} = \frac{1\cdot 2 + (-4)\cdot(-2) + 1\cdot(-1)}{\sqrt{1^2+(-4)^2+1^2}\sqrt{2^2+(-2)^2+(-1)^2}} = \frac{1}{\sqrt{2}}.$$

The desired angle is, therefore, $\theta = \pi/4$.

! **Example 1.3.4.** Find symmetric equations of the line L that passes through $(2, 1, 14)$ and perpendicularly intersects the line $L_0 : \frac{x-3}{2} = \frac{y}{1} = \frac{z-1}{1}$.

Solution. Suppose that the line L intersects L_0 at the point $P(x, y, z)$. Then the coordinates of P must have the form

$$x = 3 + 2t,\ y = t,\ \text{and}\ z = 1 + t,\ \text{for some}\ t.$$

The vector parallel to L with initial point $(2, 1, 14)$ and terminal point P is

$$\langle 3 + 2t - 2, t - 1, 1 + t - 14 \rangle = \langle 1 + 2t, t - 1, -13 + t \rangle.$$

Since the two lines intersect perpendicularly, the direction of L_0 is also perpendicular to this vector, so

$$\langle 1 + 2t, t - 1, -13 + t \rangle \cdot \langle 2, 1, 1 \rangle = 0.$$

Solving for t, we have $t = 2$. Hence, P has coordinates $(7, 2, 3)$ and a vector parallel to L is $\langle 7, 2, 3 \rangle - \langle 2, 1, 14 \rangle = \langle 5, 1, -11 \rangle$. Therefore, symmetric equations of L are

$$\frac{x-7}{5} = \frac{y-2}{1} = \frac{z-3}{-11}.$$

Note. This example shows how to find the foot of the perpendicular of a point onto a given line. This can be used to find the distance from a given point to a given line, as shown in the following example.

Example 1.3.5. Find the perpendicular distance from the point $Q(1, 2, 3)$ to the straight line with parametric equations $x = 3 + t, y = 4 - 2t, z = -2 + 2t$.

Solution. Let t be the value such that the point on the line $N(3 + t, 4 - 2t, -2 + 2t)$ is the foot of the perpendicular from the point Q to the line. The vector $\overrightarrow{NQ}$ must be perpendicular to the direction of the line, so $\overrightarrow{NQ} \cdot \langle 1, -2, 2 \rangle = 0$. This means

$$\langle 1 - (3 + t),\ 2 - (4 - 2t),\ 3 - (-2 + 2t) \rangle \cdot \langle 1, -2, 2 \rangle = 0,$$
$$\langle -2 - t,\ -2 + 2t,\ 5 - 2t \rangle \cdot \langle 1, -2, 2 \rangle = 0,$$

$$-2 - t - 2(-2 + 2t) + 2(5 - 2t) = 0,$$
$$t = \frac{4}{3}.$$

So, the foot of the perpendicular from the point Q to the line is $N(3+t, 4-2t, -2+2t) = N(\frac{13}{3}, \frac{4}{3}, \frac{2}{3})$, and the distance from the point Q to the line is

$$|NQ| = \sqrt{\left(\frac{13}{3} - 1\right)^2 + \left(\frac{4}{3} - 2\right)^2 + \left(\frac{2}{3} - 3\right)^2} = \sqrt{17}.$$

Note. The distance can also be obtained by minimizing the function $d(t) = \sqrt{|NQ|}$. Also, one can show that the distance from a point P to a line $\mathbf{r} = \mathbf{r}_0 + \mathbf{v}t$ is

$$\text{distance from } P \text{ to a line} = \frac{|\overrightarrow{MP} \times \mathbf{v}|}{|\mathbf{v}|}, \text{ where } M \text{ is any point on the line.} \quad (1.17)$$

1.3.2 Planes

A *plane* is a surface that is determined by a point $M_0(x_0, y_0, z_0)$ and a normal vector $\mathbf{n}$. That is, there is a unique plane that passes through the given point M_0 and is perpendicular to a given direction $\mathbf{n}$. How do you find an equation for this plane? Assume that $M(x, y, z)$ is a point in space. Then M is in the plane if and only if the vector $\overrightarrow{M_0M}$ is orthogonal to the normal vector $\mathbf{n}$ (see Figure 1.13), that is,

$$\mathbf{n} \cdot \overrightarrow{M_0M} = 0 \quad \text{or} \quad \mathbf{n} \cdot (\overrightarrow{OM} - \overrightarrow{OM_0}) = 0.$$

If $\mathbf{n} = \langle a, b, c \rangle$, then expanding the dot product gives

$$\mathbf{n} \cdot (\overrightarrow{OM} - \overrightarrow{OM_0}) = \langle a, b, c \rangle \cdot (\langle x, y, z \rangle - \langle x_0, y_0, z_0 \rangle).$$

Thus,

$$a(x - x_0) + b(y - y_0) + c(z - z_0) = 0. \quad (1.18)$$

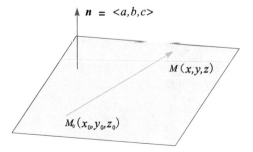

Figure 1.13: Planes in space.

This is called the *Cartesian equation/linear equation of the plane through* $M_0(x_0, y_0, z_0)$ with normal vector $\mathbf{n} = \langle a, b, c \rangle$. By collecting terms in the equation, we can write the equation as

$$ax + by + cz + d = 0, \tag{1.19}$$

where $d = -(ax_0 + by_0 + cz_0)$. A point (x, y, z) is in the plane if and only if it satisfies this equation.

Example 1.3.6. The plane $x = 0$ is the *yz*-coordinate plane, the plane $y = 0$ is the *xz*-coordinate plane, and the plane $z = 0$ is the *xy*-coordinate plane; $z = 3$ is the plane parallel to the *xy*-plane with distance 3 units from it.

Example 1.3.7. Find an equation of the plane that passes through the point $(2, 2, -1)$ with normal vector $\vec{n} = \langle 1, 2, 3 \rangle$. Also, find the intercepts of the plane with the three coordinate axes and then sketch the plane.

Solution. Plug $a = 1$, $b = 2$, $c = 3$ and $x_0 = 2$, $y_0 = 2$, $z_0 = -1$ into the equation (1.18). We get an equation of the plane

$$1(x - 2) + 2(y - 2) + 3(z + 1) = 0,$$

or

$$x + 2y + 3z = 3.$$

In order to find the *x*-intercept, we set $y = z = 0$ in this equation and solve for x to get $x = 3$. Similarly, the *y*-intercept is $3/2$ and the *z*-intercept is 1. The plane is shown in Figure 1.14(a).

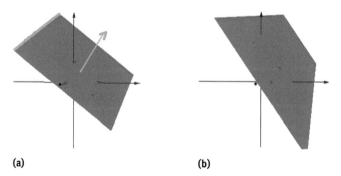

(a)   (b)

Figure 1.14: Planes, Examples 1.3.7 and 1.3.8.

Example 1.3.8. Find an equation of the plane through the three points $P(-1,-3,2)$, $Q(0,-1,7)$, and $R(3,2,-1)$.

Solution. The vectors $\vec{PQ}$ and $\vec{PR}$ are

$$\vec{PQ} = \langle 0,-1,7\rangle - \langle -1,-3,2\rangle = \langle 1,2,5\rangle$$

and

$$\vec{PR} = \langle 3,2,-1\rangle - \langle -1,-3,2\rangle = \langle 4,5,-3\rangle.$$

Their cross product $\vec{PQ} \times \vec{PR}$ is orthogonal to the desired plane and, thus, $\vec{n} = \vec{PQ} \times \vec{PR}$ is a normal vector to the plane. Hence, an equation of the plane is

$$\vec{PM} \cdot \vec{n} = \vec{PM} \cdot (\vec{PQ} \times \vec{PR}) = 0,$$

where $M(x,y,z)$ is an arbitrary point in the plane. Using the triple product formula gives

$$\begin{vmatrix} x-(-1) & y-(-3) & z-2 \\ 1 & 2 & 5 \\ 4 & 5 & -3 \end{vmatrix} = 0.$$

Simplifying this, we obtain

$$23y - 31x - 3z + 44 = 0.$$

The graph of the plane is shown in Figure 1.14(b).

In general, an equation of the plane passing through three points $P_1(x_1,y_1,z_1)$, $P_2(x_2,y_2,z_2)$, and $P_3(x_3,y_3,z_3)$ is

$$\begin{vmatrix} x-x_1 & y-y_1 & z-z_1 \\ x_2-x_1 & y_2-y_1 & z_2-z_1 \\ x_3-x_1 & y_3-y_1 & z_3-z_1 \end{vmatrix} = 0.$$

In particular, if the three points are three intercepts with the x-, y-, and z-axes given by $P_1(a,0,0)$, $P_2(0,b,0)$, and $P_3(0,0,c)$, then an equation of the plane is

$$\begin{vmatrix} x-a & y & z \\ -a & b & 0 \\ -a & 0 & c \end{vmatrix} = 0,$$

and this simplifies to (provided, a, b, and c are all nonzero)

$$\frac{x}{a} + \frac{y}{b} + \frac{z}{c} = 1.$$

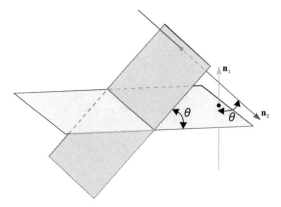

Figure 1.15: Angle between two planes.

We can define the angle between two planes using their normal vectors as shown in Figure 1.15.

Definition 1.3.1. The **angle between two planes** is defined as the acute angle between the normal vectors of the two planes. Two planes are considered to be perpendicular if their normal vectors are orthogonal.

Example 1.3.9. Find the angle between the planes $x - y - 2z = 1$ and $x + y - 3z = 1$.

Solution. The normal vectors of these two planes are
$$\vec{n}_1 = \langle 1, -1, -2 \rangle \quad \text{and} \quad \vec{n}_2 = \langle 1, 1, -3 \rangle,$$
respectively. Let θ be the angle between the two planes. Then
$$\cos\theta = \frac{\vec{n}_1 \cdot \vec{n}_2}{|\vec{n}_1||\vec{n}_2|} = \frac{1(1) + (-1)(1) + (-2)(-3)}{\sqrt{1^2 + (-1)^2 + (-2)^2}\sqrt{1^2 + 1^2 + (-3)^2}} \approx 0.73855.$$
So the acute angle between the given planes is $\cos^{-1}(0.73855) \approx 42°$.

Example 1.3.10. Find a formula for the perpendicular distance D from the point $P(x_0, y_0, z_0)$ to the plane $ax + by + cz + d = 0$.

Solution. Let $P_1(x_1, y_1, z_1)$ be any point in the given plane. Then
$$\overrightarrow{P_1P} = \langle x_0 - x_1, y_0 - y_1, z_0 - z_1 \rangle.$$
The vector $\vec{n} = \langle a, b, c \rangle$ is a normal vector of the plane. Then, as shown in Figure 1.16, the distance D from P to the plane is
$$D = |\,|\overrightarrow{P_1P}|\cos\theta\,|.$$

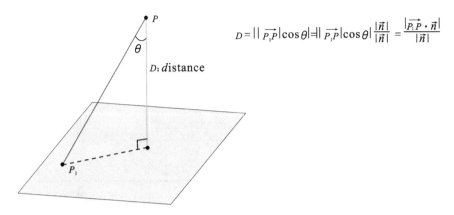

Figure 1.16: Distance from a point P to a plane $ax + by + cz + d = 0$.

Thus,

$$\begin{aligned}
D &= ||\overrightarrow{P_1P}|\cos\theta| = \left|\overrightarrow{P_1P}|\cdot\cos\theta\cdot\frac{|\vec{n}|}{|\vec{n}|}\right| \\
&= \frac{1}{|\vec{n}|}||\overrightarrow{P_1P}|\cdot\cos\theta\cdot|\vec{n}|| = \frac{|\overrightarrow{P_1P}\cdot\vec{n}|}{|\vec{n}|} \\
&= \frac{|a(x_0-x_1)+b(y_0-y_1)+c(z_0-z_1)|}{\sqrt{a^2+b^2+c^2}} \\
&= \frac{|ax_0+by_0+cz_0-(ax_1+by_1+cz_1)|}{\sqrt{a^2+b^2+c^2}} \\
&= \frac{|ax_0+by_0+cz_0+d|}{\sqrt{a^2+b^2+c^2}}, \quad (1.20)
\end{aligned}$$

since $-(ax_1 + by_1 + cz_1) = d$.

Example 1.3.11. Find the distance between the two parallel planes $x + 2y - 2z = 5$ and $2x + 4y - 4z = 3$.

Solution. The two planes are parallel to each other since their normal vectors $\langle 1, 2, -2\rangle$ and $\langle 2, 4, -4\rangle$ are parallel. In order to find the distance D between the two planes, we can, instead, find the distance from any point in one plane to the other plane. For example, we can put $y = z = 0$ in the equation of the first plane, to get $x = 5$, so $(5, 0, 0)$ is a point in the first plane. Using formula (1.20) from Example 1.3.10,

$$D = \left|\frac{2(5)+4(0)-4(0)-3}{\sqrt{2^2+4^2+(-4)^2}}\right| = \frac{7}{6}.$$

So the distance between the two planes is $7/6$.

The intersection of two planes that are not parallel is of course a line. So a line L can be described as the line of intersection of two planes in the form

$$L : \begin{cases} A_1 x + B_1 y + C_1 z = D_1, \\ A_2 x + B_2 y + C_2 z = D_2. \end{cases} \tag{1.21}$$

This is a *general equation* of the line L. The symmetric equations of a line are an example of this form. There will, of course, be infinitely many possible choices for the two planes that intersect in a given line L.

Example 1.3.12. Rewrite the line L determined by the equations below in the form of parametric equations and then in the form of symmetric equations:

$$\begin{cases} x + y - z = 1, \\ 2x + y + 3z = 4. \end{cases}$$

Solution. First of all, we find a point on the line by choosing $z = 0$ and solving the equations for x and y,

$$\begin{cases} x + y = 1, \\ 2x + y = 4, \end{cases}$$

obtaining $x = 3$ and $y = -2$. Therefore, the point $(3, -2, 0)$ lies on line L. Note that the direction vector $\mathbf{v}$ of line L is perpendicular to both normal vectors of the given planes, so it is given by the cross product

$$\mathbf{v} = \mathbf{n}_1 \times \mathbf{n}_2 = \begin{vmatrix} \vec{i} & \vec{j} & \vec{k} \\ 1 & 1 & -1 \\ 2 & 1 & 3 \end{vmatrix} = 4\vec{i} - 5\vec{j} - \vec{k}.$$

Hence, parametric equations of line L are $x = 3 + 4t$, $y = -2 - 5t$, $z = -t$. Symmetric equations of line L are

$$\frac{x-3}{4} = \frac{y+2}{-5} = \frac{z}{-1}.$$

Vector equations of planes

Suppose two nonparallel vectors $\mathbf{a}$ and $\mathbf{b}$ lie in the plane with their tails at $M_0(x_0, y_0, z_0)$. Then, any vector with its head a point in the plane and tail at M_0 can be given by a linear combination of the two vectors $\mathbf{a}, \mathbf{b}$. Assume M_0 is the head of the position vector $\mathbf{r}_0$. Then, for any position vector $\mathbf{r}$ with head at a point in the plane (see Figure 1.17), there must be two scalars λ and u such that

$$\mathbf{r} - \mathbf{r}_0 = \lambda \mathbf{a} + u \mathbf{b}.$$

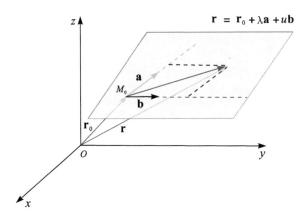

Figure 1.17: Vector equations of a plane.

Therefore, the vector equation

$$\mathbf{r} = \mathbf{r}_0 + \lambda \mathbf{a} + u \mathbf{b} \tag{1.22}$$

describes a plane in space. Suppose $\mathbf{r} = \langle x, y, z \rangle$, $\mathbf{a} = \langle a_1, a_2, a_3 \rangle$, and $\mathbf{b} = \langle b_1, b_2, b_3 \rangle$. Then

$$\begin{pmatrix} x \\ y \\ z \end{pmatrix} = \begin{pmatrix} x_0 \\ y_0 \\ z_0 \end{pmatrix} + \lambda \begin{pmatrix} a_1 \\ a_2 \\ a_3 \end{pmatrix} + u \begin{pmatrix} b_1 \\ b_2 \\ b_3 \end{pmatrix} \tag{1.23}$$

is a vector equation of the plane. This can be rewritten as

$$\begin{cases} x = x_0 + \lambda a_1 + u b_1, \\ y = y_0 + \lambda a_2 + u b_2, \\ z = z_0 + \lambda a_3 + u b_3. \end{cases} \tag{1.24}$$

These are *parametric equations* of the plane. Note that this can be written in the form

$$\mathbf{r}(\lambda, u) = \langle x(\lambda, u), y(\lambda, u), z(\lambda, u) \rangle, \tag{1.25}$$

which is a vector-valued function with two parameters.

Example 1.3.13. Rewrite the equation of the plane $2x - y - 3z = 10$ in a vector form $\vec{r} = \vec{r}_0 + \lambda \vec{a} + \mu \vec{b}$.

Solution. We must find a position vector $\vec{r}_0$ whose terminal point is a point in the plane and two nonparallel vectors $\vec{a}$ and $\vec{b}$ which are both parallel to the plane. To find three such vectors, we find three points in the plane. It is easy to check that $(5, 0, 0)$, $(0, -10, 0)$, and $(2, 0, -2)$ are three points in the plane and, therefore,

$$\vec{a} = \begin{pmatrix} 0 \\ -10 \\ 0 \end{pmatrix} - \begin{pmatrix} 5 \\ 0 \\ 0 \end{pmatrix} = \begin{pmatrix} -5 \\ -10 \\ 0 \end{pmatrix} \quad \text{and} \quad \vec{b} = \begin{pmatrix} 2 \\ 0 \\ -2 \end{pmatrix} - \begin{pmatrix} 5 \\ 0 \\ 0 \end{pmatrix} = \begin{pmatrix} -3 \\ 0 \\ -2 \end{pmatrix}$$

are two vectors parallel the plane, and $\vec{a} \ne \vec{b}$, thus, a vector equation of the plane is given by

$$\vec{r} = \begin{pmatrix} 5 \\ 0 \\ 0 \end{pmatrix} + \lambda \begin{pmatrix} -5 \\ -10 \\ 0 \end{pmatrix} + \mu \begin{pmatrix} -3 \\ 0 \\ -2 \end{pmatrix}.$$

Also, we can solve the equation $2x - y - 3z = 10$ to find a general solution. For example, we let $y = \lambda$ and $z = u$ be two free variables. Then a general solution to the equation

$$\begin{pmatrix} x \\ y \\ z \end{pmatrix} = \begin{pmatrix} \frac{10+\lambda+3u}{2} \\ \lambda \\ u \end{pmatrix} = \begin{pmatrix} 5 \\ 0 \\ 0 \end{pmatrix} + \lambda \begin{pmatrix} \frac{1}{2} \\ 1 \\ 0 \end{pmatrix} + u \begin{pmatrix} \frac{3}{2} \\ 0 \\ 1 \end{pmatrix}$$

is in the desired vector form.

1.4 Curves and vector-valued functions

A line is a special curve in space. As seen from the previous section, a line in space has parametric equations

$$\begin{cases} x = x_0 + mt, \\ y = y_0 + nt, \\ z = z_0 + pt, \end{cases}$$

where (x_0, y_0, z_0) is a point on the line and $\langle m, n, p \rangle$ is the direction of the line. We can rewrite this in a vector form

$$\mathbf{r} = \mathbf{r}_0 + \mathbf{v}t$$

with $\mathbf{r} = \langle x, y, z \rangle$, $\mathbf{r}_0 = \langle x_0, y_0, z_0 \rangle$, and $\mathbf{v} = \langle m, n, p \rangle$ is the direction vector. This can be written as

$$\mathbf{r}(t) = \langle x_0 + mt, y_0 + nt, z_0 + pt \rangle,$$

which is a vector-valued function of t with each component being a linear function of t. In general, the graph of a vector-valued function $\mathbf{r}(t) = \langle x(t), y(t), z(t) \rangle$ is a curve in space. Its parametric form is $x = x(t)$, $y = y(t)$, and $z = z(t)$. You can imagine that this curve is the trajectory of a moving object: at each specific time t, its position vector is $\mathbf{r}(t)$.

Example 1.4.1 (A helix). The graph of the vector-valued function $r(t) = 2\cos t\mathbf{i} + 2\sin t\mathbf{j} + 0.5t\mathbf{k}$, $t \ge 0$, is called a helix. The curve is shown in Figure 1.18.

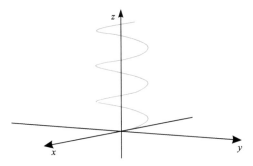

Figure 1.18: A picture of a helix.

Example 1.4.2 (Slinky curve). A slinky curve is defined as $r(t) = \langle a(t)\cos t, a(t)\sin t, 1.2\sin 20t\rangle$. The graph of the curve when $a(t) = 5 + \cos 20t$ and $0 \le t \le 2\pi$ is shown in Figure 1.19.

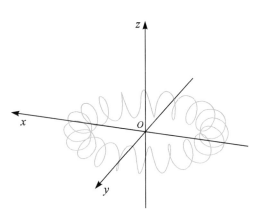

Figure 1.19: A picture of a slinky curve.

Sometimes it is helpful to visualize a curve in space by projecting the curve onto one of the coordinate planes. If a curve has the vector equation $\mathbf{r}(t) = \langle x(t), y(t), z(t)\rangle$, then its view from above is its projection onto the xy-plane, and when it is projected, its x- and y-coordinates remain unchanged, but the z-coordinate becomes 0. Thus, the projection of the curve onto the xy-plane has the equation

$$\mathbf{r}(t) = \langle x(t), y(t), 0\rangle.$$

In Example 1.4.1, the projection of the helix onto the xy-plane has the equation

$$\mathbf{r}(t) = \langle 2\cos t, 2\sin t, 0\rangle$$

or $x^2 + y^2 = 4$ and $z = 0$. It is a circle in the xy-plane.

Similarly, to obtain an equation for the projection of the curve $\mathbf{r}(t) = \langle x(t), y(t), z(t)\rangle$ onto the xz-plane, we set the y-coordinate to be 0. To obtain an equation for the projection curve of the curve $\mathbf{r}(t) = \langle x(t), y(t), z(t)\rangle$ onto the yz-plane, we set the x-coordinate to be 0. For instance, the projection of the curve $\mathbf{r}(t) = \langle 2\cos t, 2\sin t, 0.5t\rangle$ onto the yz-plane has an x-coordinate equal to 0, giving

$$\mathbf{r}(t) = \langle 0, 2\sin t, 0.5t\rangle$$

or $x = 0$ and $y = 2\sin(2z)$. It is the graph of $y = 2\sin(2z)$ in the plane $x = 0$.

1.5 Calculus of vector-valued functions

1.5.1 Limits, derivatives, and tangent vectors

We can also define the limit of a vector-valued function $\mathbf{r}(t) = \langle x(t), y(t), z(t)\rangle$ at a point t_0. Similar to a scalar function, if $t \to t_0$ implies $\mathbf{r}(t) \to \mathbf{L}$, then we say $\lim_{t \to t_0} \mathbf{r}(t) = \mathbf{L}$, where $\mathbf{L} = \langle a, b, c\rangle$ is a constant vector. More precisely, it is defined as follows.

Definition 1.5.1. Let $\mathbf{L} = \langle a, b, c\rangle$ be a constant vector and $\mathbf{r}(t) = \langle x(t), y(t), z(t)\rangle$ be a vector-valued function. Then $\lim_{t \to t_0} \mathbf{r}(t) = \mathbf{L}$ if and only if for any given $\varepsilon > 0$, there is a number $\delta > 0$ such that

$$|\mathbf{r}(t) - \mathbf{L}| < \varepsilon \text{ whenever } 0 < |t - t_0| < \delta.$$

Since $\mathbf{r}(t) = x(t)\mathbf{i} + y(t)\mathbf{j} + z(t)\mathbf{k}$ and

$$|\mathbf{r}(t) - \mathbf{L}| = \sqrt{(x(t) - a)^2 + (y(t) - b)^2 + (z(t) - c)^2} < \varepsilon,$$

using the above definition and applying the limit laws for scalar functions, we have the following theorem.

Theorem 1.5.1. *Let $\mathbf{L} = \langle a, b, c\rangle$ be a constant vector and let $\mathbf{r}(t) = \langle x(t), y(t), z(t)\rangle$ be a vector-valued function. Then*

$$\lim_{t \to t_0} \mathbf{r}(t) = \mathbf{L} \iff \lim_{t \to t_0} x(t) = a, \quad \lim_{t \to t_0} y(t) = b, \quad \text{and} \quad \lim_{t \to t_0} z(t) = c.$$

Therefore, $\lim_{t \to t_0} \mathbf{r}(t) = \langle \lim_{t \to t_0} x(t), \lim_{t \to t_0} y(t), \lim_{t \to t_0} z(t)\rangle$. That is, to evaluate the limit of a vector-valued function, we evaluate the limit of each component of the function, given that all limits exist.

Example 1.5.1. Given the vector-valued function $\mathbf{r}(t) = \langle \frac{1-\cos t}{t^2}, e^{-t}, \tan^{-1} t\rangle$, evaluate the limits (a) $\lim_{t \to 0} \mathbf{r}(t)$ and (b) $\lim_{t \to \infty} \mathbf{r}(t)$.

Solution.
(a) We have $\lim_{t \to 0} \mathbf{r}(t) = \langle \lim_{t \to 0} \frac{1-\cos t}{t^2}, \lim_{t \to 0} e^{-t}, \lim_{t \to 0} \tan^{-1} t \rangle = \langle \lim_{t \to 0} \frac{\sin t}{2t}, 1, 0 \rangle = \langle \frac{1}{2}, 1, 0 \rangle$.

(b) We have $\lim_{t \to \infty} \mathbf{r}(t) = \langle \lim_{t \to \infty} \frac{1-\cos t}{t^2}, \lim_{t \to \infty} e^{-t}, \lim_{t \to \infty} \tan^{-1} t \rangle = \langle 0, 0, \frac{\pi}{2} \rangle$.

Intuitively, we know that if each component of $\mathbf{r}(t)$ is continuous, then the curve $\mathbf{r}(t)$ must be continuous, which means that you can draw the curve continuously, without lifting your pencil. The formal definition of continuity is given below.

Definition 1.5.2. A vector-valued function $\mathbf{r}(t)$ is continuous at t_0 if and only if $\lim_{t \to t_0} \mathbf{r}(t) = \mathbf{r}(t_0)$.

This means that each component of $\mathbf{r}(t)$ must be continuous at $t = t_0$.

We now consider the trajectory of a moving object, where at any instant its position vector is given by $\mathbf{r}(t)$. Its displacement over the time Δt is $\mathbf{r}(t) - \mathbf{r}(t_0) = \mathbf{r}(t_0 + \Delta t) - \mathbf{r}(t_0)$. Therefore,

$$\frac{\Delta \mathbf{r}}{\Delta t} = \frac{\mathbf{r}(t_0 + \Delta t) - \mathbf{r}(t_0)}{\Delta t}$$

is the average velocity of the object during this time interval. The limit as $\Delta t \to 0$, if it exists, is the instantaneous velocity at that t_0. This is the definition of a derivative. Thus,

$$\mathbf{r}'(t_0) = \lim_{\Delta t \to 0} \frac{\Delta \mathbf{r}}{\Delta t} = \lim_{\Delta t \to 0} \frac{\mathbf{r}(t_0 + \Delta t) - \mathbf{r}(t_0)}{\Delta t}.$$

If $x(t)$, $y(t)$, and $z(t)$ are differentiable one-variable functions, then

$$\mathbf{r}'(t_0) = \lim_{\Delta t \to 0} \frac{\mathbf{r}(t_0 + \Delta t) - \mathbf{r}(t_0)}{\Delta t}$$

$$= \lim_{\Delta t \to 0} \frac{(x(t_0 + \Delta t) - x(t_0))\mathbf{i} + (y(t_0 + \Delta t) - y(t_0))\mathbf{j} + (z(t_0 + \Delta t) - z(t_0))\mathbf{k}}{\Delta t}$$

$$= \lim_{\Delta t \to 0} \frac{(x(t_0 + \Delta t) - x(t_0))\mathbf{i}}{\Delta t} + \lim_{\Delta t \to 0} \frac{(y(t_0 + \Delta t) - x(t_0))\mathbf{j}}{\Delta t} + \lim_{\Delta t \to 0} \frac{(y(t_0 + \Delta t) - y(t_0))\mathbf{k}}{\Delta t}$$

$$= x'(t_0)\mathbf{i} + y'(t_0)\mathbf{j} + z'(t_0)\mathbf{k}.$$

Or this can be written as $\mathbf{r}'(t_0) = \langle x'(t_0), y'(t_0), z'(t_0) \rangle$. If this vector is not $\mathbf{0}$, then it is a vector tangent (tangent vector) to the curve $\mathbf{r}(t)$ at t_0. Figure 1.20(a) illustrates this idea. We can extend this idea to define the derivative as a function of t as follows.

Definition 1.5.3. If $x(t)$, $y(t)$, and $z(t)$ are three differentiable functions on the interval (a, b), then the derivative of the vector-valued function $\mathbf{r}(t) = \langle x(t), y(t), z(t) \rangle$ is

$$\mathbf{r}'(t) = \langle x'(t), y'(t), z'(t) \rangle.$$

If this vector is not $\mathbf{0}$, then it is a vector tangent to the curve $\mathbf{r}(t)$.

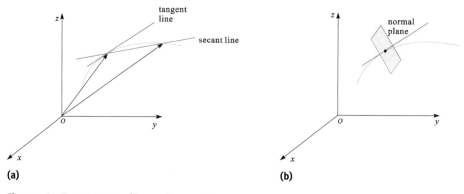

Figure 1.20: Tangent vector/line and normal plane.

In light of the above definition, we are now able to derive an equation for the tangent line to the curve $\mathbf{r}(t)$ at any point $t = t_0$. Since the curve at point $(x(t_0), y(t_0), z(t_0))$ has tangent vector $\mathbf{r}'(t_0) = \langle x'(t_0), y'(t_0), z'(t_0) \rangle$, the symmetric equations of the tangent line, provided $\mathbf{r}'(t_0) \neq \mathbf{0}$, are

$$\frac{x - x(t_0)}{x'(t_0)} = \frac{y - y(t_0)}{y'(t_0)} = \frac{z - z(t_0)}{z'(t_0)}. \tag{1.26}$$

Parametric equations of the tangent line at $t = t_0$ are

$$\begin{cases} x = x(t_0) + x'(t_0)t, \\ y = y(t_0) + y'(t_0)t, \\ z = z(t_0) + z'(t_0)t, \end{cases} \tag{1.27}$$

and a vector equation of the tangent line at $t = t_0$ is

$$\mathbf{r}(t) = \mathbf{r}(t_0) + \mathbf{r}'(t_0)t, \tag{1.28}$$

where $\mathbf{r}'(t_0)$ is a tangent vector. The *unit tangent vector* at $t = t_0$ is $\mathbf{T} = \frac{\mathbf{r}'(t_0)}{|\mathbf{r}'(t_0)|}$.

Note that the plane passing through the curve at $t = t_0$ with a normal vector parallel to the tangent vector to the curve at $t = t_0$ is the *normal plane* to the curve at $t = t_0$, as shown in Figure 1.20(b). The normal plane to the curve at $t = t_0$ has the equation

$$x'(t_0)(x - x_0) + y'(t_0)(y - y_0) + z'(t_0)(z - z_0) = 0. \tag{1.29}$$

Example 1.5.2. Find an equation for the tangent line and normal plane to the curve $\mathbf{r}(t) = \langle \sin t, \cos t, \sin 2t \rangle$ at $t = \pi/6$.

Solution. The point is $(\sin \pi/6, \cos \pi/6, \sin(2 \times \pi/6)) = (1/2, \sqrt{3}/2, \sqrt{3}/2)$, and since $\mathbf{r}'(t) = \langle \cos t, -\sin t, 2\cos 2t \rangle$,

$$\mathbf{r}'(\pi/6) = \langle \cos(\pi/6), -\sin(\pi/6), 2\cos(2 \cdot \pi/6) \rangle = \langle \sqrt{3}/2, -1/2, 1 \rangle.$$

So, the parametric equations for the desired tangent line are

$$\begin{cases} x = \frac{1}{2} + \frac{\sqrt{3}}{2}t, \\ y = \frac{\sqrt{3}}{2} - \frac{1}{2}t, \\ z = \frac{\sqrt{3}}{2} + t. \end{cases}$$

An equation for the normal plane at $t = \pi/6$ is

$$\frac{\sqrt{3}}{2}\left(x - \frac{1}{2}\right) - \frac{1}{2}\left(y - \frac{\sqrt{3}}{2}\right) + \left(z - \frac{\sqrt{3}}{2}\right) = 0.$$

Figure 1.21 shows the tangent line and normal plane at $t = \pi/6$.

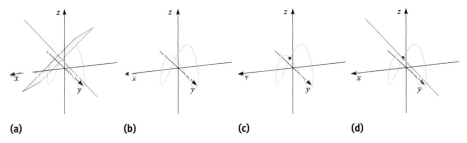

(a)   (b)   (c)   (d)

Figure 1.21: Tangent line and normal plane, Example 1.5.2.

By using the above definition of the derivative for a vector-valued function, we can deduce the following theorem, the proof of which is omitted here.

Theorem 1.5.2. *Let $u(t)$ and $v(t)$ be two differentiable vector-valued functions and $f(t)$ be a differentiable scalar-valued function over $a < t < b$. Let c be a constant vector. Then at any t in (a, b), we have:*

1. $\frac{d}{dt}(c) = 0$,
2. $\frac{d}{dt}(u(t) \pm v(t)) = \frac{d}{dt}u(t) \pm \frac{d}{dt}v(t)$ *(sum or difference rule)*,
3. $\frac{d}{dt}(f(t)u(t)) = (\frac{d}{dt}f(t))u(t) + f(t)\frac{d}{dt}u(t)$ *(constant multiple rule)*,
4. $\frac{d}{dt}u(f(t)) = u'(f(t))f'(t)$ *(chain rule)*,
5. $\frac{d}{dt}(u(t) \cdot v(t)) = u'(t) \cdot v(t) + u(t) \cdot v'(t)$ *(dot product rule)*,
6. $\frac{d}{dt}(u(t) \times v(t)) = u'(t) \times v(t) + u(t) \times v'(t)$ *(cross product rule)*.

1.5.2 Antiderivatives and definite integrals

Similar to scalar functions, if $R'(t) = r(t)$, then we say $R(t)$ is an antiderivative of $r(t)$, and we write the *indefinite integral* of $r(t)$ as

$$\int r(t)dt = R(t) + C,$$

where **C** is an arbitrary constant vector. For definite integrals, we write $\int_a^b \mathbf{r}(t)dt = \mathbf{R}(b) - \mathbf{R}(a)$. In light of the previous definition for derivative, we have the following formal definition using the components of $\mathbf{r}(t)$.

Definition 1.5.4. If $\mathbf{r}(t) = \langle x(t), y(t), z(t) \rangle$ is continuous for $a \leq t \leq b$, then we define

$$\int \mathbf{r}(t)dt = \left\langle \int x(t)dt, \int y(t)dt, \int z(t)dt \right\rangle \text{ and}$$

$$\int_a^b \mathbf{r}(t)dt = \left\langle \int_a^b x(t)dt, \int_a^b y(t)dt, \int_a^b z(t)dt \right\rangle.$$

Example 1.5.3. If $\mathbf{r}'(t) = e^{2t}\mathbf{i} + \sec^2 t\mathbf{j} + \sin t\mathbf{k}$,
1. find $\mathbf{r}(t)$.
2. furthermore, if $\mathbf{r}(0) = \langle 1, 1, 2 \rangle$, determine $\mathbf{r}(t)$.
3. find $\int_0^{\pi/4} \mathbf{r}(t)dt$.

Solution.
1. Since $\mathbf{r}'(t) = \langle e^{2t}, \sec^2 t, \sin t \rangle$,

$$\mathbf{r}(t) = \int \mathbf{r}'(t)dt = \left\langle \int e^{2t}dt, \int \sec^2 t\, dt, \int \sin t\, dt \right\rangle$$

$$= \left\langle \frac{1}{2}e^{2t} + c_1, \tan t + c_2, -\cos t + c_3 \right\rangle$$

$$= \left\langle \frac{1}{2}e^{2t}, \tan t, -\cos t \right\rangle + \langle c_1, c_2, c_3 \rangle.$$

2. Since $\mathbf{r}(0) = \langle 1, 1, 2 \rangle$, we have

$$\langle 1, 1, 2 \rangle = \left\langle \frac{1}{2}e^0, \tan 0, -\cos 0 \right\rangle + \langle c_1, c_2, c_3 \rangle$$

so $\langle c_1, c_2, c_2 \rangle = \left\langle \frac{1}{2}, 1, 3 \right\rangle$.

Then, $\mathbf{r}(t) = \langle \frac{1}{2}e^{2t} + \frac{1}{2}, \tan t + 1, -\cos t + 3 \rangle$.

3. By definition

$$\int_0^{\pi/4} \mathbf{r}(t)dt = \left\langle \int_0^{\pi/4}\left(\frac{1}{2}e^{2t} + \frac{1}{2}\right)dt, \int_0^{\pi/4}(\tan t + 1)dt, \int_0^{\pi/4}(-\cos t + 3)dt \right\rangle$$

$$= \left\langle \frac{1}{4}e^{2t} + \frac{t}{2}\Big|_0^{\pi/4}, -\ln|\cos t|\Big|_0^{\pi/4} + \pi/4, -\sin t + 3t\Big|_0^{\pi/4} \right\rangle$$

$$= \left\langle \left(\frac{e^{\pi/2}}{4} + \frac{\pi}{8} - \frac{1}{4}\right), \frac{\ln 2}{2} + \frac{\pi}{4}, -\frac{\sqrt{2}}{2} + \frac{3\pi}{4} \right\rangle.$$

1.5.3 Length of curves, curvatures, TNB frame

As seen before, s, the arc length, or length of a plane curve $\langle x(t), y(t) \rangle$ for $a \leq t \leq b$, is

$$s = \int_a^b \sqrt{[x'(t)]^2 + [y'(t)]^2}\, dt.$$

The analog for a curve in space is the length of a curve $\mathbf{r}(t) = \langle x(t), y(t), z(t) \rangle$ for $a \leq t \leq b$, which is

$$s = \int_a^b \sqrt{[x'(t)]^2 + [y'(t)]^2 + [z'(t)]^2}\, dt,$$

provided that the integrand is integrable. The integrand is always integrable when the curve is smooth, that is, $x'(t)$, $y'(t)$, and $z'(t)$ are continuous on $[a, b]$.

Again, thinking of a moving object along the curve, the length of the curve is indeed the distance traveled by the object over time interval $[u, b]$. Since the derivative of position with respect to time is the velocity, $\mathbf{v}(t)$, and the derivative of distance traveled with respect the time t is the speed, we have

$$\mathbf{v}(t) = \mathbf{r}'(t) \quad \text{and} \quad \frac{ds}{dt} = |\mathbf{v}(t)|.$$

It is not a surprise that the length of the curve is $s = \int_a^b |\mathbf{v}(t)|\, dt$.
We conclude this in the following definition.

Definition 1.5.5. If $\mathbf{r}'(t)$ is continuous, the curve $\mathbf{r}(t)$ is a smooth curve, and the length of this curve for $a \leq t \leq b$ is defined as

$$\int_a^b |\mathbf{r}'(t)|\, dt = \int_a^b \sqrt{[x'(t)]^2 + [y'(t)]^2 + [z'(t)]^2}\, dt.$$

Example 1.5.4. Find the length of the curve $\mathbf{r}(t) = \langle 3\cos t, 4\cos t, 5\sin t \rangle$ for $0 \leq t \leq 2\pi$.

Solution. The length of the curve s is given by the integral

$$s = \int_0^{2\pi} \sqrt{((3\cos t)')^2 + ((4\cos t)')^2 + ((5\sin t)')^2}\, dt$$

$$= \int_0^{2\pi} \sqrt{9\sin^2 t + 16\sin^2 t + 25\cos^2 t}\, dt = \int_0^{2\pi} 5\, dt = 10\pi.$$

Parameterization by arc length

Now we consider a vector-valued function $\mathbf{r}(t) = \langle x(t), y(t) \rangle$ with the following parametric equations representations:

(a) $\begin{cases} x = R\cos t, \\ y = R\sin t, \end{cases}$ $0 \le t \le 2\pi$, (b) $\begin{cases} x = R\cos \frac{u}{2}, \\ y = R\sin \frac{u}{2}, \end{cases}$ $0 \le u \le 4\pi$,

(c) $\begin{cases} x = R\cos 2t, \\ y = R\sin 2t, \end{cases}$ $0 \le t \le \pi$, (d) $\begin{cases} x = R\sin 3\theta, \\ y = R\cos 3\theta, \end{cases}$ $0 \le \theta \le \frac{2\pi}{3}$.

They actually describe the same curve. In this case, it is a circle centered at $(0,0)$ with radius R. The name of a parameter, of course, does not matter. However, how the curve evolves as the parameter increases does make a difference. For example, in (a), the circle is formed counterclockwise while in (d) it is formed clockwise. The *positive orientation* of a curve is the direction in which the curve is generated as the parameter increases. So the positive orientation of (a) is counterclockwise, while the positive orientation of (d) is clockwise.

A curve may be parameterized in many ways, as shown above. In some ways, the parameter may have a nice geometric interpretation. For example, in (a), at each point on the circle, the corresponding value of the parameter t is exactly the angle (measured in radians) formed by the corresponding radius and the positive x-axis. We now introduce a very natural way for describing a curve where its parameter represents the arc length. We first investigate the following curve:

$$\begin{cases} x = 2\cos \frac{t}{2}, \\ y = 2\sin \frac{t}{2}, \end{cases} \text{ for } 0 \le t \le 4\pi.$$

The initial point is $(2, 0)$. When $t = \pi$, the corresponding arc length is also π. When $t = 2\pi$, the corresponding arc length is also 2π. In general, the length of the interval $[0, t]$ is equal to the length of the curve generated. We say that the curve $\mathbf{r}(t) = \langle 2\cos \frac{t}{2}, 2\sin \frac{t}{2} \rangle$ is parameterized by arc length. In this case we also write

$$\mathbf{r}(s) = \left\langle 2\cos \frac{s}{2}, 2\sin \frac{s}{2} \right\rangle$$

with s being the arc length parameter.

But how do we know whether a curve $\mathbf{r}(t)$ is parameterized by arc length? First, we note that by definition

$$s = \int_a^t |\mathbf{r}'(t)|\,dt,$$

and so $\frac{ds}{dt} = |\mathbf{r}'(t)|$. This means $ds = |\mathbf{r}'(t)|\,dt$. Therefore, the change in t is equal to the change in s if and only of if $|\mathbf{r}'(t)| = 1$. In particular, if the curve starts at $\mathbf{r}(a)$ and $|\mathbf{r}'(t)| = 1$ for all t, then when $t = a$, we have $s = 0$, and when $t \ne a$, we have $s = t - a$.

Example 1.5.5. Determine whether the curves

(a) $r(t) = \langle \sin t, 1, \cos t \rangle$ for $t \geq 1$ and (b) $r(t) = \langle t, t+1, 6t \rangle$ for $0 \leq t \leq 12$

use arc length as a parameter. If not, find a description that uses arc length as a parameter.

Solution.
1. For (a), $\mathbf{r}'(t) = \langle \cos t, 0, -\sin t \rangle$, so $|\mathbf{r}'(t)| = \sqrt{(\cos t)^2 + 0^2 + (-\sin t)^2} = 1$. Yes, it uses arc length as a parameter.
2. For (b), $\mathbf{r}'(t) = \langle 1, 1, 6 \rangle$, so $|\mathbf{r}'(t)| = \sqrt{1^2 + 1^2 + 6^2} = \sqrt{38} \neq 1$. No, it does not use arc length as a parameter. Since

$$s = \int_0^t |\mathbf{r}'(t)| dt = \int_0^t \sqrt{38} dt = \sqrt{38} t,$$

if we replace t by $\frac{s}{\sqrt{38}}$, the parameterized curve

$$\mathbf{r}_1(s) = \left\langle \frac{s}{\sqrt{38}}, \frac{s}{\sqrt{38}} + 1, \frac{6s}{\sqrt{38}} \right\rangle$$

uses arc length as a parameter.

Curvature, normal vector, and the TNB frame

If you observe the two curves in Figure 1.22(a,b) which are both bending downward, you will notice a difference. One curve bends more shapely than the other. To measure the sharpness that a curve bends, we need the concept of curvature. Also observing the unit tangent vectors **T** of the curve which bends more shapely, you will see that the change (in direction) of the unit tangent vector with respect to arc length is quicker than the one that bends less sharply. In other words, over a certain length of curve, the unit tangent vector changes more in direction if the curve bends more sharply. Therefore, we have the following definition.

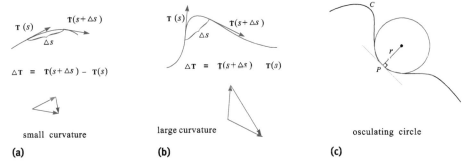

(a) small curvature (b) large curvature (c) osculating circle

Figure 1.22: Curvatures, osculating circles.

> **Definition 1.5.6** (Curvature). If r(t) is a smooth curve and T is its unit tangent vector, then the curvature κ of a smooth curve r(t) is defined as
> $$\kappa = \left|\frac{d\mathbf{T}}{ds}\right|.$$

Because $\frac{ds}{dt} = |\mathbf{r}'(t)|$, by using the chain rule, we have

$$\kappa = \left|\frac{d\mathbf{T}}{ds}\right| = \left|\frac{d\mathbf{T}}{dt}\right|\frac{1}{\left|\frac{ds}{dt}\right|} = \frac{1}{|\mathbf{r}'(t)|}\left|\frac{d\mathbf{T}}{dt}\right|.$$

Intuitively speaking, since curvature measures the degree that a curve bends, a straight line must have 0 curvature. At all points on a circle, the curvature would be the same constant, and a smaller circle should have larger curvature. Let us use Definition 1.5.6 to verify this understanding.

> **Example 1.5.6.** Find the curvature for the straight line $\mathbf{r}(t) = \langle x_0 + mt, y_0 + nt, z_0 + pt\rangle$.

Solution. Since $\frac{d\mathbf{r}}{dt} = \langle m, n, p\rangle$ is constant, $\mathbf{T} = \frac{1}{\sqrt{m^2+n^2+p^2}}\langle m, n, p\rangle$. Then,

$$\frac{d\mathbf{T}}{dt} = \frac{d}{dt}\left(\frac{1}{\sqrt{m^2+n^2+p^2}}\langle m, n, p\rangle\right) = \mathbf{0}.$$

Therefore, the curvature at any point on the line is $\kappa = \frac{1}{|\mathbf{r}'(t)|}\left|\frac{d\mathbf{T}}{dt}\right| = 0$. This agrees with our intuition.

Example 1.5.7. Find the curvature for the circle $\mathbf{r}(t) = \langle R\cos t, R\sin t\rangle$.

Solution. Since $\frac{d\mathbf{r}}{dt} = \langle -R\sin t, R\cos t\rangle$,

$$\mathbf{T} = \frac{1}{\sqrt{(-R\sin t)^2 + (R\cos t)^2}}\langle -R\sin t, R\cos t\rangle = \langle -\sin t, \cos t\rangle.$$

Then the curvature is

$$\kappa = \frac{1}{|\mathbf{r}'(t)|}\left|\frac{d\mathbf{T}}{dt}\right| = \frac{1}{\sqrt{(-R\sin t)^2 + (R\cos t)^2}}\left|\frac{d\mathbf{T}}{dt}\right|$$
$$= \frac{1}{R}\left|\frac{d}{dt}\langle -\sin t, \cos t\rangle\right| = \frac{1}{R}|\langle -\cos t, -\sin t\rangle| = \frac{1}{R}\sqrt{(-\cos t)^2 + (-\sin t)^2} = \frac{1}{R}.$$

The curvature is the same at each point on a circle, and a larger circle has a smaller curvature.

In general, calculating the curvature using the definition involves many steps. However, sometimes it is easier to calculate the curvature of a curve by using the following theorem, which can be derived using Theorem 1.5.2.

Theorem 1.5.3. *Let r(t) be a twice differentiable smooth curve. The curvature of r(t) is then*

$$\kappa = \frac{|\mathbf{r}'(t) \times \mathbf{r}''(t)|}{|\mathbf{r}'(t)|^3}.$$

This theorem allows us to evaluate the curvature of a parameterized curve by evaluating its first- and second-order derivatives at a point.

Example 1.5.8. Find the curvature of the curve $y = x^2$ at the point with greatest curvature.

Solution. Let $x = t$, $y = t^2$, and $z = 0$. Then the curve is $\mathbf{r}(t) = \langle t, t^2, 0 \rangle$. At each t,

$$\mathbf{r}'(t) = \langle 1, 2t, 0 \rangle, \quad \text{and} \quad \mathbf{r}''(t) = \langle 0, 2, 0 \rangle$$
$$\mathbf{r}'(t) \times \mathbf{r}''(t) = \langle 1, 2t, 0 \rangle \times \langle 0, 2, 0 \rangle = \langle 0, 0, 2 \rangle.$$

So the curvature is

$$\kappa = \frac{|\langle 0, 0, 2 \rangle|}{|\langle 1, 2t, 0 \rangle|^3} = \frac{2}{\sqrt{1 + 4t^2}^3}.$$

So at $t = 0$, which is the origin, the parabola $y = x^2$ has the greatest curvature, which is 2.

Note. If a curve $\mathbf{r}(t)$ is parameterized by arc length, then $|\mathbf{r}'(t)| = 1$, $ds = dt$, and so $\frac{d\mathbf{r}}{ds} = \mathbf{T}$, and we have

$$\kappa = \frac{1}{|\mathbf{r}'(s)|}\left|\frac{d\mathbf{T}}{ds}\right| = \left|\frac{d\mathbf{T}}{ds}\right| = \left|\frac{d^2\mathbf{r}}{ds^2}\right| \quad \text{or} \quad \kappa = |\mathbf{r}''(s)|.$$

The curvature tells how fast a curve turns. But in which direction does it turn? The *principal unit normal vector* determines this.

Definition 1.5.7. *Let r be a smooth curve. If $\frac{d\mathbf{T}}{dt}$ is not 0, then the principal unit normal vector N at a point on the curve is defined to be*

$$\mathbf{N} = \frac{\frac{d\mathbf{T}}{dt}}{|\frac{d\mathbf{T}}{dt}|}.$$

If the curve is parameterized by arc length, then we have

$$\mathbf{N} = \frac{\frac{d\mathbf{T}}{ds}}{|\frac{d\mathbf{T}}{ds}|} = \frac{1}{\kappa}\frac{d\mathbf{T}}{ds} \quad \text{or} \quad \frac{d\mathbf{T}}{ds} = \kappa\mathbf{N}.$$

Because $1 = \mathbf{T} \cdot \mathbf{T}$, differentiation with respect to s shows that

$$0 = \frac{d\mathbf{T}}{ds} \cdot \mathbf{T} + \mathbf{T} \cdot \frac{d\mathbf{T}}{ds} = 2\kappa \mathbf{N} \cdot \mathbf{T},$$

meaning that $\mathbf{T}$ and $\mathbf{N}$ are orthogonal at all points of the curve. Also, $\mathbf{N}$ points to the inside of the curve in the direction where the curve is turning.

There is one more aspect of the curve that we need to consider: the curve might twist out of the plane determined by $\mathbf{T}$ and $\mathbf{N}$. We define the *unit binormal vector* $\mathbf{B}$ to be $\mathbf{T} \times \mathbf{N}$. Then

$$\frac{d\mathbf{B}}{ds} = \frac{d}{ds}(\mathbf{T} \times \mathbf{N}) = \frac{d\mathbf{T}}{ds} \times \mathbf{N} + \mathbf{T} \times \frac{d\mathbf{N}}{ds} = \mathbf{T} \times \frac{d\mathbf{N}}{ds}.$$

Note that $\frac{d\mathbf{T}}{ds} \times \mathbf{N} = \mathbf{0}$, since $\frac{d\mathbf{T}}{ds}$ and $\mathbf{N}$ are parallel to each other. So, we know that $\frac{d\mathbf{B}}{ds}$ is orthogonal to $\mathbf{T}$. On the other hand, since $1 = \mathbf{B} \cdot \mathbf{B}$, we have

$$0 = \frac{d\mathbf{B} \cdot \mathbf{B}}{ds} = \frac{d\mathbf{B}}{ds} \cdot \mathbf{B} + \mathbf{B} \cdot \frac{d\mathbf{B}}{ds}.$$

This means that $\frac{d\mathbf{B}}{ds}$ is also perpendicular to $\mathbf{B}$. Therefore, $\frac{d\mathbf{B}}{ds}$ must be parallel to $\mathbf{B} \times \mathbf{T} = \mathbf{N}$. Then there is a scalar τ such that

$$\frac{d\mathbf{B}}{ds} = -\tau \mathbf{N},$$

where τ is the *torsion*, whose magnitude is the rate at which the curve twists out of the TN plane. Furthermore, we can derive $\frac{d\mathbf{N}}{ds} = -\kappa \mathbf{T} + \tau \mathbf{B}$ (left as an exercise).

In summary, the formulas

$$\begin{array}{rl} \frac{d\mathbf{T}}{ds} = & \kappa \mathbf{N} \\ \frac{d\mathbf{N}}{ds} = & -\kappa \mathbf{T} \quad +\tau \mathbf{B} \\ \frac{d\mathbf{B}}{ds} = & -\tau \mathbf{N} \end{array} \quad \text{or} \quad \begin{pmatrix} \mathbf{T}' \\ \mathbf{N}' \\ \mathbf{B}' \end{pmatrix} = \begin{pmatrix} 0 & \kappa & 0 \\ -\kappa & 0 & \tau \\ 0 & -\tau & 0 \end{pmatrix} \begin{pmatrix} \mathbf{T} \\ \mathbf{N} \\ \mathbf{B} \end{pmatrix},$$

are called Frenet–Serret formulas or Frenet–Serret theorem, named after two French mathematicians. The TNB frame, as shown in Figure 1.23, is useful when it is impossible or hard to assign a natural coordinate system for a trajectory as in relative theory or models of microbial motion. It is also called the Frenet–Serret frame.

1.6 Surfaces in space

1.6.1 Graph of an equation $F(x, y, z) = 0$

As seen before, a plane, which is a simple surface in space, has equation

$$a(x - x_0) + b(y - y_0) + c(z - z_0) = 0 \quad \text{or} \quad ax + by + cz + d = 0.$$

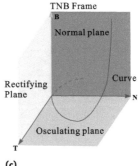

Figure 1.23: TNB frame.

All the solutions to the equation are points in the plane. Also, for any point in the plane, its coordinates must satisfy the equation. In general, for an equation of three variables,

$$F(x,y,z) = 0,$$

all its solutions are the set of points in space that form a surface, which is called the graph of the equation.

Example 1.6.1. Find an equation for the sphere with radius R centered at $P_0(x_0, y_0, z_0)$.

Solution. Suppose $P(x, y, z)$ is a point on the sphere. The distance between P and P_0 must be R, that is,

$$|PP_0| = R,$$
$$\sqrt{(x - x_0)^2 + (y - y_0)^2 + (z - z_0)^2} = R,$$

and so, this sphere has an equation

$$(x - x_0)^2 + (y - y_0)^2 + (z - z_0)^2 = R^2.$$

Example 1.6.2. Find the locus of points with equal distance from the two points $A(1, 2, 3)$ and $B(2, 1, -4)$.

Solution. If $P(x, y, z)$ is any point with equal distance from A and from B, then $|PA| = |PB|$, and this becomes

$$\sqrt{(x - 1)^2 + (y - 2)^2 + (z - 3)^2} = \sqrt{(x - 2)^2 + (y - 1)^2 + (z + 4)^2}.$$

Squaring both sides and simplifying this expression gives

$$2x - 2y - 14z - 7 = 0.$$

The locus form a plane.

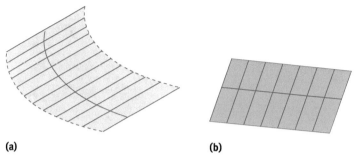

(a) (b)

Figure 1.24: Definition of a cylinder.

Note that squaring both sides of an equation, as above, can introduce extra solutions, because $A = B$ and $A = -B$ both square to $A^2 = B^2$. However, this does not happen here because the square roots on both sides must be nonnegative, and this does not allow one side to be negative.

1.6.2 Cylinder

Consider a plane in space again. A plane can be considered as the surface which is formed by all lines that are parallel to a given line and pass through a given curve. Or, in other words, the plane is formed by moving a line along a curve. This type of surface is called a cylinder, as shown in Figure 1.24.

Definition 1.6.1. A **cylinder** is defined as a surface that consists of all lines (called **rulings**) that are parallel to a given line and pass through a given curve.

We first consider the cases where all rulings are parallel to one of the coordinate axes.

Example 1.6.3 (Parabolic cylinder). Sketch the graph of the surface $y^2 = 2x$ in three-dimensional space and show that it is a cylinder.

Solution. Note that the equation of the graph $y^2 = 2x$ does not involve z. This means that for any x_0 and y_0 satisfying this equation, there is a line of solutions (x_0, y_0, z) for every possible z-value (that is, z is unrestricted by the equation). Furthermore, any horizontal plane with equation $z = k$ (parallel to the xy-plane) intersects the graph in the same curve with equation $y^2 = 2x$, a parabola. Figure 1.25(a) shows how the graph is formed by moving a line parallel to the z-axis along the parabola $y^2 = 2x$ in the xy-plane. This surface is called a *parabolic cylinder*. The graph can also be formed by infinitely many shifted copies of the same parabola $y^2 = 2x$ along the z-axis.

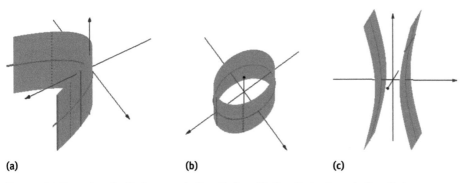

(a) (b) (c)

Figure 1.25: Examples of cylinders: parabolic cylinder, elliptic cylinder, hyperbolic cylinder.

Note.
1. In general, if one of the variables x, y, or z is missing from the equation of a surface, then the surface is a cylinder with rulings parallel to the axis of the missing variable.
2. It is useful to sketch surfaces in space by using the *traces* which are the intersection curves of the surface and planes parallel to one of the coordinate planes.

Example 1.6.4 (Elliptic cylinder). Identify and sketch in three-dimensional space the surfaces (a) $x^2 + 2y^2 = R^2$ and (b) $x^2 + z^2 = R^2$.

Solution. (a) Since z is missing, this must be a cylinder with rulings parallel to the z-axis. The graph of the equation $x^2 + 2y^2 = R^2$, for $z = k$ (a constant), is an ellipse in the plane $z = k$. Hence, the surface $x^2 + 2y^2 = R^2$ is an elliptic cylinder whose rulings are parallel to the z-axis and, so, are vertical (see Figure 1.25(b)).

(b) Similarly, $x^2 + z^2 = R^2$ is a circular cylinder whose rulings are parallel to the y-axis and thus they are horizontal.

Example 1.6.5 (Hyperbolic cylinder). Identify and sketch the surface $y^2 - \frac{z^2}{9} = 1$.

Solution. Since x is missing, this is a cylinder with rulings parallel to the x-axis. All the traces for constant x are hyperbolas. This is a hyperbolic cylinder whose graph is shown in Figure 1.25(c).

Example 1.6.6. Describe the surfaces of (a) $x = \sin y$ and (b) $z = \ln x$.

Solution. (a) It is a cylinder with rulings parallel to the z-axis. The trace with $z = 0$ is the curve $x = \sin y$ in the xy-plane. The cylinder is generated by moving this curve up and down along the z-axis.

(b) It is a cylinder with rulings parallel to the y-axis. The trace with $y = 0$ is the curve $z = \ln x$ in the xz-plane. The cylinder is generated by moving this curve left and right along the y-axis.

1.6.3 Quadric surfaces

A quadric surface is the graph of a second-degree polynomial equation with three variables, x, y, and z. The most general such equation is

$$Ax^2 + By^2 + Cz^2 + Dxy + Eyz + Fxz + Gx + Hy + Iz + J = 0,$$

where $A, B, C, \ldots, J$ are constants. By translation and rotation of the axes (in algebra: completing the square and making a linear transformation) it is possible to bring this equation into one of the two standard forms,

$$Ax^2 + By^2 + Cz^2 + D = 0 \quad \text{or} \quad Ax^2 + By^2 + Iz = 0,$$

where A, B, C, and I are nonzero (otherwise, the graphs are cylinders). The signs (positive or negative) of these constants and whether D is zero lead to the following list of the types of quadric surfaces in three-dimensional space:

1. **elliptic cone** $\frac{z^2}{c^2} = \frac{x^2}{a^2} + \frac{y^2}{b^2}$ $(D = 0)$,
2. **ellipsoid** $\frac{x^2}{a^2} + \frac{y^2}{b^2} + \frac{z^2}{c^2} = 1$ $(D \neq 0)$,
3. **hyperboloid of one sheet** $\frac{x^2}{a^2} + \frac{y^2}{b^2} - \frac{z^2}{c^2} = 1$ $(D \neq 0)$,
4. **hyperboloid of two sheets** $\frac{x^2}{a^2} - \frac{y^2}{b^2} - \frac{z^2}{c^2} = 1$ $(D \neq 0)$,
5. **elliptic paraboloid** $z = \frac{x^2}{a^2} + \frac{y^2}{b^2}$,
6. **hyperbolic paraboloid** $z = \frac{x^2}{a^2} - \frac{y^2}{b^2}$.

One needs to be aware that the same surfaces with different orientations are obtained when the roles of the variables are interchanged.

Like conic sections in two-dimensional space, quadric surfaces admit similar geometric and physical properties, which makes them useful in designing satellite dishes, headlamps, mirrors in telescopes, cooling towers for nuclear power plants, water tanks, and so forth.

Using traces, it is not hard to sketch the graph of these quadric surfaces. They are summarized in Figure 1.26.

1.6.4 Surface of revolution

One type of special surface in space is obtained by revolving a curve about a line. For example, if we revolve the plane curve $y = x^2$ about the x axis, we obtain a surface of

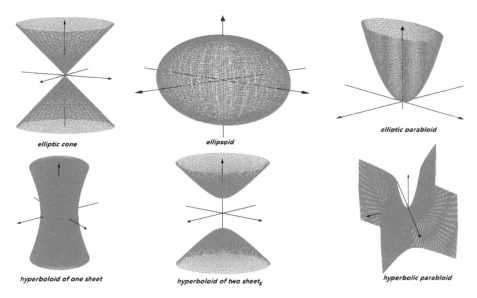

Figure 1.26: Quadric surfaces.

revolution in space. How do we find an equation for this surface? We consider a more general case. Suppose we have a curve $f(y,z) = 0$ in the yz-plane (this means $x = 0$), and we rotate this curve about the z-axis, as shown in Figure 1.27. To find an equation for the surface, we consider a point $P(x,y,z)$ on the surface. The point P is obtained by revolving the point $P_0(y_0, z_0)$ on the original curve $f(y,z) = 0$ about the z-axis. Note that P and P_0 actually have the same height above the xy-plane and the same distance from the z-axis; therefore, we have

$$z_0 = z \quad \text{and} \quad |y_0| = \sqrt{x^2 + y^2}$$

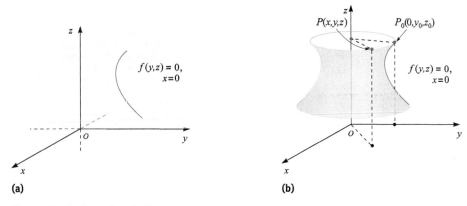

Figure 1.27: Surface of revolution.

and, thus, $(\pm\sqrt{x^2+y^2},z)$ must satisfy the equation $f(y,z)=0$. Therefore, we have

$$f(\pm\sqrt{x^2+y^2},z)=0.$$

This equation has three variables and is exactly an equation of the surface obtained by rotating the curve $f(y,z)=0$ in the yz-plane about the z-axis.

Example 1.6.7. Find an equation of the surface of revolution formed by revolving a straight line L in the yz-plane with equation $z=ay$ about the z-axis.

Solution. As discussed above, an equation of the surface is

$$z=\pm a\sqrt{x^2+y^2}.$$

That gives

$$z^2=a^2(x^2+y^2).$$

This type of surface is called a *circular cone*. The graph of a circular cone is shown in Figure 1.28.

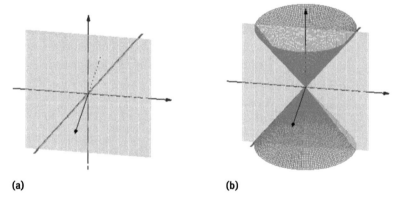

(a)   (b)

Figure 1.28: Surface of revolution, Example 1.6.7.

Using similar ideas, one can determine that

$$f(y,\pm\sqrt{x^2+z^2})=0$$

is the equation of the surface obtained by revolving the curve $f(y,z)=0$ in the yz-plane about the y-axis. Also, an equation for the surface obtained by revolving a curve in a coordinate plane other than the yz-plane about one of the axes can be determined in a similar manner.

Example 1.6.8. Find an equation for each surface of revolution obtained by revolving the curve $x^2 + \frac{4}{9}y^2 = 1$ in the xy-plane about (a) the x-axis and (b) the y-axis.

Solution. For (a), a rotation about the x-axis, keep x unchanged and replace y with $\pm\sqrt{y^2 + z^2}$, yielding

$$x^2 + \frac{4}{9}\left(\pm\sqrt{y^2 + z^2}\right)^2 = 1.$$

This simplifies to

$$x^2 + \frac{4}{9}y^2 + \frac{4}{9}z^2 = 1,$$

which is an equation for the desired surface.

For (b), similarly, we keep y unchanged, and we replace x by $\pm\sqrt{x^2 + z^2}$ in the equation of the curve. We obtain

$$x^2 + z^2 + \frac{4}{9}y^2 = 1,$$

which is an equation of that surface of revolution. The graphs of these surfaces of revolution are shown in Figure 1.29.

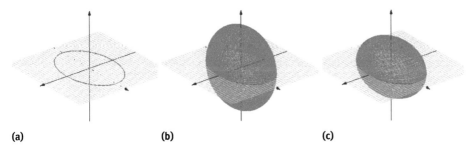

(a) (b) (c)

Figure 1.29: Surface of revolution, Example 1.6.8.

1.7 Parameterized surfaces

As seen in Section 1.3.2, a plane with an equation $ax + by + cz + d = 0$ also has a vector equation

$$\mathbf{r}(u, v) = \mathbf{r}_0 + u\mathbf{a} + v\mathbf{b}.$$

In general, the graph of the vector-valued function $\mathbf{r}(u, v)$ with two independent parameters u and v is a surface in space. Its parametric form is

$$x = x(u, v), \quad y = y(u, v), \quad \text{and} \quad z = z(u, v).$$

Example 1.7.1. Find a parameterization for each of the following surfaces:

$$(1) 2x + 4y - z = 5, \quad (2) x^2 + 2y^2 = 1, \quad (3) x^2 + y^2 + z^2 = 4, \quad (4) z^2 = x^2 + 4y^2.$$

Solution.
1. Let $x = u$ and $y = v$. Then $z = 2u + 4v - 5$. Or

$$\begin{cases} x = u, \\ y = v, \\ z = 2u + 4v - 5. \end{cases}$$

2. Let $x = \cos u$ and $y = \frac{1}{\sqrt{2}} \sin u$. Since the equation does not involve the variable z, z could be any real number. Thus,

$$\begin{cases} x = \cos u, \\ y = \frac{1}{\sqrt{2}} \sin u, \\ z = v \end{cases}$$

is a parameterization for the cylinder $x^2 + 2y^2 = 1$.

3. For the sphere, one can check that

$$x = 2\cos\phi\cos\theta, \quad y = 2\cos\phi\sin\theta, \quad \text{and} \quad z = 2\sin\phi$$

is a parameterization.

4. For the cone, let $x = u\cos v$, $y = \frac{1}{2}u\sin v$, and $z^2 = u^2$. So

$$x = u\cos v, \quad y = \frac{1}{2}u\sin v, \quad \text{and} \quad z = u$$

is a parameterization.

1.8 Intersecting surfaces and projection curves

Besides using a vector-valued function, there is another way of interpreting a curve in space. As we have seen before, a line is the curve of intersection of two planes, and a trace is, in fact, the curve of intersection of a plane and a surface in space.

Example 1.8.1. For each of the following curves, find two surfaces so that the curve is their intersection curve:
1. the line $r(t) = \langle 2 - 3t, 4 + t, -2 - 5t \rangle$,
2. the helix $r(t) = \langle 2\cos t, 2\sin t, 0.5t \rangle$.

Solution.
1. The line has symmetric equations

$$\frac{x-2}{-3} = \frac{y-4}{1} = \frac{z+2}{-5},$$

so

$$\begin{cases} \frac{x-2}{-3} = \frac{y-4}{1}, \\ \frac{y-4}{1} = \frac{z+2}{-5}, \end{cases} \text{ which simplifies to } \begin{cases} x + 3y - 14 = 0, \\ 5y + z - 18 = 0. \end{cases}$$

So the line is the line of intersection of the two planes $x+3y-14 = 0$ and $5y+z-18 = 0$.

2. The helix has parametric equations

$$x = 2\cos t, \quad y = 2\sin t, \quad z = 0.5t.$$

Therefore, $x^2 + y^2 = 4$ and $t = \frac{z}{0.5} = 2z$, so $x = 2\cos(2z)$. Both of them are cylinders. Therefore, the helix is the curve of intersection of the two cylinders, and we have

$$\begin{cases} x^2 + y^2 = 4, \\ x = 2\cos(2z). \end{cases}$$

Figure 1.30 shows the graphs of the cylinder $x^2 + y^2 = 4$ and the cylinder $x = 2\cos(2z)$.

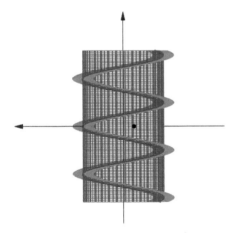

Figure 1.30: A helix is the curve of intersection of two cylinders.

In general, the graphs $F(x, y, z) = 0$ and $G(x, y, z) = 0$ are surfaces in space, and

$$\begin{cases} F(x, y, z) = 0, \\ G(x, y, z) = 0 \end{cases}$$

describes the curve of intersection of the two surfaces. We call it a general equation of the curve. Intuitively, if the system of equation has just one independent variable, say, x, then y and z are dependent variables. Therefore, we have the parametric equations $x = x$, $y = y(x)$, and $z = z(x)$, which describe a curve in space. For example,

$$\begin{cases} z = x^2 + 2y^2, \\ z = 3 \end{cases}$$

describes an ellipse in the $z = 3$ plane. It has parametric equations $x = \sqrt{3}\cos t$, $y = \sqrt{\frac{3}{2}}\sin t$, and $z = 3$.

Example 1.8.2. Describe the curve given by the equations $\begin{cases} z = x^2 + y^2, \\ x^2 + y^2 + z^2 = 2. \end{cases}$

Solution. Since $z = x^2 + y^2$ and $x^2 + y^2 + z^2 = 2$,

$$z^2 + z - 2 = 0.$$

This means that either $z = -2$ (rejected, as $z \geq 0$) or $z = 1$. Thus,

$$\begin{cases} z = 1, \\ x^2 + y^2 = 1. \end{cases}$$

This curve is a circle in the $z = 1$ plane. It has parametric equations $x = \cos t$, $y = \sin t$, and $z = 1$. This curve is the intersection curve of a paraboloid and a sphere with center at the origin and radius $\sqrt{2}$, as shown in Figure 1.31(b).

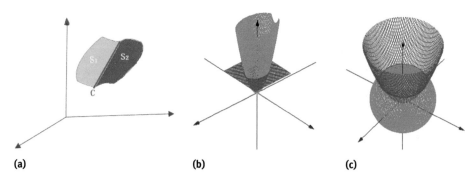

(a) (b) (c)

Figure 1.31: Curves as intersections of surfaces.

We were fortunate that in the previous example, there exist nice equations to describe curves in space. However, in some cases, it might be hard to find a simple equation for a space curve as the intersection curve of two surfaces. For example, consider

$$\begin{cases} x + y + 2z = 0, \\ x^2 + y^2 + z^2 = 4 \end{cases} \quad \text{or} \quad \begin{cases} z = x^2 + 2y^2, \\ x^2 + y^2 = 1. \end{cases}$$

We know that the first example is the intersection curve of a plane and a sphere. Intuitively, we know this is a circle in space. But for the second one, it might be hard to visualize it. To study a curve like this, it would be helpful to view it from the top or side. This means that we can study its projection curves onto one of the coordinate planes. How can we find equations for those projection curves? We first consider the case where we project the curve

$$\begin{cases} F(x,y,z) = 0, \\ G(x,y,z) = 0 \end{cases}$$

onto the xy-plane.

Well, the xy-plane has the feature that $z = 0$. So, if we try to eliminate the variable z from the simultaneous equations, we might obtain an equation $H(x,y) = 0$. This is actually a cylinder parallel to the z-axis, and this cylinder contains the curve of intersection! This is because for any point on the curve of intersection, its coordinates must satisfy both $F(x,y,z) = 0$ and $G(x,y,z) = 0$, and, therefore, satisfy $H(x,y) = 0$. Since this cylinder is vertical, if we view it from above, then the projection curve will be

$$\begin{cases} H(x,y) = 0, \\ z = 0. \end{cases}$$

Similarly, if we want the projection curve on the xz- or yz-coordinate planes, we simply eliminate the variable y or x from the simultaneous equations. This gives a cylinder parallel to the y- or x-axis, respectively. The curve of intersection of the cylinder with the xz- or yz-coordinate planes is the desired projection curve. So, finding projection curves is not hard now.

Example 1.8.3. Find an equation for the projection curve of the intersecting curve of the plane $x + y + 2z = 0$ and the sphere $x^2 + y^2 + z^2 = 4$:
1. onto the xy-plane,
2. onto the yz-plane.

Solution.
1. Onto the xy-plane, we eliminate the variable z to obtain

$$x^2 + y^2 + \left(-\frac{x+y}{2}\right)^2 = 4,$$

which simplifies to $5x^2 + 5y^2 + 2xy = 0$. This is an elliptic cylinder, and the projection curve

$$\begin{cases} 5x^2 + 5y^2 + 2xy = 16, \\ z = 0 \end{cases}$$

is an ellipse in the xy-plane.

2. Onto the yz-plane, we eliminate the variable x to obtain

$$(-y - 2z)^2 + y^2 + z^2 = 4,$$

which simplifies to $2y^2+5z^2+4yz = 4$. This is an elliptic cylinder, and the projection curve

$$\begin{cases} 2y^2 + 5z^2 + 4yz = 4, \\ x = 0 \end{cases}$$

is also an ellipse in the yz-plane. Figure 1.32 shows the graphs and projections.

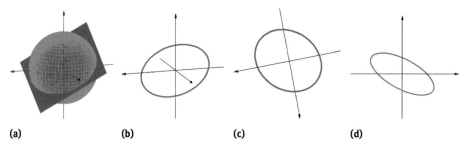

(a) (b) (c) (d)

Figure 1.32: Curves as intersections of surfaces.

Example 1.8.4 (Viviani curve). The following curve C is given:

$$\begin{cases} x^2 + y^2 + z^2 = 4, \\ (x-1)^2 + y^2 = 1. \end{cases}$$

Find the projection curve onto the xy-plane.

Solution. The curve is the intersection curve of a sphere and a circular cylinder. The variable z is missing in the second equation, so there is no need to eliminate z because the second equation is already a cylinder containing the curve C. So the projection curve onto the xy-plane is a circle centered at $(1, 0)$ with radius 1 (as in Figure 1.33) given by

$$\begin{cases} z = 0, \\ (x-1)^2 + y^2 = 1. \end{cases}$$

Note. This curve has nice parametric equations. If $x = 1 + \cos t$ and $y = \sin t$, then

$$z^2 = 4 - (1 + \cos t)^2 - (\sin t)^2$$
$$= 4 - 1 - 2\cos t - (\cos t)^2 - (\sin t)^2$$
$$= 2 - 2\cos t$$

1.8 Intersecting surfaces and projection curves — 55

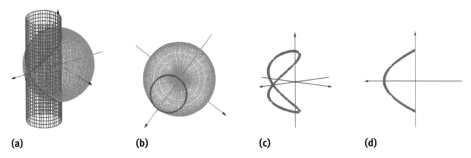

(a) (b) (c) (d)

Figure 1.33: Viviani curve, top view.

$$= 2(1 - \cos t)$$
$$= 4\sin^2 \frac{t}{2}.$$

Thus, we can take $z = 2\sin \frac{t}{2}$. A vector equation for this curve is, therefore,

$$\mathbf{r}(t) = \left\langle 1 + \cos t, \sin t, \sin \frac{t}{2} \right\rangle$$

Setting the z-component to 0, we have the projection curve onto the xy-plane, i.e.,

$$\mathbf{r}(t) = \langle 1 + \cos t, \sin t, 0 \rangle,$$

which is the same as $(x-1)^2 + y^2 = 1$. The intersection curve of the cylinder $(x-a)^2 + y^2 = a^2$ and the sphere $x^2 + y^2 + z^2 = 4a^2$ is called a Viviani curve, named after an Italian mathematician. Figure 1.33 shows the top and side views of a Viviani curve.

In some cases, we can also find projection curves onto some plane which is not parallel to any of the coordinate planes.

Example 1.8.5. Find the projection line of the line L: $\begin{cases} 2x-y+z=0, \\ x-y-2z+10=0 \end{cases}$ onto the plane $y + 2z + 2 = 0$.   !

Solution. It is easy to see that both planes $2x - y + z = 0$ and $x - y - 2z + 10 = 0$ contain L, and any linear combination of these equations gives a plane containing L. We have

$$(2x - y + z) + \lambda(x - y - 2z + 10) = 0,$$

where λ is any fixed number. Among all these planes that contain L, there must be exactly one plane which contains the projection line of L onto the plane $y + 2z + 2 = 0$. This plane is the one which is perpendicular to the plane $y + 2z + 2 = 0$. This means that the normal vectors of the two planes are perpendicular, i.e.,

$$\langle 0, 1, 2 \rangle \cdot \langle 2 + \lambda, -1 - \lambda, 1 - 2\lambda \rangle = 0.$$

Therefore, we have $0(2 + \lambda) + 1(-1 - \lambda) + 2(1 - 2\lambda) = 0$ with solution $\lambda = \frac{1}{5}$. Hence, the plane containing L and the projection line is

$$(2x - y + z) + \frac{1}{5}(x - y - 2z + 10) = 0, \quad \text{or}$$
$$11x - 6y + 3z + 10 = 0.$$

The projection line L is the line of intersection of this plane and the plane $y + 2z + 2 = 0$. So the projection line is

$$\begin{cases} y + 2z + 2 = 0, \\ 11x - 6y + 3z + 10 = 0. \end{cases}$$

Figure 1.34 shows these planes and the projection line onto the plane $y + 2z + 2 = 0$ in blue.

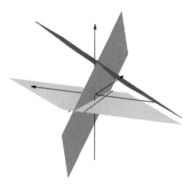

Figure 1.34: Projection line.

1.9 Regions bounded by surfaces

Suppose a region in space is bounded by two surfaces $F(x, y, z) = 0$ and $G(x, y, z) = 0$. How do you find its projection region onto one of the three coordinate planes? For example, the region R is bounded by the paraboloid $z = x^2 + y^2$ and the top semisphere $z = \sqrt{2 - x^2 - y^2}$. The intersecting curve of the two surfaces is

$$\begin{cases} z = x^2 + y^2, \\ z = \sqrt{2 - x^2 - y^2}. \end{cases}$$

Projecting this curve onto the xy-plane to obtain an equation of the projection curve gives

$$z = 0 \quad \text{and} \quad x^2 + y^2 = 1,$$

and the projection region of R onto the xy-plane is, therefore,

$$z = 0 \quad \text{and} \quad x^2 + y^2 \leq 1.$$

To project the region onto the xz-plane, we first note that the projection region of $z = x^2 + y^2$ onto the xz-plane is

$$y = 0 \quad \text{and} \quad z \geq x^2.$$

The projection region of $z = \sqrt{2 - x^2 - y^2}$ onto the xz-plane is a half-circle,

$$y = 0 \quad \text{and} \quad z \leq \sqrt{2 - x^2}.$$

Therefore, the projection region of R onto the xz-plane is the region bounded by $z = x^2$ and $z = \sqrt{2 - x^2}$ in the plane $y = 0$.

1.10 Review

The main concepts discussed in this chapter are listed below.
1. Vector operations of addition, subtraction, and scalar multiplication, and the dot product and cross product:

$$\mathbf{a} \cdot \mathbf{b} = a_1 b_1 + a_2 b_2 + a_3 b_3, \quad \mathbf{a} \times \mathbf{b} = \begin{vmatrix} \mathbf{i} & \mathbf{j} & \mathbf{k} \\ a_1 & a_2 & a_3 \\ b_1 & b_2 & b_3 \end{vmatrix}.$$

2. Equations of lines and planes in space:

$$\text{lines}: \mathbf{r} = \mathbf{r}_0 + t\mathbf{v}, \quad \mathbf{r} = \langle x(t), y(t), z(t) \rangle, \quad \text{or} \quad \begin{cases} a_1 x + b_1 y + c_1 z = d_1, \\ a_2 x + b_2 y + c_2 z = d_2, \end{cases}$$

or $\dfrac{x - x_0}{m} = \dfrac{y - y_0}{n} = \dfrac{z - z_0}{p}$

or $x = x_0 + mt$, $y = y_0 + nt$ and $z = z_0 + pt$,

planes: $a(x - x_0) + b(y - y_0) + c(z - z_0) = 0$ or $ax + by + cz + d = 0$.

3. Distance from a point $P(x_0, y_0, z_0)$ to a plane $ax + by + cz + d = 0$,

$$\text{distance} = \left| \frac{\overrightarrow{P_1 P} \cdot \mathbf{n}}{|\mathbf{n}|} \right| = \frac{|ax_0 + by_0 + cz_0 + d|}{\sqrt{a^2 + b^2 + c^2}},$$

where P_1 is any point in the plane and $\mathbf{n}$ is a normal vector of the plane.

4. Distance from a point $P(x_0, y_0, z_0)$ to a line $\mathbf{r} = \mathbf{r}_0 + t\mathbf{v}$,

$$\text{distance} = \frac{|\overrightarrow{P_1 P} \times \mathbf{v}|}{|\mathbf{v}|},$$

where P_1 is any point in the plane and $\mathbf{v}$ is a direction vector of the line.

5. Distance between two skew lines $\mathbf{r} = \mathbf{r}_1 + t\mathbf{v}_1$ and $\mathbf{r} = \mathbf{r}_2 + t\mathbf{v}_2$,

$$\text{distance} = \frac{|\overrightarrow{P_1P_2} \cdot (\mathbf{v}_1 \times \mathbf{v}_2)|}{|\mathbf{v}_1 \times \mathbf{v}_2|},$$

where P_1 is any point on the first line and P_2 is any point on the second line.

6. For a vector-valued function $\mathbf{r}(t)$:
   (a) $\mathbf{r}'(t) = \langle x'(t), y'(t), z'(t) \rangle$, $\int \mathbf{r}(t)dt = \langle \int x(t)dt, \int y(t)dt, \int z(t)dt \rangle$,
   (b) tangent line at $t = t_0$: $\mathbf{r} = \mathbf{r}(t_0) + t\mathbf{r}'(t_0)$,
   (c) normal plane at $t = t_0$:

$$x'(t_0)(x - x(t_0)) + y'(t_0)(y - y(t_0)) + z'(t_0)(z - z(t_0)) = 0,$$

   (d) length of curves: $s = \int_a^b |\mathbf{r}'(t)|dt$, for $a \leq t \leq b$,
   (e) curvature: $\kappa = \frac{d\mathbf{T}}{ds} = \frac{1}{|\mathbf{r}'(t)|}|\frac{d\mathbf{T}}{dt}| = \frac{|\mathbf{r}'(t) \times \mathbf{r}''(t)|}{|\mathbf{r}''(t)|^3}$,
   (f) the principal unit normal vector: $\mathbf{N} = \frac{d\mathbf{T}}{dt}/|\frac{d\mathbf{T}}{dt}|$,
   (g) the unit binormal vector: $\mathbf{B} = \mathbf{T} \times \mathbf{N}$,
   (h) torsion: $\tau = -\mathbf{N} \cdot \frac{d\mathbf{B}}{ds}$.
7. The cylinders parallel to one of the axes: $F(x,y) = 0$, $G(x,z) = 0$, and $H(y,z) = 0$.
8. Quadric surfaces:

$\frac{z^2}{a^2} = \frac{x^2}{b^2} + \frac{y^2}{c^2}$ elliptic cone $\qquad \frac{x^2}{a^2} + \frac{y^2}{b^2} + \frac{z^2}{c^2} = 1$ ellipsoid

$\frac{x^2}{a^2} + \frac{y^2}{b^2} - \frac{z^2}{c^2} = 1$ hyperboloid of one sheet $\qquad z = \frac{x^2}{a^2} + \frac{y^2}{b^2}$ elliptic paraboloid

$\frac{x^2}{a^2} - \frac{y^2}{b^2} - \frac{z^2}{c^2} = 1$ hyperboloid of two sheets $\qquad z = \frac{x^2}{a^2} - \frac{y^2}{b^2}$ hyperbolic paraboloid

9. Surface of revolution:

$$\begin{cases} f(y,z) = 0, \\ x = 0, \end{cases} \qquad \text{about the } z\text{-axis } f(\pm\sqrt{x^2 + y^2}, z) = 0,$$

$$\text{about the } y\text{-axis } f(y, \pm\sqrt{x^2 + z^2}) = 0.$$

Similar results hold for curves in other coordinate planes rotated about one of the axes.

10. Vector form of a plane: $\mathbf{r} = \mathbf{r}_0 + u\mathbf{a} + v\mathbf{b}$.
11. Surfaces with vector parametric forms: $\mathbf{r} = \langle x(u,v), y(u,v), z(u,v) \rangle$.
12. Finding the projections of curves by eliminating one of the variables x, y, or z.

1.11 Exercises

1.11.1 Vectors

1. Sketch the following points in a three-dimensional coordinate system:
$$(1,2,3),\ (-1,0,2),\ (0,2,0),\ (0,-1,1),\ (2,-1,2),\ (2,0,0).$$

2. Find the three points that are symmetrical to $M_0(x_0, y_0, z_0)$ about the x-axis, the xz-plane, and the origin, respectively.

3. If $|\vec{AB}| = 11$, $A(4,-7,1)$, and $B(6,2,z)$, find z.

4. If $\vec{a} = \langle 5,7,8 \rangle$, $\vec{b} = \langle 3,-4,6 \rangle$, and $\vec{c} = \langle -6,-9,-5 \rangle$, then find the length and direction angles of the vector $\vec{a} + \vec{b} + \vec{c}$.

5. If $\mathbf{r} = \mathbf{i} - 2\mathbf{j} - 2\mathbf{k}$, then find the unit vector in the direction of $\mathbf{r}$. Also, find the three direction cosines.

6. Which of the following expressions make sense? Which do not make sense? Explain your answers.
   (1) $(\vec{a} \cdot \vec{b}) \cdot \vec{c}$,   (2) $(\vec{a} \cdot \vec{b})\vec{c}$,   (3) $|\vec{a}|(\vec{b} \cdot \vec{c})$,
   (4) $(\vec{a} \mid \vec{b}) \cdot \vec{c}$,   (5) $\vec{a} + \vec{b} \cdot \vec{c}$,   (6) $|\vec{a}|(\vec{b} + \vec{c})$.

7. Which of the following identities are true? Which are false? Explain your answers.
   (1) $|\vec{a}|\vec{a} = \vec{a} \cdot \vec{a}$,   (2) $(\vec{a} \cdot \vec{b})(\vec{a} \cdot \vec{b}) = (\vec{a} \cdot \vec{a})(\vec{b} \cdot \vec{b})$,
   (3) $(\vec{a} \cdot \vec{b})\vec{c} = \vec{a}(\vec{b} \cdot \vec{c})$,   (4) $(\mathbf{a} + \mathbf{b}) \cdot (\mathbf{a} + \mathbf{b}) = \mathbf{a} \cdot \mathbf{a} + 2\mathbf{a} \cdot \mathbf{b} + \mathbf{b} \cdot \mathbf{b}$.

8. Simplify the following expressions:
   (1) $\vec{i} \times (\vec{j} + \vec{k}) - \vec{j} \times (\vec{i} + \vec{k}) + \vec{k} \times (\vec{i} + \vec{j} + \vec{k})$,   (2) $(2\mathbf{a} + \mathbf{b}) \times (\mathbf{c} - \mathbf{a}) + (\mathbf{b} - \mathbf{c}) \times (\mathbf{a} \mid \mathbf{b})$.

9. Prove Theorem 1.2.4.

10. If $\vec{a} = 3\vec{i} - \vec{j} - 2\vec{k}$, $\vec{b} = \vec{i} + 2\vec{j} - \vec{k}$, and $\vec{c} = \vec{i} + \vec{j}$, find
    (1) $\vec{a} \cdot \vec{b}$ and $\vec{a} \times \vec{b}$,   (2) $\text{Proj}_{\vec{b}}\vec{a}$,   (3) $(-2\vec{a}) \cdot 3\vec{b}$,
    (4) the angle between $\vec{a}$ and $\vec{b}$,   (5) $(\vec{a} \times \vec{b}) \cdot \vec{c}$.

11. Prove that $\mathbf{a} \cdot (\mathbf{b} \times \mathbf{c}) = (\mathbf{a} \times \mathbf{b}) \cdot \mathbf{c}$.

12. Prove that the four points $A(2,-1,-2)$, $B(1,2,1)$, $C(2,3,0)$, and $D(-1,5,4)$ are not coplanar.

13. Prove the Cauchy-Schwarz inequality for three-dimensional vectors $\vec{a}$ and $\vec{b}$,
$$|\vec{a} \cdot \vec{b}| \le |\vec{a}||\vec{b}|.$$
A general form of the inequality is given by
$$\left(\sum_{i=1}^{n} a_i b_i \right)^2 \le \sum_{i=1}^{n} a_i^2 \sum_{i=1}^{n} b_i^2.$$

14. Use the projection method to show that in $\mathbb{R}^2$ the distance from a point $P(x_0, y_0)$ to the line $ax + by + c = 0$ is
$$\frac{|ax_0 + by_0 + c|}{\sqrt{a^2 + b^2}}.$$

15. If $\mathbf{r}_1$, $\mathbf{r}_2$, and $\mathbf{r}_3$ are three nonzero position vectors, show that their heads are collinear if and only if

$$\mathbf{r}_1 \times \mathbf{r}_2 + \mathbf{r}_2 \times \mathbf{r}_3 + \mathbf{r}_3 \times \mathbf{r}_1 = \mathbf{0}.$$

1.11.2 Lines and planes in space

1. Find an equation for the line that:
   (1) passes through the two points $(1, 3, 2)$ and $(-1, 3, -2)$.
   (2) passes through the point $(1, 2, -3)$ and is parallel to the vector $\mathbf{v} = \langle 0, 1, 1 \rangle$.
   (3) passes through the point $(-3, 2, 1)$ and is parallel to the x-axis.
   (4) passes through the point $(0, 3, 2)$ and is perpendicular to the xz-plane.
2. Find an equation for the plane that:
   (1) passes through the point $(6, 3, 2)$ and is parallel to another plane $3x - 7y + 5z - 12 = 0$.
   (2) passes through the points $(1, 2, 1)$, $(0, 1, 1)$, and $(1, -1, 2)$.
   (3) passes through the point $(1, 7, -3)$ and contains the line $x = 1 - 3t$, $y = t$, $z = 2 - t$.
3. Determine whether each pair of lines is parallel, perpendicular, intersecting, or skew lines.

   (1) $\ell_1 : \dfrac{x+1}{2} = \dfrac{y-2}{0} = \dfrac{z-3}{2}$ $\quad \ell_2 : x = 2 - t, y = 3 + 2t$, and $z = 4t$,

   (2) $\ell_1 : x = t, y = 0,$ and $z = -2t$ $\quad \ell_2 : \dfrac{x+1}{2} = \dfrac{y+7}{9} = \dfrac{z}{-4}$,

   (3) $\ell_1 : \mathbf{r} = \langle t, t, -3t \rangle$ $\quad \ell_2 : \mathbf{r} = \langle 1 + t, 2t, 5 - 3t \rangle$.

4. Find the shortest distance between the two lines

   $$\dfrac{x-1}{0} = \dfrac{y}{1} = \dfrac{z}{1} \quad \text{and} \quad \dfrac{x}{2} = \dfrac{y}{-1} = \dfrac{z+2}{0}.$$

5. Find the plane that passes through the point $(1, 2, 1)$ and contains the line of intersection of the planes $x - y + z = 2$ and $2x - y - 2z = 1$.
6. Find parametric equations of the line that passes through $(4, 1, 3)$ and is parallel to the line

   $$\dfrac{x-3}{2} = \dfrac{y}{1} = \dfrac{z-1}{5}.$$

7. Find parametric equations of the line through $(2, 1, 1)$ that intersects the line $\dfrac{x+1}{3} = \dfrac{y+3}{1} = \dfrac{z}{2}$ and is parallel to the plane $3x - 4y + z = 10$.
8. Find the angle between two planes $4x + 2y + 4z - 7 = 0$ and $3x - 4y = 0$.
9. Find the distance from the point $A(1, 2, 3)$ to the line $x = t, y = 4 - 3t, z = 3 - 2t$.

10. Find the acute angle between the line $\begin{cases} x+y+3z=0, \\ x-y-z=0 \end{cases}$ and the plane $x - y - z + 1 = 0$.
11. Find symmetric equations of the line that is contained in the plane $\pi : x+y+z+1 = 0$, passes through the point of intersection of the plane π and the line $L : \begin{cases} y+z+1=0, \\ x+2z=0, \end{cases}$ and is perpendicular to the line L.
12. Find the distance from the point $(1, -1, 2)$ to the plane $x - y + 2z = 2$.
13. Find a vector equation in the form $\mathbf{r} = \mathbf{r}_0 + u\mathbf{r}_1 + v\mathbf{r}_2$ for the plane $2x + 5y - z = 8$.

1.11.3 Curves and surfaces in space

1. Sketch the following curves in space:
   (1) $\mathbf{r}(t) = \cos t\mathbf{i} + 2\mathbf{j} + 2\sin t\mathbf{k}, \ 0 \le t \le 2\pi$,
   (2) $\mathbf{r}(t) = 2\cos t\mathbf{i} + 2\sin t\mathbf{j} + 3\mathbf{k}, \ 0 \le t \le 2\pi$.
2. If $\mathbf{r}(t) = \langle te^{-t}, \sin\frac{1}{t}, t\sin\frac{1}{t} \rangle$, find $\lim_{t\to 0} \mathbf{r}(t)$ and $\lim_{t\to\infty} \mathbf{r}(t)$. Is $\mathbf{r}(t)$ continuous at $t = 0$?
3. Find the point of intersection, if it exists, of the plane $x + y = 0$ and the curve $\mathbf{r}(t) = \langle t, \sin t, \cot t \rangle$.
4. Prove Theorem 1.5.2.
5. For the vector-valued function
$$\mathbf{r}(t) = \left\langle \frac{\cos t}{\sqrt{2}} + \frac{\sin t}{\sqrt{3}}, -\frac{\cos t}{\sqrt{2}} + \frac{\sin t}{\sqrt{3}}, \frac{\sin t}{\sqrt{3}} \right\rangle,$$
find:
   (a) an equation for its tangent line at $t = \frac{\pi}{2}$.
   (b) an equation for its normal plane at $t = \frac{\pi}{2}$.
   (c) an equation of the projection curve onto the xy-plane.
6. If $\mathbf{r}'(t) = \langle te^t, t\cos t^2, -\frac{2t}{\sqrt{t^2+4}} \rangle$ and $\mathbf{r}(0) = \langle 0, 2, 4 \rangle$, find $\mathbf{r}(t)$.
7. A particle is moving in space with velocity $\mathbf{v}(t) = \langle 3\cos t, 4\cos t, 5\sin t \rangle$. Initially it starts at the origin. Find its position when $t = \pi$ and the distance it traveled during $0 \le t \le \pi$.
8. Determine whether the curves
   (1) $\mathbf{r}(t) = \cos t^2\mathbf{i} + \sin t^2\mathbf{j}$, (2) $\mathbf{r}(t) = 5\cos t\mathbf{i} + 3\sin t\mathbf{j} + 4\sin t\mathbf{k}$
   use arc length as a parameter. If not, find a description that uses arc length as a parameter.
9. Prove Theorem 1.5.3.
10. For each curve

    (1) $\mathbf{r}(t) = 3\cos t\mathbf{i} + 3\sin t\mathbf{j} + 2t\mathbf{k}$ and (2) $\mathbf{r}(t) = e^t\cos t\mathbf{i} + e^t\sin t\mathbf{j} + e^t\mathbf{k}$,

    find
    (a) the unit tangent vector $\mathbf{T}$ at $t = 0$.
    (b) the principal unit normal vector $\mathbf{N}$ at $t = 0$.

(c) the curvature of the curve at $t = 0$.
(d) the unit binormal vector **B** at $t = 0$.
(e) the torsion at $t = 0$.

11. Prove that $\frac{d\mathbf{N}}{ds} = -\kappa\mathbf{T} + \tau\mathbf{B}$.

12. On a smooth curve **C**, the *kissing circle* (the circle of curvature or osculating circle, see Figure 1.22(c)) at a point P is the circle that (a) is tangent to C at P, (b) has the same curvature as C at P, and (c) lies on the same side of C as the principal normal vector **N**. The radius of this circle is $\frac{1}{\kappa}$ and is called the radius of curvature. If the curve is $y = y(x)$, then the curvature of the curve at $P(x_0, y_0)$ can be shown to be

$$\frac{|y''(x_0)|}{(1+[y'(x_0)]^2)^{3/2}},$$

and the center of this circle at $P(x_0, y_0)$ can be shown to be

$$\left(x_0 - y'(x_0)\frac{1+[y'(x_0)]^2}{y''(x_0)}, y_0 + \frac{1+[y'(x_0)]^2}{y''(x_0)}\right).$$

(a) Find an equation of the kissing circle for the parabola $y = x^2$ at $x = 2$.
(b) Find an equation of the kissing circle for the cycloid $\mathbf{r}(t) = \langle t - \sin t, 1 - \cos t\rangle$ at $t = \pi$.

13. Find an equation of the sphere that passes through the point $(2, -4, 3)$ and contains the circle $\begin{cases} x^2+y^2=5, \\ z=0. \end{cases}$

14. Identify and sketch the following surfaces:
    (1) $z = y^2$,
    (2) $x^2 + 8y^2 = 4$,
    (3) $x^2 - z^2 = 9$,
    (4) $y = x^3$,
    (5) $4x^2 + y^2 + \frac{z^2}{16} = 1$,
    (6) $z = 2x^2 + 6y^2$,
    (7) $z^2 + 2y^2 - 2x = 0$,
    (8) $z + y^2 - \frac{x^2}{4} = 0$,
    (9) $\frac{x^2}{25} + \frac{y^2}{16} - z^2 = 0$,
    (10) $\frac{x^2}{25} + \frac{y^2}{16} - z^2 = 1$,
    (11) $-\frac{x^2}{9} - \frac{y^2}{16} + z^2 = 1$.

15. Find the intersections of the given surface with the planes $x = k$, $y = k$, $z = k$ and identify the surface and sketch it.
    (1) $x^2 + y^2 + z^2 = 2az$,
    (2) $x^2 + y^2 = 2az$,
    (3) $x^2 + z^2 = 2az$,
    (4) $x^2 - y^2 = z^2$,
    (5) $x^2 = 2az$.

16. Find an equation for the surface obtained by rotating the parabola $y = x^2$, $z = 0$ about the y-axis.

17. Find an equation for the surface S obtained by rotating the curve $\begin{cases} y^2=6-z, \\ x=0 \end{cases}$ about the z-axis. Find the projection curve onto the xy-plane of the curve of intersection of the surface S and the cone $z = \sqrt{x^2 + y^2}$.

18. Find an equation of the cylinder consisting of all lines parallel to the direction $\langle 2, 1, -1\rangle$ and passing through the curve $\begin{cases} y^2-4x=0, \\ z=0. \end{cases}$

19. Find the projection curve of
    (1) $\begin{cases} z=xy, \\ x^2+2y^2=1 \end{cases}$ onto the xy-plane,
    (2) $\mathbf{r}(t) = \langle 1 + t^2, \sin t, t\rangle$ onto the xz-plane,

(3) $z = x^2 + 2y^2$ and $x^2 + y^2 = 1$ onto (a) the xy-plane and (b) the yz-plane.
20. Find the projection regions of
    (1) the paraboloid $z = x^2 + y^2$,
    (2) the solid bounded by the cone $z = \sqrt{x^2 + 2y^2}$ and the sphere $x^2 + y^2 + z^2 = 1$
    onto the three coordinate planes.
21. Find an equation of the projection line of the line $\mathbf{r}(t) = \langle 2 + t, 3 - 2t, 4t \rangle$ onto the plane $x + y - z = 1$.
22. Try to find an equation for the projection curve of the curve $\mathbf{r}(t) = \langle 2\cos t, 2\sin t, 4t \rangle$ onto the plane $x - 2y + 3z = 2$.
23. (**Ruled surfaces**) In geometry, a surface S is *ruled* (also called a *scroll*) if through every point of S there is a straight line that lies on S. For example, a plane and a circular cone are both ruled surfaces. A surface is *doubly ruled* if through every point of S there are two distinct lines that lie on the surface.
    Show that:
    (a) the cylinder $x^2 + 2z^2 = 1$ is ruled.
    (b) the hyperbolic paraboloid $z = x^2 - y^2$ and the hyperboloid of one sheet $x^2 + y^2 - z^2 = 1$ are doubly ruled surfaces.

2 Functions of multiple variables

In single-variable calculus, a function depends on only one variable. However, in the real world, physical quantities often depend on two or more variables. For example, the volume V of a circular cone depends on its base radius r and height h, so it is a function of two variables. The temperature T of a city in China depends on the time t, the longitude x, and the latitude y of the city. The temperature T here is a function of three variables. A smart person's IQ may depend on his genes, thinking skills, education, and so forth. It is a function of more than three variables. In this chapter, we study multivariable functions and apply differential calculus to such functions.

2.1 Functions of multiple variables

2.1.1 Definitions

We first start with functions of two variables. Similar to functions of one variable, a real-valued *function f of two variables* is defined as follows.

Definition 2.1.1. A real function of two real variables is a rule f (also called a mapping or correspondence) that assigns to each ordered pair of real numbers (x, y) in a set $D \subset \mathbb{R}^2$ a unique number $z \in \mathbb{R}$. We denote this rule by $z = f(x, y)$. The set D is called the *domain* of f. The set $\{f(x,y) | (x,y) \in D\}$ is called the *range* of the function f. The variables x and y are called the *independent variables* and z is called the *dependent variable*.

We usually denote a function of two variables explicitly by the equation $z = f(x, y)$ or implicitly by the equation $F(x, y, z) = 0$. Usually, we assume that the domain of a function of two variables is all possible ordered pairs (x, y) of real numbers for which $f(x, y)$ makes physical and/or mathematical sense. That is, when the domain is not otherwise specified, we take it to be

$$D = \text{domain of } f = \{(x, y) : f(x, y) \text{ is defined}\}$$

or, if f describes some real-life situation,

$$D = \text{domain of } f = \{(x, y) : f(x, y) \text{ makes physical sense}\}.$$

The domain of a function of two variables can have interior points and boundary points, and it may be an open or closed region in the xy-plane, just the way the domains of one-variable functions defined on subsets of the real number line can (see Figure 2.1(b) and (c)). We give the following important definitions.

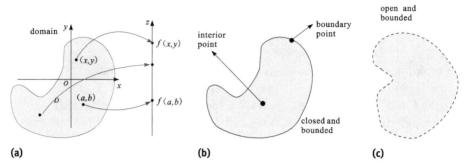

Figure 2.1: Functions of two variables, domain, range, boundary point, interior point.

Definition 2.1.2.
1. A point (x_0, y_0) in a region (set) R in the xy-plane is an **interior point** of R if there exists a disk with center (x_0, y_0) that lies entirely in R.
2. A point (x_0, y_0) is a **boundary point** of R if every disk centered at (x_0, y_0) contains points that lie outside of R as well as points that lie in R. The boundary point may or may not belong to R.
3. The interior points of a region make up the **interior** of the region.
4. The region's boundary points make up its **boundary**.
5. A region is **open** if it consists entirely of interior points.
6. A region is **closed** if it contains all its boundary points.
7. A region in the plane is **bounded** if it lies inside a disk of fixed radius.
8. A region is **unbounded** if it is not bounded.

The structures of domains of two-variable functions can vary considerably from one function to another, as shown in the following examples.

Example 2.1.1. The function f defined by

$$f(x, y) = \sqrt{(4 - x^2 - y^2)(x^2 + y^2 - 1)}$$

has a maximum domain consisting of all (x, y) satisfying $(4 - x^2 - y^2)(x^2 + y^2 - 1) \geq 0$ ensuring the square root exists. Solving this inequality gives

$$D = \{(x, y) : 1 \leq x^2 + y^2 \leq 4\}.$$

One can easily draw a picture of this domain, which is an annulus (ring), as shown in Figure 2.2(a).

Example 2.1.2. Find the domain of the function $z = \sqrt{1 - \frac{x^2}{a^2} - \frac{y^2}{b^2}} + \ln(x + y)$ ($a > 0$ and $b > 0$).

Solution. The function z is defined only for these pairs of (x, y) satisfying $1 - \frac{x^2}{a^2} - \frac{y^2}{b^2} \geq 0$ and $x + y > 0$, so the domain for this function of two variables is

$$D = \left\{ (x, y) \,\bigg|\, \frac{x^2}{a^2} + \frac{y^2}{b^2} \leq 1 \text{ and } x + y > 0 \right\}.$$

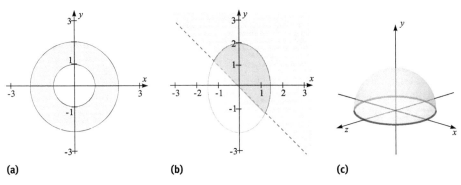

Figure 2.2: Domains of functions of two variables, Example 2.1.1, Example 2.1.2, and Example 2.1.3.

It is a region bounded by the elliptical curve and the line $y = -x$. This is neither an open nor a closed region, as shown in Figure 2.2(b).

Example 2.1.3. Find the domain and range of the function z defined by

$$z(x,y) = \sqrt{9 - x^2 - y^2}.$$

Solution. The domain D of z is found by noticing that a square root cannot take negative values, i.e.,

$$D = \{(x,y)|9 - x^2 - y^2 \geq 0\} = \{(x,y)|x^2 + y^2 \leq 9\},$$

which is the disk with center $(0,0)$ and radius 3. The range of z is

$$\{z|z = \sqrt{9 - x^2 - y^2}, \text{ for all } (x,y) \in D\}.$$

Since z is a positive square root, $z \geq 0$ and the domain restriction $0 \leq x^2 + y^2 \leq 9$ shows that

$$0 \leq \sqrt{9 - x^2 - y^2} \leq 3.$$

Hence, the range is the set of real numbers

$$\{z|0 \leq z \leq 3\} = [0,3].$$

The graph of the function $z = \sqrt{9 - x^2 - y^2}$ is shown in Figure 2.2(c).

2.1.2 Graphs and level curves

As seen in the previous chapter, the set of points (x,y,z) whose coordinates satisfy the equation $z = f(x,y)$ consist of a surface in space. This surface is called the graph of the function $z = f(x,y)$. For instance, the graph of the function $z = 3x - y + 3$ is a plane, the graph of the function $z = \sqrt{a^2 - x^2 - y^2}$ is the upper semisphere, and the graph of $z = x^2 - 3y^2$ is a hyperbolic paraboloid in space.

As seen in the previous chapter, it might be hard to visualize or sketch the graph of a function of two variables, for example, $z = x^2 - 4y^2$, which is a hyperbolic paraboloid. However, we can use the ideas of *contour curves* and *level curves* to help visualize the graph. One may already have an idea from the daily weather forecasts or topographic mappings of a mountain, as shown in Figure 2.3.

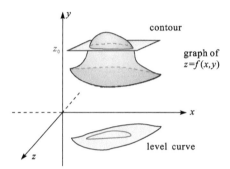

Figure 2.3: Graphs, contours, and level curves.

Assume that you are walking on the surface of a mountain, and the surface is the graph of a function $z = f(x,y)$. If you walk along a path on which your elevation remains constant, say, z_0, which is actually the height above the bottom plane of the mountain, then the path is part of a contour curve which is the intersecting curve of the surface $z = f(x,y)$ and the plane $z = z_0$. When the contour curves are projected onto the xy-plane, those projection curves are called level curves.

 Example 2.1.4. Find and sketch the level curves of the following surfaces:
1. $z = x^2 - 4y^2$,
2. $f(x,y) = x^4 + y^4 + 8xy$.

Solution.
1. The level curves are described by the equations

$$\begin{cases} z_0 = x^2 - 4y^2, \\ z = 0, \end{cases}$$

where z_0 is a constant. For $z_0 = 0$, the level curve is two straight lines $x = \pm 2y$ in the xy-plane. For all values of $z_0 \neq 0$, the level curves are hyperbolas in the xy-plane. Setting $z_0 = 1, 4, 8, -1, -4, -8$ enables us to obtain Figure 2.4.

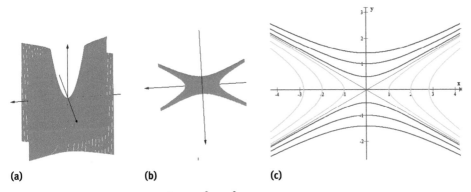

(a) (b) (c)

Figure 2.4: Graph and level curves for $z = x^2 - 4y^2$.

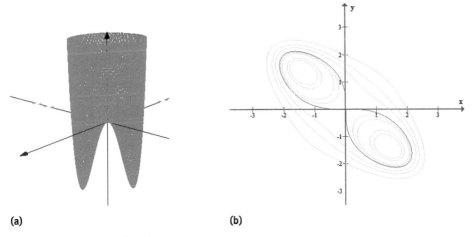

(a) (b)

Figure 2.5: Graph of $z = x^4 + y^4 + 8xy$ and its level curves.

2. We have not seen the graph of $f(x,y)$ before. We use a graphing utility to help sketch the surface and graph the level curves, as shown in Figure 2.5.

2.1.3 Functions of more than two variables

Likewise, a function f of three variables x, y, and z is a rule that assigns to each ordered triple (x, y, z) in a domain set $D \subset \mathbb{R}^3$ a unique real number u, and we write

$$u = f(x, y, z)$$

to indicate that it is a function of these three variables. If the domain of a function f of three variables x, y, and z is not specified, then it is usually taken to be the set of ordered triples (x, y, z) in $\mathbb{R}^3$ for which $f(x, y, z)$ makes physical or mathematical sense.

The graph of a function of three variables is the set of those points (x,y,z,u) in four-dimensional space where $u = f(x,y,z)$! Therefore, we cannot visualize it in three-dimensional space. To get some sense of the graph, we could use an idea similar to level curves. If we set $u_0 = f(x,y,z)$, for any constant u_0, the graph of $u_0 = f(x,y,z)$ is a *level surface* in space. For example, the level surfaces of the function $u = x^2 + y^2 + z^2$ are spheres in space.

Although most of the functions that we will work with in this textbook will be functions of two or three variables, scientists, engineers, and mathematicians often need to work with functions of four or more variables. Those functions are defined in a similar way. In general, a function f of n variables is a rule that assigns a unique number u to an n-tuple $(x_1, x_2, \ldots, x_n) \in \mathbb{R}^n$ of real numbers, and we write $u = f(x_1, x_2, \ldots, x_n)$. Sometimes we use vector notation to write such functions more compactly. That is, if the ordered n-tuple is considered to be a vector $\mathbf{x} = \langle x_1, x_2, \ldots, x_n \rangle$, then we write $f(\mathbf{x})$ in place of $f(x_1, x_2, \ldots, x_n)$, and we can then write the function compactly as $u = f(\mathbf{x})$, or $u = f(P)$, for $P \in \mathbb{R}^n$.

2.1.4 Limits

Limits for functions of two variables are required to develop the calculus of functions of two variables. They can often be interpreted in much the same way as we interpret limits of functions of one variable. For example, the statement

$$\lim_{(x,y) \to (2,-1)} 3x^2 y = -12$$

means that the value of $3x^2 y$ gets closer and closer to -12 as (x,y) gets closer and closer to $(2, -1)$. It may seem obvious that if (x,y) is close to $(2, -1)$, then x is close to 2 and y is close to -1, so the value of $3x^2 y$ is close to $3(2^2)(-1) = -12$. However, we will soon see that the limit of a function of two variables is not always this clear. Vague phrases like "closer and closer to" can be hard to interpret in some circumstances, and we can avoid most of these difficulties by defining the limit more precisely.

Definition 2.1.3. Let D be a region in the xy-plane and (a, b) be an interior point of D. Let f be a real-valued function of two variables defined on D except possibly at (a, b). We say that a real number L is the limit of f as (x, y) approaches (a, b), and we write

$$\lim_{(x,y) \to (a,b)} f(x,y) = L$$

if for every number $\varepsilon > 0$ there is a number $\delta > 0$ (the value of δ depends on ε) such that

$$|f(x,y) - L| < \varepsilon \quad \text{for all } (x,y) \in D \text{ satisfying } 0 < \sqrt{(x-a)^2 + (y-b)^2} < \delta.$$

We also write $f(x,y) \to L$ (meaning $f(x,y)$ approaches L) as $(x,y) \to (a,b)$ (meaning (x,y) approaches (a,b)).

Note that the set of points (x, y) satisfying the condition $0 < \sqrt{(x-a)^2 + (y-b)^2} < \delta$ forms a punctured open disk with center (a, b) and radius δ ("punctured" means one point, the center (a, b), is excluded and "open" means the boundary circle is not included). This punctured disk is sometimes denoted as $\mathring{U}((a, b), \delta)$. Note that we can relax the requirement that the point (a, b) be in the interior of D as long as for every $\delta > 0$ the punctured disk $\mathring{U}((a, b), \delta)$ contains elements of D where f is defined. For example, $\lim_{(x,y) \to (0,0)} \sqrt{x+y}$ exists, even though $\sqrt{x+y}$ is not defined for $x+y < 0$.

Example 2.1.5. Show that $\lim_{(x,y) \to (1,2)} (2x + 4y) = 10$.

Proof. For any given $\varepsilon > 0$, we choose $\delta = \frac{\varepsilon}{8}$. Then when

$$0 < \sqrt{(x-1)^2 + (y-2)^2} < \delta,$$

we have $|x - 1| < \frac{\varepsilon}{8}$ and $|y - 2| < \frac{\varepsilon}{8}$, and then

$$|(2x + 4y) - 10| = |2(x-1) + 4(y-2)|$$
$$\leq 2|x-1| + 4|y-2|$$
$$\leq 2\frac{\varepsilon}{8} + 4\frac{\varepsilon}{8} < \varepsilon.$$

So, by the definition, we conclude that $\lim_{(x,y) \to (1,2)} (2x + 4y) = 10$. $\square$

Example 2.1.6. Assume $f(x, y) = \frac{xy}{\sqrt{x^2+y^2}}$. Show that

$$\lim_{(x,y) \to (0,0)} f(x, y) = 0.$$

Proof. The domain of the function $f(x, y)$ is $D = \mathbb{R}^2 \setminus \{(0, 0)\}$. Since $|2xy| \leq x^2 + y^2$, it follows that

$$|f(x,y) - 0| = \left| \frac{xy}{\sqrt{x^2+y^2}} - 0 \right| = \frac{|xy|}{\sqrt{x^2+y^2}} \leq \frac{1}{2} \frac{x^2+y^2}{\sqrt{x^2+y^2}} = \frac{\sqrt{x^2+y^2}}{2}.$$

Therefore, for any $\varepsilon > 0$, we can ensure that $|f(x,y) - 0| < \varepsilon$ if we choose $\delta = 2\varepsilon$. This is because when

$$0 < \sqrt{(x-0)^2 + (y-0)^2} = \sqrt{x^2+y^2} < \delta = 2\varepsilon,$$

we have

$$|f(x,y) - 0| \leq \frac{\sqrt{x^2+y^2}}{2} < \frac{2\varepsilon}{2} = \varepsilon.$$

That is, whenever the point $P(x,y) \in \mathbb{R}^2 \setminus \{(0,0)\}$ and $|OP| < \delta = 2\varepsilon$, we always have

$$|f(x,y) - 0| < \varepsilon.$$

Thus,

$$\lim_{(x,y) \to (0,0)} f(x,y) = 0. \qquad \square$$

Note. The function $f(x,y)$ is not defined when $(x,y) = (0,0)$; however, a limit can exist at a point where the function is not defined.

The definitions and properties of limits of functions of two variables are very similar to those of one-variable functions, and can be extended in a very similar way to functions of more variables.

Theorem 2.1.1 (Limit laws). *Suppose that* $\lim_{(x,y) \to (a,b)} f(x,y) = L$, $\lim_{(x,y) \to (a,b)} g(x,y) = M$, *where L and M are real numbers. Then:*
1. $\lim_{(x,y) \to (a,b)} (f(x,y) \pm g(x,y)) = L \pm M$ *(sum/difference rule),*
2. $\lim_{(x,y) \to (a,b)} f(x,y) g(x,y) = LM$ *(product rule),*
3. $\lim_{(x,y) \to (a,b)} \frac{f(x,y)}{g(x,y)} = \frac{L}{M}$ *(quotient rule, given that $M \neq 0$).*

Example 2.1.7. Show that $\lim_{(x,y) \to (0,2)} \frac{\sin xy}{x} = 2$.

Solution. We know the one-variable limit $\lim_{x \to 0} \frac{\sin x}{x} = 1$, and we can use this here by considering the product xy to be a single variable such that $xy \to 0$ as $(x,y) \to (0,2)$, as follows:

$$\lim_{(x,y) \to (0,2)} \frac{\sin(xy)}{x} = \lim_{(x,y) \to (0,2)} \left[\frac{\sin(xy)}{xy} \cdot y\right] = \lim_{xy \to 0} \frac{\sin(xy)}{xy} \cdot \lim_{y \to 2} y = 1 \cdot 2 = 2.$$

Example 2.1.8. Evaluate $\lim_{(x,y) \to (0,1)} \frac{3 - \sqrt{xy+9}}{xy}$.

Solution. Let $t = xy$. Then $t \to 0$. We have

$$\lim_{(x,y) \to (0,1)} \frac{3 - \sqrt{xy+9}}{xy} = \lim_{t \to 0} \frac{3 - \sqrt{t+9}}{t} = \lim_{t \to 0} \frac{(3 - \sqrt{t+9})(3 + \sqrt{t+9})}{t(3 + \sqrt{t+9})}$$

$$= \lim_{t \to 0} \frac{-t}{t(3 + \sqrt{t+9})} = \lim_{t \to 0} \frac{-1}{3 + \sqrt{t+9}} = -\frac{1}{6}.$$

For limits of functions of a single variable, when we let x approach a number a, there are only two possible directions of approach: from the left or from the right. If the left-hand limit and right-hand limit differ, $\lim_{x \to a^-} f(x) \neq \lim_{x \to a^+} f(x)$, then $\lim_{x \to a} f(x)$ does not exist.

For functions of two variables, the situation is different. This is because we can let (x,y) approach (a,b) from an infinite number of directions in any manner so long as (x,y) stays within the domain of f, as shown in Figure 2.6(b). The existence of the limit $\lim_{(x,y)\to(a,b)} f(x,y)$ means that $f(x,y)$ approaches the same value no matter in what direction (x,y) approaches (a,b). Therefore, if there are two different routes for $(x,y) \to (a,b)$ along which the function $f(x,y)$ approaches different values, then we can conclude that the limit $\lim_{(x,y)\to(a,b)} f(x,y)$ does not exist.

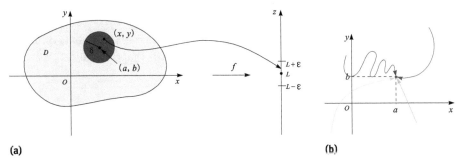

(a) (b)

Figure 2.6: Limits of functions of two variables.

Example 2.1.9. Investigate whether or not the limit $\lim_{(x,y)\to(0,0)} f(x,y)$ exists when f is defined in two parts by

$$f(x,y) = \begin{cases} \dfrac{xy}{x^2+y^2} & \text{when } (x,y) \neq (0,0), \\ 0 & \text{when } (x,y) = (0,0). \end{cases}$$

Solution. It is easy to check that if $(x,y) \to (0,0)$ along the x-axis, then $f(x,y) \to 0$. If $(x,y) \to (0,0)$ along the y-axis, then $f(x,y) \to 0$. Now, we have obtained identical limits along the two axes. However, this does not show that the given limit is 0. If we let (x,y) approach $(0,0)$ along the line $y = kx$, we have

$$\lim_{(x,y)\to(0,0), y=kx} \frac{xy}{x^2+y^2} = \lim_{x\to 0} \frac{kx^2}{x^2+k^2x^2} = \frac{k}{1+k^2}.$$

Obviously, this limit varies with different values of k. So the limit $\lim_{(x,y)\to(0,0)} f(x,y)$ does not exist.

Example 2.1.10. Investigate

$$\lim_{(x,y)\to(0,0)} \frac{x^2 y}{x^4 + y^2}.$$

Solution. Allowing (x, y) to approach $(0, 0)$ along any line $y = kx$, the limit is

$$\lim_{(x,y) \to (0,0)} \frac{x^2 y}{x^4 + y^2} = \lim_{x \to 0} \frac{kx^3}{x^4 + k^2 x^2} = \lim_{x \to 0} \frac{kx}{x^2 + k^2} = 0.$$

However, this is not sufficient to prove the existence of the limit, even though we have infinitely many paths along which the limits are all 0. Let us see what happens when the path is a parabola, say, $y = mx^2$. Then the limit is

$$\lim_{(x,y) \to (0,0)} \frac{x^2 y}{x^4 + y^2} = \lim_{x \to 0} \frac{mx^4}{x^4 + m^2 x^4} = \frac{m}{1 + m^2}.$$

The limit depends on m! This means that when (x, y) approaches $(0, 0)$ along different parabolas, we have different limits. Therefore, we conclude that the limit

$$\lim_{(x,y) \to (0,0)} \frac{x^2 y}{x^4 + y^2}$$

does not exist.

Iterated limits

Now we consider the two iterated limits

$$\lim_{x \to a} \lim_{y \to b} f(x, y) \quad \text{and} \quad \lim_{y \to b} \lim_{x \to a} f(x, y).$$

The limit $\lim_{x \to a} \lim_{y \to b} f(x, y)$ means we evaluate the one-variable limit $\lim_{y \to b} f(x, y)$ first holding x as a constant, and then evaluate the one-variable limit $\lim_{x \to a} (\lim_{y \to b} f(x, y))$ letting $x \to a$. Note that the two iterated limits are actually two specific paths by which a point (x, y) approaches (a, b). Therefore, we have the following theorem.

Theorem 2.1.2. *If* $\lim_{(x,y) \to (a,b)} f(x, y)$ *exists, then both* $\lim_{x \to a} \lim_{y \to b} f(x, y)$ *and* $\lim_{y \to b} \lim_{x \to a} f(x, y)$ *exist and*

$$\lim_{(x,y) \to (a,b)} f(x, y) = \lim_{x \to a} \lim_{y \to b} f(x, y) = \lim_{y \to b} \lim_{x \to a} f(x, y).$$

This theorem indicates a way to evaluate the limit of a function of two variables, that is, to evaluate two one-variable limits given that all limits involved exist. For instance,

$$\lim_{(x,y) \to (1,2)} (2x + 4y) = \lim_{x \to 1} \left(\lim_{y \to 2} (2x + 4y) \right) = \lim_{x \to 1} (2x + 8) = 10.$$

However, the converse is not true. For instance,

$$\lim_{x \to 0} \lim_{y \to 0} \frac{xy}{x^2 + y^2} = \lim_{y \to 0} \lim_{x \to 0} \frac{xy}{x^2 + y^2} = 0,$$

but $\lim_{(x,y) \to (0,0)} \frac{xy}{x^2+y^2}$ does not exist.

For functions of more than two variables, say, n variables, we can define the limit at a point $P_0 \in \mathbb{R}^n$ in a similar manner. In a compact notation, $\lim_{P \to P_0} f(P) = L$ means that for any given $\varepsilon > 0$, there is a $\delta > 0$ such that whenever $0 < |PP_0| < \delta$, we have $|f(P) - L| < \varepsilon$. Also, the limit laws apply to limits of functions of more than two variables as well.

Example 2.1.11. Find $\lim_{(x,y,z) \to (1,1,1)} \frac{\sqrt{xy} + \sqrt{yz} - \sqrt{xz} - z}{\sqrt{xz} + \sqrt{yz} - \sqrt{xy} - y}$.

Solution. This limit requires (x, y, z) to approach $(1, 1, 1)$, which is a boundary point of the domain of the function. We can assume all x, y, and z are positive and try factorization. We obtain

$$\lim_{(x,y,z) \to (1,1,1)} \frac{\sqrt{xy} + \sqrt{yz} - \sqrt{xz} - z}{\sqrt{xz} + \sqrt{yz} - \sqrt{xy} - y} = \lim_{(x,y,z) \to (1,1,1)} \frac{\sqrt{x}(\sqrt{y} - \sqrt{z}) + \sqrt{z}(\sqrt{y} - \sqrt{z})}{\sqrt{x}(\sqrt{z} - \sqrt{y}) + \sqrt{y}(\sqrt{z} - \sqrt{y})}$$

$$= \lim_{(x,y,z) \to (1,1,1)} \frac{(\sqrt{x} + \sqrt{z})(\sqrt{y} - \sqrt{z})}{(\sqrt{x} + \sqrt{y})(\sqrt{z} - \sqrt{y})}$$

$$= \lim_{(x,y,z) \to (1,1,1)} -\frac{(\sqrt{x} + \sqrt{z})}{(\sqrt{x} + \sqrt{y})} = -1.$$

2.1.5 Continuity

Recall that evaluating limits of a continuous function f of a single variable is easy because f is defined to be continuous at $x = a$ when $\lim_{x \to a} f(x) = f(a)$. That is, evaluation of the limit can be accomplished by direct substitution of $x = a$ in $f(x)$. This definition includes three separate conditions: (a) the limit must exist as $x \to a$, (b) the function value $f(a)$ must exist, and finally (c) the limit and this function value must be the same. Similarly, we can define the continuity for functions of two variables.

Definition 2.1.4. Let f be a function of two variables, and let (a, b) be in its domain $\mathcal{D}$. We say that f is **continuous** at (a, b) if

$$\lim_{(x,y) \to (a,b)} f(x, y) = f(a, b).$$

If f is not continuous at (a, b), then we say that f is **discontinuous** at (a, b) and that f has a **discontinuity** at (a, b). If f is continuous at every point in $\mathcal{D}$, then we say that f is **continuous on** $\mathcal{D}$.

For example, the function

$$f(x, y) = \begin{cases} \frac{xy}{x^2 + y^2} & \text{when } x^2 + y^2 \neq 0, \\ 0 & \text{when } x^2 + y^2 = 0 \end{cases}$$

is discontinuous at $(0, 0)$ because the limit does not even exist there, as shown in an earlier example.

All the points on the circle $C = \{(x,y)|x^2+y^2 = 1\}$ are discontinuities of the function $f(x,y) = \sin\frac{1}{x^2+y^2-1}$ because the function is not defined at any point of this circle. It is also possible to define a function f that is discontinuous at (a,b) such that the limit of f exists as $(x,y) \to (a,b)$ and $f(a,b)$ exists, but the two values are different.

Similar to the continuity of functions of one variable, all elementary functions of two variables are continuous on their natural domains. That is, the limit of an elementary function f of two variables at point (a,b) in its domain is given by

$$\lim_{(x,y)\to(a,b)} f(x,y) = f(a,b).$$

Example 2.1.12. Find $\lim_{(x,y)\to(1,\pi)} \frac{\cos(xy)+\sin(xy)}{x+y}$.

Solution. The function $\frac{\cos(xy)+\sin(xy)}{x+y}$ is an elementary function and its domain is

$$D = \{(x,y)|x+y \neq 0\}.$$

Since $(1,\pi)$ lies in its domain, this function is continuous at the point $(1,\pi)$ and

$$\lim_{(x,y)\to(1,\pi)} \frac{\cos(xy)+\sin(xy)}{x+y} = \frac{\cos\pi + \sin\pi}{1+\pi} = -\frac{1}{1+\pi} \approx -0.241.$$

The continuity of functions of more than two variables is defined similarly using the compact notation.

Definition 2.1.5. A function f of n variables is continuous at P_0, if and only if $\lim_{P\to P_0} f(P) = f(P_0)$, where P and P_0 are points in $\mathbb{R}^n$.

Finally, the *extreme value theorem*, *intermediate value theorem*, and *uniform continuity theorem* also hold for continuous functions f of n variables defined on a closed, bounded region in $\mathbb{R}^n$.

2.2 Partial derivatives

2.2.1 Definition

Suppose that a pollution index Q depends on two factors, x and y, which are outputs of pollutants from two factories. Now, you have a little money in hand and can invest it in only one of the factories to reduce the pollution index. Which factory will you put your money on? Of course, one would like to choose the factory whose small change in output of pollutants will result in the greatest drop in Q. This involves the idea of partial derivatives in which we hold one of the variables constant, and try to find the rate of change of a function with respect to the other variable. So a partial derivative is simply a one-variable differentiation applied to a two- (or more) variable function. That is, suppose f is a function of two variables, x and y, but we let only x vary while holding $y = b$ constant. We have then converted f into a function of a single variable, x, i.e., $g(x) = f(x, b)$. If $g(x) = f(x, b)$ has a derivative at a, then we call it the partial derivative of f with respect to x at (a, b) and denote it by $f_x(a, b)$ or $\frac{\partial f(a,b)}{\partial x}$. One-variable derivatives were defined in terms of a limit. The definition of partial derivatives for functions of two variables is also defined using limits in a similar way.

Definition 2.2.1. Let $f(x, y)$ be a real-valued function with domain $D \subset \mathbb{R}^2$, and let (a, b) be an interior point of D. The **partial derivative of f with respect to x at (a, b)** is denoted and defined by

$$\frac{\partial f(a,b)}{\partial x} = \lim_{\Delta x \to 0} \frac{f(a + \Delta x, b) - f(a, b)}{\Delta x}.$$

The **partial derivative of f with respect to y at (a, b)** is denoted and defined as

$$\frac{\partial f(a,b)}{\partial y} = \lim_{\Delta y \to 0} \frac{f(a, b + \Delta y) - f(a, b)}{\Delta y}.$$

Note.
1. The derivative $f'(x)$ is interpreted as rate of change. Partial derivatives are also rates of change. If $z = f(x, y)$, then $\partial f / \partial x$ represents the rate of change of z with respect to x when y is held constant. Similarly, $\partial f / \partial y$ represents the rate of change of z with respect to y when x is held constant.
2. Partial derivatives are sometimes called partials.
3. Sometimes, we use h in the above limits in place of Δx or Δy.

If $z = f(x, y)$ has a partial derivative with respect to x at all points $(x, y) \in D$, then $f_x(x, y)$ is also a function of x and y with domain D. We call it the partial derivative of $f(x, y)$ with respect to x and denote it by any of the following:

$$\frac{\partial z}{\partial x}, \frac{\partial f}{\partial x}, \frac{\partial f(x, y)}{\partial x}, z_x, \text{ or } f_x(x, y).$$

Similarly, we denote the partial derivative of $f(x, y)$ with respect to y by

$$\frac{\partial z}{\partial y}, \frac{\partial f}{\partial y}, \frac{\partial f(x, y)}{\partial y}, z_y, \text{ or } f_y(x, y).$$

Sometimes the subscript "x" is replaced by the number 1, "y" by 2, and so on, so that

$$\frac{\partial f}{\partial x} = f_1, \quad \frac{\partial f}{\partial y} = f_2.$$

So, to compute the partial derivative f_x or f_y, all we have to remember is that it is just the ordinary one-variable derivative where we regard y or x as a constant, and we can, therefore, apply all the derivative laws for functions of a single variable when finding f_x or f_y.

Example 2.2.1. If $f(x,y) = \cos(\frac{x}{2+y})$, compute $\frac{\partial f}{\partial x}$ and $\frac{\partial f}{\partial y}$.

Solution. To compute $\frac{\partial f}{\partial x}$, we regard y as a constant. Using the chain rule for functions of one variable, we have

$$\frac{\partial f}{\partial x} = -\sin\left(\frac{x}{2+y}\right) \cdot \frac{\partial}{\partial x}\left(\frac{x}{2+y}\right) = -\sin\left(\frac{x}{2+y}\right) \cdot \frac{1}{2+y}.$$

Similarly, we compute $\frac{\partial f}{\partial y}$ as follows:

$$\frac{\partial f}{\partial y} = -\sin\left(\frac{x}{2+y}\right) \cdot \frac{\partial}{\partial y}\left(\frac{x}{2+y}\right) = \sin\left(\frac{x}{2+y}\right) \cdot \frac{x}{(2+y)^2}.$$

Example 2.2.2. If $f(x,y) = 2x^2 y - 3xy^2 + 2x - y^2 + 3$, then find $\frac{\partial f}{\partial x}(2,-3)$ and $\frac{\partial f}{\partial y}(2,-3)$ by using one-variable differentiation formulas and a second time by using the definition as a limit.

Solution. Method 1: We have

$$\frac{\partial f(x,y)}{\partial x} = \frac{\partial}{\partial x}(2x^2 y - 3xy^2 + 2x - y^2 + 3)$$
$$= 4xy - 3y^2 + 2.$$

So

$$\frac{\partial f}{\partial x}(2,-3) = 4xy - 3y^2 + 2\Big|_{\substack{x=2\\y=-3}} = 4 \cdot 2 \cdot (-3) - 3 \cdot (-3)^2 + 2 = -49.$$

Method 2: We have

$$\frac{\partial f}{\partial x}\Big|_{x=2,y=-3} = \lim_{\Delta x \to 0} \frac{f(2+\Delta x, -3) - f(2,-3)}{\Delta x}$$
$$= \lim_{\Delta x \to 0} \frac{(-6(2+\Delta x)^2 - 25(2+\Delta x) - 6) + 80}{\Delta x}$$
$$= \lim_{\Delta x \to 0} \frac{-6\Delta x^2 - 49\Delta x}{\Delta x}$$
$$= -49.$$

Method 3: We can plug in $y = -3$ first to get

$$f(x,-3) = 2x^2(-3) - 3x(-3)^2 + 2x - (-3)^2 + 3$$
$$= -6x^2 - 25x - 6.$$

Then $f_x(x,-3) = -12x - 25$ and so $f_x(2,-3) = -12(2) - 25 = -49$.
Similarly, we get $\frac{\partial f(2,-3)}{\partial y} = 50$.

Example 2.2.3. The **Cobb–Douglas production function** is defined as

$$P(L,C) = bL^\alpha C^{1-\alpha},$$

where P is the total production (the monetary value of all goods produced in a period), L is the amount of labor (some measure of the total labor used in that period), and C is the amount of capital invested (the monetary worth of all machinery, equipment, and buildings); b and α are constants. Find the partial derivatives P_L and P_C.

Solution. We have

$$P_L(L,C) = \alpha b L^{\alpha-1} C^{1-\alpha} \quad \text{and} \quad P_C = b(1-\alpha) L^\alpha C^{-\alpha}.$$

Note. In 1928 Charles Cobb and Paul Douglas used this function to model the growth of American economy during the period 1899–1922. Their model turned out to be remarkably accurate even though there were many factors affecting economic performance.

Example 2.2.4. The ideal gas equation is given by $pV = RT$, where R is a constant. Show that

$$\frac{\partial p}{\partial V} \cdot \frac{\partial V}{\partial T} \cdot \frac{\partial T}{\partial p} = -1.$$

Proof. From

$$p = \frac{RT}{V} \implies \frac{\partial p}{\partial V} = -\frac{RT}{V^2},$$

$$V = \frac{RT}{p} \implies \frac{\partial V}{\partial T} = \frac{R}{p},$$

$$T = \frac{pV}{R} \implies \frac{\partial T}{\partial p} = \frac{V}{R},$$

it follows that

$$\frac{\partial p}{\partial V} \cdot \frac{\partial V}{\partial T} \cdot \frac{\partial T}{\partial p} = -\frac{RT}{V^2} \cdot \frac{R}{p} \cdot \frac{V}{R} = -\frac{RT}{pV} = -1. \qquad \square$$

Note. This example also shows that partial derivatives cannot be interpreted as ratios of differentials, as otherwise $\frac{\partial p}{\partial V} \cdot \frac{\partial V}{\partial T} \cdot \frac{\partial T}{\partial p}$ would be equal to 1.

Example 2.2.5. Investigate $\frac{\partial f}{\partial x}$ and $\frac{\partial f}{\partial y}$ at $(0,0)$ for $f(x,y) = \sqrt{x^2 + y^2}$.

Solution. By applying the definition of partial differentiation, we have

$$\frac{\partial f(0,0)}{\partial x} = \lim_{\Delta x \to 0} \frac{f(0+\Delta x, 0) - f(0,0)}{\Delta x}$$

$$= \lim_{\Delta x \to 0} \frac{\sqrt{(0+\Delta x)^2 + 0^2} - \sqrt{0^2 + 0^2}}{\Delta x}$$

$$= \lim_{\Delta x \to 0} \frac{|\Delta x|}{\Delta x}.$$

We know that this limit does not exist as the two one-sided limits $\Delta x \to 0^+$ and $\Delta x \to 0^-$ do not match. Therefore, $\frac{\partial f(0,0)}{\partial x}$ does not exist. Similarly, $\frac{\partial f(0,0)}{\partial y}$ does not exist either.

Now, we have seen that for some functions the partial derivatives exist, and for some functions they do not. Besides using the definition, we shall try to interpret partials geometrically.

2.2.2 Interpretations of partial derivatives

There is a geometric interpretation of partial derivatives $f_x(a, b)$ and $f_y(a, b)$. When we compute $\frac{\partial f}{\partial x}$, we keep y fixed, say, $y = b$. Therefore, we only consider the points (x, b, z). So

$$z = f(x, y) \quad \text{and} \quad y = b,$$

and this is actually the intersection curve C of the plane $y = b$ and the surface S, the graph of $z = f(x, y)$. The derivative $f_x(a, b)$ is therefore the slope of the tangent line to the curve C at $(a, b, f(a, b))$ on S. Similarly, $f_y(a, b)$ is the slope of the tangent line to the curve $z = f(a, y)$ in the $x = a$ plane at the point $(a, b, f(a, b))$. Figure 2.7(a) illustrates the geometric interpretation of partial derivatives.

Now, we can explain why the function $z = \sqrt{x^2 + y^2}$ has no partials at $(0, 0)$. The graph of this function is an upper cone with its vertex at the origin. The plane $y = 0$

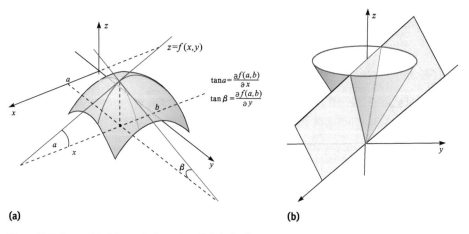

Figure 2.7: Geometric interpretation of partial derivatives.

intersects the cone in two lines $z = x$ and $z = -x$ in the xz-plane. The origin is a corner of the intersection lines, and, therefore, it has no derivative there, as shown in Figure 2.7(b).

We already know that when a function of one variable has a derivative at a point P, it must also be continuous at the point P. However, continuity does not follow for a function of two variables just because it has partial derivatives at a point. Take, for example, the function $f(x,y)$ equal to $x+y$ when either x or y equals 0, but with $f(x,y) = 4$ at all other points. This function has partial derivatives with respect to x and y equal to 1 at $(0,0)$, but the function is clearly not continuous at $(0,0)$. The next example shows that a function can have partial derivatives at every point yet still be a discontinuous function.

Example 2.2.6. Find the partial derivatives of the function

$$f(x,y) = \begin{cases} \frac{xy}{x^2+y^2} & \text{when } x^2+y^2 \neq 0, \\ 0 & \text{when } x^2+y^2 = 0. \end{cases}$$

Solution. If $(x,y) \neq (0,0)$, the one-variable quotient rule gives

$$\frac{\partial f(x,y)}{\partial x} = \left(\frac{xy}{x^2+y^2}\right)_x = \frac{y(x^2+y^2) - xy \cdot 2x}{(x^2+y^2)^2} = \frac{y(y^2-x^2)}{(x^2+y^2)^2}.$$

If $(x,y) = (0,0)$, then using the definition of partial differentiation we find

$$\frac{\partial f(0,0)}{\partial x} = \lim_{\Delta x \to 0} \frac{f(0+\Delta x, 0) - f(0,0)}{\Delta x} = \lim_{\Delta x \to 0} \frac{0-0}{\Delta x} = 0.$$

Similarly, when $(x,y) \neq (0,0)$,

$$\frac{\partial f(x,y)}{\partial y} = \frac{x(x^2-y^2)}{(x^2+y^2)^2}$$

and

$$\frac{\partial f(0,0)}{\partial y} = 0.$$

So, this function has a partial derivative at every point (x,y); however, we have already seen in a previous example that this function is not continuous at the point $(0,0)$.

The concept of partial derivatives can be extended to functions of more than two variables in a natural way. For instance, a function $u = f(x,y,z)$ generally has three partial derivatives and the partial derivative of the function with respect to x is defined as

$$f_x(x,y,z) = \lim_{\Delta x \to 0} \frac{f(x+\Delta x, y, z) - f(x,y,z)}{\Delta x},$$

where (x, y, z) is an interior point in the domain of u. However, there is no nice geometric interpretation for f_x as a slope of some visible tangent line. To find the partial derivative f_x we hold y and z constant and use the one-variable derivative rules to find f_x. In a similar manner, we can find f_y and f_z.

Example 2.2.7. Find the partial derivatives of the function $r = \sqrt{x^2 + y^2 + z^2}$.

Solution. To find $\frac{\partial r}{\partial x}$, we regard y and z as constants. Then we find

$$\frac{\partial r}{\partial x} = \frac{2x}{2\sqrt{x^2 + y^2 + z^2}} = \frac{x}{\sqrt{x^2 + y^2 + z^2}} = \frac{x}{r}.$$

By symmetry, $\frac{\partial r}{\partial y} = \frac{y}{r}$ and $\frac{\partial r}{\partial z} = \frac{z}{r}$.

2.2.3 Partial derivatives of higher order

For a function $z = f(x, y)$, the partial derivatives $f_x(x, y)$ and $f_y(x, y)$ can themselves be differentiated, giving four more derivatives, i.e., $(f_x)_x$, $(f_x)_y$, $(f_y)_x$, $(f_y)_y$, called *second derivatives* or *second-order partial derivatives*. The standard notation for these second-order partial derivatives are similar to the notations y'' and d^2y/dx^2 for the second derivatives of a function $y = f(x)$ of a single variable. We write

$$(f_x)_x = f_{xx} = \frac{\partial}{\partial x}\left(\frac{\partial f}{\partial x}\right) = \frac{\partial^2 f}{\partial x^2} = \frac{\partial^2 z}{\partial x^2},$$

$$(f_x)_y = f_{xy} = \frac{\partial}{\partial y}\left(\frac{\partial f}{\partial x}\right) = \frac{\partial^2 f}{\partial y \partial x} = \frac{\partial^2 z}{\partial y \partial x},$$

$$(f_y)_x = f_{yx} = \frac{\partial}{\partial x}\left(\frac{\partial f}{\partial y}\right) = \frac{\partial^2 f}{\partial x \partial y} = \frac{\partial^2 z}{\partial x \partial y},$$

$$(f_y)_y = f_{yy} = \frac{\partial}{\partial y}\left(\frac{\partial f}{\partial y}\right) = \frac{\partial^2 f}{\partial y^2} = \frac{\partial^2 z}{\partial y^2}.$$

The notation f_{xy} or $\frac{\partial^2 f}{\partial y \partial x}$ means that we first differentiate with respect to x (keeping y constant) and then differentiate with respect y (keeping x constant), whereas in computing $f_{yx} = \frac{\partial^2 f}{\partial x \partial y}$, the differentiation order is reversed.

Example 2.2.8. Find the second-order partial derivatives of

$$f(x, y) = x^2 e^y + y \cos x.$$

Solution. We find the partials f_x and f_y, and we differentiate each of these functions with respect to each of x and y to give f_{xx}, f_{xy}, f_{yx}, and f_{yy}. Then we obtain the following:

$$f_x = 2xe^y - y \sin x,$$

$$f_y = x^2 e^y + \cos x,$$
$$f_{xx} = 2e^y - y\cos x,$$
$$f_{xy} = 2xe^y - \sin x,$$
$$f_{yx} = 2xe^y - \sin x,$$
$$f_{yy} = x^2 e^y.$$

We note that for this function, whose second-order partial derivatives are continuous, the "mixed partials" f_{xy} and f_{yx} are equal. This result holds generally for functions with continuous second-order derivatives.

Theorem 2.2.1 (Clairaut's theorem: equality of mixed partial derivatives). *If $z = f(x, y)$ is defined and has continuous second-order partial derivatives throughout a domain $\mathcal{D}$, then the two functions f_{xy} and f_{yx} are identical at any interior point of $\mathcal{D}$.*

Notations for third- and higher-order partial derivatives are defined in a similar way.

Example 2.2.9. Calculate f_{xxyz} when $f(x, y, z) = \sin(3x + 2yz)$.

Solution. We have
$$f_x = 3\cos(3x + 2yz),$$
$$f_{xx} = -9\sin(3x + 2yz),$$
$$f_{xxy} = -18z\cos(3x + 2yz),$$
$$f_{xxyz} = -18\cos(3x + 2yz) + 36yz\sin(3x + 2yz).$$

2.3 Total differential

2.3.1 Linearization and differentiability

In one-variable calculus, if a function $y = f(x)$ is differentiable at $x = a$ this means that the change Δy in y can be written as

$$\Delta y = A\Delta x + o(\Delta x),$$

where A is a constant that only depends on the point a, not the change Δx in x. It turns out that the constant A is exactly $f'(a)$, the derivative at $x = a$. Thus,

$$\Delta y = f'(a)\Delta x + o(\Delta x).$$

When Δx is small $\Delta y = y - f(a) \approx f'(a)\Delta x$, which can be rewritten as

$$y \approx f(a) + f'(a)(x - a).$$

The function $L(x) = f(a) + f'(a)(x - a)$ is the local linearization of $f(x)$ at $x = a$, which is, in fact, the tangent line approximation of f at $x = a$.

A similar approximation can be made for a function of two variables, $z = f(x,y)$. Consider a small change Δz in z at a point (a,b), caused by changes Δx in x and Δy in y:

$$\Delta z = f(a + \Delta x, b + \Delta y) - f(x,y).$$

The increment Δz represents the change in the value of f when (a,b) changes from (a,b) to $(a + \Delta x, b + \Delta y)$. In general, the exact increment Δz in z is hard to find. For example, for $z = x^y$ at $(1,1)$ with $\Delta x = 0.09$ and $\Delta y = -0.02$, $\Delta z = 1.09^{0.98} - 1^1$. However, even though z is not a linear function, Δz can be very close to a linear expression of Δx and Δy, and the difference is negligible as $\Delta x \to 0$ and $\Delta y \to 0$. When this happens, we say that the function $z = f(x,y)$ is differentiable at the point (a,b). The formal definition is given below.

Definition 2.3.1 (Differentiability of a real-valued function $z = f(x,y)$ **of two variables**). Assume that f is a real-valued function with domain $D \subset \mathbb{R}^2$ and that (a,b) is an interior point of D. The function f is **differentiable** at (a,b) if there exist constants A and B such that

$$\Delta z = A\Delta x + B\Delta y + o(\rho),$$

where A and B depend on a and b but are independent of Δx and Δy, and $\rho = \sqrt{(\Delta x)^2 + (\Delta y)^2}$. If $z = f(x,y)$ is differentiable at every point in D, then we say that z is *differentiable on D*.

Note that if $z = f(x,y)$ is differentiable at (a,b), then $\Delta z \approx A\Delta x + B\Delta y$ for some constants A and B. This expression can be rewritten as

$$f(x,y) \approx f(a,b) + A(x - a) + B(y - b).$$

The formula $L(x,y) = f(a,b) + A(x-a) + B(y-b)$ gives the local linearization of $z = f(x,y)$ at (a,b), which is the equation of a plane that approximates the surface well at points near $(a,b,f(a,b))$. This plane, as we will see later, is indeed the *tangent plane* to the graph of $z = f(x,y)$ at $(a,b,f(a,b))$. So, intuitively speaking, if a function $z = f(x,y)$ is differentiable at a point, then its graph must be continuous and "smooth" at that point so that there exists a plane that could nicely touch (be tangent to) the surface at that point. This is indeed the case, as shown in the following two theorems.

Theorem 2.3.1. *If $z = f(x,y)$ is differentiable at (a,b), then it must be continuous at the point (a,b).*

Proof. In fact, from the above definition, if $z = f(x,y)$ is differentiable at (a,b), then

$$\Delta z = A\Delta x + B\Delta y + o(\sqrt{(\Delta x)^2 + (\Delta y)^2}),$$

$$\lim_{(\Delta x, \Delta y) \to (0,0)} \Delta z = \lim_{(\Delta x, \Delta y) \to (0,0)} \left(A\Delta x + B\Delta y + o(\sqrt{(\Delta x)^2 + (\Delta y)^2}) \right) = 0.$$

But $\Delta z = f(a + \Delta x, b + \Delta y) - f(a, b) = f(x, y) - f(a, b)$; therefore,

$$\lim_{(\Delta x, \Delta y) \to (0,0)} \Delta z = \lim_{(x,y) \to (a,b)} (f(x,y) - f(a,b)) = 0,$$

$$\lim_{(x,y) \to (a,b)} f(x,y) = f(a,b),$$

which means $z = f(x, y)$ is continuous at (a, b). □

By the above theorem, we know the following.

If a function is not continuous at (a, b), then it is not differentiable at (a, b).

Therefore, functions such as

$$f(x, y) = \begin{cases} \frac{xy}{x^2+y^2} & (x, y) \neq (0, 0), \\ 0 & (x, y) = (0, 0) \end{cases} \quad \text{and} \quad f(x, y) = \frac{xy^2}{x^4 + y^4}$$

are not differentiable at $(0, 0)$ since they are not continuous there.

Also, as in one-variable calculus, if $y = f(x)$ is differentiable at $x = a$, then dy/dx exists at $x = a$. For a function of two variables, we claim that if it is differentiable, then it has partial derivatives.

Theorem 2.3.2. *If $z = f(x, y)$ is differentiable at point (a, b), then the partial derivatives $\frac{\partial z(a,b)}{\partial x}$ and $\frac{\partial z(a,b)}{\partial y}$ exist. Furthermore,*

$$\Delta z = \frac{\partial z(a, b)}{\partial x} \Delta x + \frac{\partial z(a, b)}{\partial y} \Delta y + o(\rho).$$

Proof. If the function $z = f(x, y)$ is differentiable at a point (a, b), then there exist A and B, independent of Δx and Δy, such that

$$\Delta z = A \Delta x + B \Delta y + o(\rho), \quad \text{where } \rho = \sqrt{(\Delta x)^2 + (\Delta y)^2}.$$

In particular, when $\Delta y = 0$ (that is, $y = b$), then $\rho = |\Delta x|$ and

$$\Delta z = A \Delta x + o(|\Delta x|).$$

Dividing both sides by Δx gives

$$\frac{\Delta z}{\Delta x} = A \frac{\Delta x}{\Delta x} + \frac{o(|\Delta x|)}{\Delta x}.$$

Note that in the case of $\Delta y = 0$, $\Delta z = f(a + \Delta x, b) - f(a, b)$. Then, when taking the limit of both sides as $\Delta x \to 0$, we obtain

$$\frac{\partial z(a, b)}{\partial x} = A.$$

Similarly, $\frac{\partial z(a,b)}{\partial y} = B$. This completes the proof. □

Note. By the above theorem, we conclude the following.

If either $f_x(a, b)$ or $f_y(a, b)$ does not exist, then $f(x, y)$ is not differentiable at (a, b).

If one replaces the point (a, b) by a general point (x, y), then when a function $z = f(x, y)$ is differentiable at (x, y), we have

$$\Delta z = \frac{\partial z}{\partial x}\Delta x + \frac{\partial z}{\partial y}\Delta y + o(\rho).$$

Example 2.3.1. Determine whether the following functions are differentiable at $(0, 0)$:

(1) $f(x, y) = \sin\frac{1}{x^2 + y^2}$, (2) $g(x, y) = \begin{cases} \frac{xy}{\sqrt{x^2+y^2}} & (x, y) \neq (0, 0), \\ 0 & (x, y) = (0, 0), \end{cases}$ and

(3) $h(x, y) = \begin{cases} (x^2 + y^2)\sin\frac{1}{x^2+y^2} & (x, y) \neq (0, 0), \\ 0 & (x, y) = (0, 0). \end{cases}$

Solution.
1. Since $f(x, y) = \sin\frac{1}{x^2+y^2}$ is undefined at $(0, 0)$, it is not continuous at $(0, 0)$. Therefore, it is not differentiable at $(0, 0)$.
2. As seen in a previous example, $\lim_{(x,y)\to(0,0)} \frac{xy}{\sqrt{x^2+y^2}} = 0 = g(0, 0)$. Therefore, $g(x, y)$ is continuous at $(0, 0)$. Computing $g_x(0, 0)$, we have

$$g_x(0, 0) = \lim_{\Delta x \to 0} \frac{g(0 + \Delta x, 0) - g(0, 0)}{\Delta x} = \lim_{\Delta x \to 0} \frac{\frac{\Delta x \cdot 0}{\sqrt{(\Delta x)^2 + 0^2}} - 0}{\Delta x} = 0.$$

Similarly, $g_y(0, 0) = 0$. Thus, if $g(x, y)$ is differentiable at $(0, 0)$, we must have

$$\Delta g(x, y) = g_x(0, 0)\Delta x + g_y(0, 0)\Delta y + o(\rho),$$
$$g(0 + \Delta x, 0 + \Delta y) - g(0, 0) = g_x(0, 0)\Delta x + g_y(0, 0)\Delta y + o(\rho),$$
$$\frac{\Delta x \Delta y}{\sqrt{(\Delta x)^2 + (\Delta y)^2}} = 0 + 0 + o(\rho), \quad \text{where } \rho = \sqrt{(\Delta x)^2 + (\Delta y)^2}.$$

That is, $\frac{\Delta x \Delta y}{\sqrt{(\Delta x)^2+(\Delta y)^2}}$ is negligible with respect to $\sqrt{(\Delta x)^2 + (\Delta y)^2}$, as $(\Delta x, \Delta y) \to (0, 0)$, i.e.,

$$\lim_{(\Delta x, \Delta y) \to (0,0)} \frac{\frac{\Delta x \Delta y}{\sqrt{(\Delta x)^2+(\Delta y)^2}}}{\sqrt{(\Delta x)^2 + (\Delta y)^2}} = 0.$$

However,

$$\lim_{(\Delta x, \Delta y) \to (0,0)} \frac{\frac{\Delta x \Delta y}{\sqrt{(\Delta x)^2+(\Delta y)^2}}}{\sqrt{(\Delta x)^2 + (\Delta y)^2}} = \lim_{(\Delta x, \Delta y) \to (0,0)} \frac{\Delta x \Delta y}{(\Delta x)^2 + (\Delta y)^2}.$$

This limit does not exist as if $\Delta y = k\Delta x$, the limit depends on k. So, we reached a contradiction. So $g(x,y)$ is not differentiable at $(0,0)$.

3. For $h(x,y)$, it is easy to see that $\lim_{(x,y)\to(0,0)} h(x,y) = 0 = h(0,0)$, so $h(x,y)$ is continuous at $(0,0)$. We now compute

$$h_x(0,0) = \lim_{\Delta x \to 0} \frac{h(0+\Delta x, 0) - h(0,0)}{\Delta x} = \lim_{\Delta x \to 0} \frac{((\Delta x)^2 + 0^2)\sin \frac{1}{(\Delta x)^2 + 0^2} - 0}{\Delta x} = 0.$$

Similarly, $h_y(0,0) = 0$. We need to check whether the difference between Δh and the linearization $h_x(0,0)\Delta x + h_y(0,0)\Delta y$ is negligible with respect to $\sqrt{(\Delta x)^2 + (\Delta y)^2}$ as $(\Delta x, \Delta y) \to (0,0)$. We write

$$\lim_{(\Delta x, \Delta y) \to (0,0)} \frac{\Delta h - h_x(0,0)\Delta x - h_y(0,0)\Delta y}{\sqrt{(\Delta x)^2 + (\Delta y)^2}}$$

$$= \lim_{(\Delta x, \Delta y) \to (0,0)} \frac{((\Delta x)^2 + (\Delta y)^2)\sin \frac{1}{(\Delta x)^2 + (\Delta y)^2} - 0}{\sqrt{(\Delta x)^2 + (\Delta y)^2}}$$

$$= \lim_{(\Delta x, \Delta y) \to (0,0)} \sqrt{(\Delta x)^2 + (\Delta y)^2} \sin \frac{1}{(\Delta x)^2 + (\Delta y)^2} = 0.$$

So $\Delta h - h_x(0,0)\Delta x - h_y(0,0)\Delta y = o(\sqrt{(\Delta x)^2 + (\Delta y)^2})$. That is,

$$\Delta h = h_x(0,0)\Delta x + h_y(0,0)\Delta y + o(\sqrt{(\Delta x)^2 + (\Delta y)^2}),$$

and the function $h(x,y)$ is differentiable at $(0,0)$.

Now, we have seen some nice properties of a differentiable function of two variables. However, how can we determine whether a function is differentiable? In one-variable calculus, we know that as long as $f'(a)$ exists, $y = f(x)$ is differentiable at $x = a$. But in multivariable calculus, this is not the case. The previous theorem shows that existence of the partial derivatives is a necessary condition for differentiability, but it is not a sufficient condition for differentiability. In Example 2.2.6, we saw that the function has partials at $(0,0)$ but is not continuous there; therefore, it is not differentiable there. For a sufficient condition, we have the following theorem.

Theorem 2.3.3 (Test for differentiability of a real-valued function). *Let $z = f(x,y)$ be a real-valued function of two variables and (a,b) an interior point of its domain D. If $\frac{\partial z}{\partial x}$ and $\frac{\partial z}{\partial y}$ are continuous at (a,b), then f is differentiable at (a,b).*

Proof. Recall the Lagrange mean value theorem for differentiable functions of one variable, $y = f(x)$,

$$f(x) - f(a) = f'(\xi)(x-a), \quad \text{where } \xi \text{ is some number between } a \text{ and } x, \text{ or}$$

$$f(a + \Delta x) - f(a) = f'(a + \theta \Delta x)\Delta x, \quad \text{where } 0 < \theta < 1.$$

Therefore,

$$\Delta z = f(a + \Delta x, b + \Delta y) - f(a, b)$$
$$= f(a + \Delta x, b) - f(a, b) + f(a + \Delta x, b + \Delta y) - f(a + \Delta x, b)$$
$$= f_x(a + \theta_1 \Delta x, b)\Delta x + f_y(a + \Delta x, b + \theta_2 \Delta y)\Delta y, \quad \text{where } 0 < \theta_1, \theta_2 < 1.$$

Since $\frac{\partial z}{\partial x} = f_x(x, y)$ and $\frac{\partial z}{\partial y} = f_y(a, b)$ are both continuous at (a, b), we have

$$\lim_{(\Delta x, \Delta y) \to (0,0)} f_x(a + \theta_1 \Delta x, b) = f_x(a, b) \quad \text{and} \quad \lim_{(\Delta x, \Delta y) \to (0,0)} f_y(a + \Delta x, b + \theta_2 \Delta y) = f_y(a, b).$$

Thus, there exist ε_1 and ε_2 such that $\varepsilon_1 \to 0$, $\varepsilon_2 \to 0$ as $\Delta x \to 0$ and $\Delta y \to 0$, and

$$f_x(a + \theta_1 \Delta x, b) = f_x(a, b) + \varepsilon_1 \quad \text{and} \quad f_y(a + \Delta x, b + \theta_2 \Delta y) = f_y(a, b) + \varepsilon_2.$$

So,

$$\Delta z = f_x(a + \theta_1 \Delta x, b)\Delta x + f_y(a + \Delta x, b + \theta_2 \Delta y)\Delta y$$
$$= (f_x(a, b) + \varepsilon_1)\Delta x + (f_y(a, b) + \varepsilon_2)\Delta y$$
$$= f_x(a, b)\Delta x + f_y(a, b)\Delta y + \varepsilon_1 \Delta x + \varepsilon_2 \Delta y.$$

Furthermore,

$$\lim_{(\Delta x, \Delta y) \to (0,0)} \frac{\Delta z - f_x(a, b)\Delta x - f_y(a, b)\Delta y}{\sqrt{(\Delta x)^2 + (\Delta y)^2}} = \lim_{(\Delta x, \Delta y) \to (0,0)} \frac{\varepsilon_1 \Delta x + \varepsilon_2 \Delta y}{\sqrt{(\Delta x)^2 + (\Delta y)^2}}$$
$$= \lim_{(\Delta x, \Delta y) \to (0,0)} \varepsilon_1 \left(\frac{\Delta x}{\sqrt{(\Delta x)^2 + (\Delta y)^2}} \right) + \varepsilon_2 \left(\frac{\Delta y}{\sqrt{(\Delta x)^2 + (\Delta y)^2}} \right)$$
$$= 0 + 0 = 0.$$

Therefore,

$$\Delta z - f_x(a, b)\Delta x - f_y(a, b)\Delta y = o(\sqrt{(\Delta x)^2 + (\Delta y)^2}) \quad \text{or}$$
$$\Delta z = f_x(a, b)\Delta x + f_y(a, b)\Delta y + o(\sqrt{(\Delta x)^2 + (\Delta y)^2}),$$

which proves that $z = f(x, y)$ is differentiable at (a, b). □

The converse of this theorem is not true. A counterexample is the following:

$$f(x, y) = \begin{cases} (x^2 + y^2) \sin \frac{1}{x^2 + y^2} & \text{when } x^2 + y^2 \neq 0, \\ 0 & \text{when } x^2 + y^2 = 0. \end{cases}$$

One can show that the partial derivatives of $f(x, y)$ exist, and f is differentiable, but the partial derivatives are not continuous at $(0, 0)$.

2.3.2 The total differential

In one-variable calculus, we defined the differential dy to be the function of the differential dx by

$$dy = f'(x)dx,$$

and when $dx = \Delta x$, we have $dy = f'(x)\Delta x$. Similarly, we define the differential dz as follows.

Definition 2.3.2. If $z = f(x, y)$ is differentiable at (a, b), the differential dz (sometimes called the total differential) at (a, b) is the function defined by

$$dz = \frac{\partial z(a, b)}{\partial x}dx + \frac{\partial z(a, b)}{\partial y}dy,$$

where the two independent variables are the differentials dx and dy.

Example 2.3.2. When $z = e^{xy}$, find $dz(2, 1)$.

Solution. Since

$$\frac{\partial z}{\partial x} = f_x = ye^{xy} \quad \text{and} \quad \frac{\partial z}{\partial y} = f_y = xe^{xy}$$

are both continuous at $(2, 1)$, by the definition of dz at a point (a, b), we have

$$dz = \frac{\partial z(2, 1)}{\partial x}dx + \frac{\partial z(2, 1)}{\partial y}dy$$
$$= e^2 dx + 2e^2 dy.$$

If the total differential of $z = f(x, y)$ exists for all points in a region D, then dz becomes a function of $(x, y) \in D$ and $(dx, dy) \in \mathbb{R}^2$.

Example 2.3.3. Find the differential $dz = df(x, y)$ when f is the function defined by

$$z = f(x, y) = 1 + \ln(3 - x^2 - y^2) \quad \text{for } (x, y) \in D = [-1, 1] \times [-1, 1].$$

Solution. Since

$$\frac{\partial z}{\partial x} = f_x = \frac{-2x}{3 - x^2 - y^2} \quad \text{and} \quad \frac{\partial z}{\partial y} = f_y = \frac{-2y}{3 - x^2 - y^2}$$

are both continuous on D, by the definition of dz, we have

$$dz = \frac{\partial f}{\partial x}dx + \frac{\partial f}{\partial y}dy$$

$$= (1 + \ln(3 - x^2 - y^2))_x dx + (1 + \ln(3 - x^2 - y^2))_y dy$$

$$= \frac{-2x}{3 - x^2 - y^2} dx + \frac{-2y}{3 - x^2 - y^2} dy.$$

The definition for a total differential can be extended naturally to differentiable functions of more than two variables. For instance, if $u = f(x, y, z)$ is differentiable, then the total differential du is defined as

$$du = \frac{\partial u}{\partial x} dx + \frac{\partial u}{\partial y} dy + \frac{\partial u}{\partial z} dz.$$

Example 2.3.4. Find du if $u = x + \sin\frac{y}{2} + e^{yz}$.

Solution. Since

$$\frac{\partial u}{\partial x} = 1, \quad \frac{\partial u}{\partial y} = \frac{1}{2}\cos\frac{y}{2} + ze^{yz}, \quad \frac{\partial u}{\partial z} = ye^{yz},$$

it follows that

$$du = \frac{\partial u}{\partial x} dx + \frac{\partial u}{\partial y} dy + \frac{\partial u}{\partial z} dz = dx + \left(\frac{1}{2}\cos\frac{y}{2} + ze^{yz}\right) dy + ye^{yz} dz.$$

2.3.3 The linear/differential approximation

In one-variable calculus, when $dx = \Delta x$, then $dy = f'(a)\Delta x$ and $\Delta y \approx dy$. So the linearization is essentially the differential approximation. Similarly, for a differentiable function $z = f(x, y)$ of two variables, when $dx = \Delta x$ and $dy = \Delta y$, then at (a, b) we have

$$dz = \frac{\partial z(a, b)}{\partial x} \Delta x + \frac{\partial z(a, b)}{\partial y} \Delta y.$$

Thus,

$$\Delta z = dz + o(\sqrt{(\Delta x)^2 + (\Delta y)^2}).$$

When Δx and Δy are both small, this gives the approximation

$$\Delta z = f(x, y) - f(a, b) \approx dz = \frac{\partial z(a, b)}{\partial x} \Delta x + \frac{\partial z(a, b)}{\partial y} \Delta y, \tag{2.1}$$

or, equivalently,

$$f(x, y) \approx f(a, b) + \frac{\partial z(a, b)}{\partial x} \Delta x + \frac{\partial z(a, b)}{\partial y} \Delta y. \tag{2.2}$$

The function
$$L(x,y) = f(a,b) + \frac{\partial z(a,b)}{\partial x}(x-a) + \frac{\partial z(a,b)}{\partial y}(y-b)$$
is called the *local linearization* of the function $z = f(x,y)$ at the point (a,b). This is essentially the total differential approximation.

Example 2.3.5. Find an approximation for $(1.04)^{2.02}$ using a suitable local linearization.

Solution. Let $f(x,y) = x^y$. Then $1.04^{2.02} = f(1.04, 2.02)$ and this is close to $f(1,2) = 1^2 = 1$, so we can use the differential approximation to approximate the change from this known value. Since

$$\frac{\partial z}{\partial x} = yx^{y-1} \quad \text{and} \quad \frac{\partial z}{\partial y} = x^y \ln x,$$

their values at $(1,2)$ are

$$\frac{\partial z}{\partial x}(1,2) = 2 \times 1^{2-1} = 2 \quad \text{and} \quad \frac{\partial z}{\partial y}(1,2) = 1^2 \ln 1 = 0.$$

Changing $x : 1 \to 1.04$ gives $\Delta x = 0.04$ and $y : 2 \to 2.02$ gives $\Delta y = 0.02$. Thus,

$$f(x,y) \approx f(1,2) + \frac{\partial z(1,2)}{\partial x}\Delta x + \frac{\partial z(1,2)}{\partial y}\Delta y.$$

By substitution,

$$f(1.04, 2.02) \approx f(1,2) + \frac{\partial z}{\partial x}(1,2)0.04 + \frac{\partial z}{\partial y}(1,2)0.02$$
$$= 1 + 2 \times 0.04 + 0 \times 0.02 = 1.08.$$

Note. A calculator gives a better approximation, $(1.04)^{2.02} = 1.082448\ldots$, but our error is less than 3×10^{-3}.

Example 2.3.6. Use the total differential to estimate of the change of the function $z = \sqrt{20 - 7x^2 - y^2}$ when (x,y) changes from $(1,2)$ to $(0.98, 2.03)$.

Solution. Since

$$\frac{\partial z(1,2)}{\partial x} = \frac{-14x}{2\sqrt{20-7x^2-y^2}}\bigg|_{(1,2)} = \frac{-14 \times 1}{2\sqrt{20-7(1)^2-2^2}} = -\frac{7}{3} \quad \text{and}$$

$$\frac{\partial z(1,2)}{\partial y} = \frac{-2y}{2\sqrt{20-7x^2-y^2}}\bigg|_{(1,2)} = \frac{-2 \times 2}{2\sqrt{20-7(1)^2-2^2}} = -\frac{2}{3},$$

the differential is
$$dz = \frac{\partial z(1,2)}{\partial x}\Delta x + \frac{\partial z(1,2)}{\partial y}\Delta y = -\frac{7}{3}(0.98-1) - \frac{2}{3}(2.03-2) = 0.0026667.$$

Since $\Delta z = f(0.98, 2.03) - f(1,2) \approx dz$, the change in z is approximately 0.00266667. Using a calculator,
$$\sqrt{20 - 7 \times 0.98^2 - 2.03^2} - \sqrt{20 - 7 \times 1^2 - 2^2} \approx 0.0025938.$$

The two values are very close, but dz is much easier to evaluate when a calculator is not available.

Note. The linearization for a function of two variables at a point is essentially a plane approximation. For this function, it is
$$L(x,y) = 3 - \frac{7}{3}(x-1) - \frac{2}{3}(y-2),$$
which is the equation of a plane. This is the tangent plane at $(1,2)$ as we will see later in this chapter.

Linear approximations can be used for differentiable functions of more than two variables. For example, if $u = f(x,y,z)$ is differentiable at (a,b,c), then
$$\Delta u \approx dz \quad \text{and}$$
$$f(x,y,z) \approx f(a,b,c) + f_x(a,b,c)\Delta x + f_y(a,b,c)\Delta y + f_z(a,b,c)\Delta z.$$

There are some exercises involving differential approximation for functions of more than two variables in the end of this chapter.

2.4 The chain rule

2.4.1 The chain rule with one independent variable

In single-variable calculus, we have found that the chain rule is useful for differentiating a composite function: if $y = f(x)$ and $x = x(t)$ are both differentiable, then the composite function y is a differentiable function of t, and the derivative of y with respect to t is
$$\frac{dy}{dt} = \frac{dy}{dx}\frac{dx}{dt}. \tag{2.3}$$

For functions of several variables, there are several versions of the chain rules. We first consider $z = f(x,y)$, where each variable $x = \phi(t)$ and $y = \psi(t)$ is, in turn, a differentiable function of a variable t. This means $z = f(\phi(t), \psi(t))$ is a function of the variable t. When f_x and f_y are both continuous, we are able to differentiate z with respect to t to get $\frac{dz}{dt}$, as seen in the following theorem.

Theorem 2.4.1. *If $x = \phi(t)$ and $y = \psi(t)$ are two differentiable functions of t, and $z = f(x, y)$ is a differentiable function of x and y, then the composite function $z = f(x, y) = f(\phi(t), \psi(t))$ is a differentiable function of t and*

$$\frac{dz}{dt} = \frac{\partial z}{\partial x}\frac{dx}{dt} + \frac{\partial z}{\partial y}\frac{dy}{dt}. \tag{2.4}$$

Proof. A small change Δt in t causes changes of Δx in x and Δy in y. So, there is a change of Δz in z, and since z is differentiable, we have

$$\Delta z = \frac{\partial z}{\partial x}\Delta x + \frac{\partial z}{\partial y}\Delta y + o(\sqrt{(\Delta x)^2 + (\Delta y)^2}).$$

Dividing both sides of this equation by Δt, we have

$$\frac{\Delta z}{\Delta t} = \frac{\partial z}{\partial x}\frac{\Delta x}{\Delta t} + \frac{\partial z}{\partial y}\frac{\Delta y}{\Delta t} + \frac{o(\sqrt{(\Delta x)^2 + (\Delta y)^2})}{\Delta t}$$

$$= \frac{\partial z}{\partial x}\frac{\Delta x}{\Delta t} + \frac{\partial z}{\partial y}\frac{\Delta y}{\Delta t} + \frac{o(\sqrt{(\Delta x)^2 + (\Delta y)^2})}{\sqrt{(\Delta x)^2 + (\Delta y)^2}} \cdot \frac{\sqrt{(\Delta x)^2 + (\Delta y)^2}}{\Delta t}$$

$$= \frac{\partial z}{\partial x}\frac{\Delta x}{\Delta t} + \frac{\partial z}{\partial y}\frac{\Delta y}{\Delta t} + \frac{o(\sqrt{(\Delta x)^2 + (\Delta y)^2})}{\sqrt{(\Delta x)^2 + (\Delta y)^2}} \cdot \sqrt{\left(\frac{\Delta x}{\Delta t}\right)^2 + \left(\frac{\Delta y}{\Delta t}\right)^2}.$$

Since x is differentiable, it is continuous. So, if we now let $\Delta t \to 0$, then $\Delta x \to 0$. Similarly, we also have $\Delta y \to 0$. This in turn means that

$$\frac{dz}{dt} = \lim_{\Delta t \to 0} \frac{\Delta z}{\Delta t} = \frac{\partial z}{\partial x} \lim_{\Delta t \to 0} \frac{\Delta x}{\Delta t} + \frac{\partial z}{\partial y} \lim_{\Delta t \to 0} \frac{\Delta y}{\Delta t}$$

$$+ \lim_{\Delta x \to 0, \Delta y \to 0} \frac{o(\sqrt{(\Delta x)^2 + (\Delta y)^2})}{\sqrt{(\Delta x)^2 + (\Delta y)^2}} \lim_{\Delta t \to 0} \sqrt{\left(\frac{\Delta x}{\Delta t}\right)^2 + \left(\frac{\Delta y}{\Delta t}\right)^2}$$

$$= \frac{\partial z}{\partial x}\frac{dx}{dt} + \frac{\partial z}{\partial y}\frac{dy}{dt} + 0 \cdot \sqrt{\left(\frac{dx}{dt}\right)^2 + \left(\frac{dy}{dt}\right)^2}$$

$$= \frac{\partial z}{\partial x}\frac{dx}{dt} + \frac{\partial z}{\partial y}\frac{dy}{dt}. \qquad \square$$

It is helpful to use a tree diagram to remember this chain rule (and other chain rules as well), as shown in Figure 2.8. For the function in the previous theorem, since z is a function of x and y, we draw branches from the dependent variable z to the intermediate variables x and y. Then, we draw branches from x and y to the independent variable t, since both x and y are functions of t. Then, on each branch, we write the corresponding derivatives. To find $\frac{dz}{dt}$, we multiply the derivatives along each path from z to t, and then add these products to get

$$\frac{dz}{dt} = \frac{\partial z}{\partial x}\frac{dx}{dt} + \frac{\partial z}{\partial y}\frac{dy}{dt}.$$

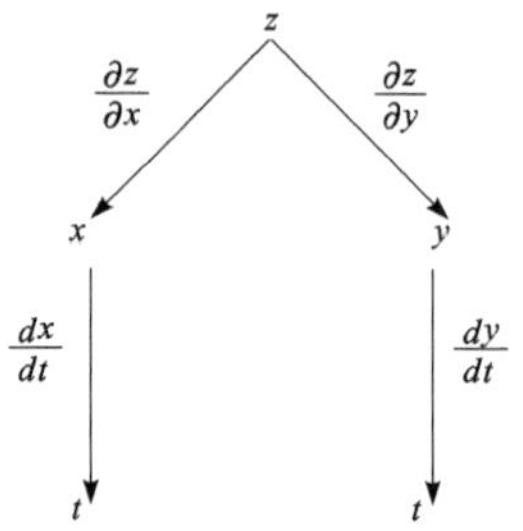

Figure 2.8: Chain rule: one independent variable.

> **Example 2.4.1.** Suppose $z = f(x,y) = x^2 - y^2$, $x = \sin t$, $y = \cos t$. Find $\frac{dz}{dt}$ using the chain rule.

Solution. Applying the chain rule, we obtain

$$\frac{dz}{dt} = \frac{\partial z}{\partial x}\frac{dx}{dt} + \frac{\partial z}{\partial y}\frac{dy}{dt}$$
$$= 2x \cdot \cos t + (-2y) \cdot (-\sin t)$$
$$= 2\sin t \cos t + 2\cos t \sin t = 2\sin 2t.$$

In this example, we can eliminate the intermediate variables x and y to obtain

$$z = x^2 - y^2 = (\sin t)^2 - (\cos t)^2 = -\cos 2t$$

by using the double-angle formula. So $dz/dt = (-\cos 2t)' = 2\sin 2t$, which agrees with the answer we obtained by using the chain rule.

2.4.2 The chain rule with more than one independent variable

We now consider another case, $z = f(x,y)$, but where each of x and y is a function of two variables s and t, i.e., $x = \phi(s,t)$, $y = \psi(s,t)$. Then $z = f(\phi(s,t), \psi(s,t))$ is indirectly a function of s and t. We can apply Theorem 2.4.1 to find $\frac{\partial z}{\partial t}$ and $\frac{\partial z}{\partial s}$. That is, if we hold s fixed, then we can compute $\frac{\partial z}{\partial t}$ by using Theorem 2.4.1 to get

$$\frac{\partial z}{\partial t} = \frac{\partial z}{\partial x}\frac{\partial x}{\partial t} + \frac{\partial z}{\partial y}\frac{\partial y}{\partial t}.$$

Similarly, we can compute $\frac{\partial z}{\partial s}$ if we hold t fixed. We summarize this in the following theorem.

> **Theorem 2.4.2.** Suppose that $z = f(x,y)$ is a differentiable function of x and y, where $x = \phi(s,t)$ and $y = \psi(s,t)$ are differentiable functions of s and t. Then
>
> $$\frac{\partial z}{\partial s} = \frac{\partial z}{\partial x}\frac{\partial x}{\partial s} + \frac{\partial z}{\partial y}\frac{\partial y}{\partial s} \quad \text{and} \quad \frac{\partial z}{\partial t} = \frac{\partial z}{\partial x}\frac{\partial x}{\partial t} + \frac{\partial z}{\partial y}\frac{\partial y}{\partial t}.$$

In this version of the chain rule, there are three types of variables, i. e., s and t are *independent variables*, x and y are called *intermediate variables*, and z is the *dependent* variable. Like the one-independent-variable case, to remember this chain rule (or any other one), it is helpful to draw a tree diagram representation of the function relationships, as shown in Figure 2.9. This time, we have two more branches. To find $\frac{\partial z}{\partial s}$ we multiply the partial derivatives along each path from z to s, and then add these products, i. e.,

$$\frac{\partial z}{\partial s} = \frac{\partial z}{\partial x}\frac{\partial x}{\partial s} + \frac{\partial z}{\partial y}\frac{\partial y}{\partial s}.$$

Similarly, we find

$$\frac{\partial z}{\partial t} = \frac{\partial z}{\partial x}\frac{\partial x}{\partial t} + \frac{\partial z}{\partial y}\frac{\partial y}{\partial t}$$

by using the paths from z to t.

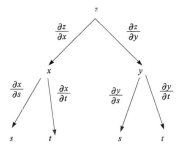

Figure 2.9: Chain rule: two independent variables.

Example 2.4.2. Using the chain rule as an aid, find the partial derivatives of the function $z = e^x \sin y$, where $x = st$ and $y = s + t$.

Solution. The tree diagram representation of function relationships is the same as in Figure 2.9, so we have

$$\begin{aligned}\frac{\partial z}{\partial s} &= \frac{\partial z}{\partial x}\frac{\partial x}{\partial s} + \frac{\partial z}{\partial y}\frac{\partial y}{\partial s} \\ &= e^x \cdot \sin y \cdot t + e^x \cos y \cdot 1 \\ &= e^{st} \sin(s+t)t + e^{st} \cos(s+t) \\ &= e^{st}(t\sin(s+t) + \cos(s+t)),\end{aligned}$$

and, similarly,

$$\frac{\partial z}{\partial t} = \frac{\partial z}{\partial x}\frac{\partial x}{\partial t} + \frac{\partial z}{\partial y}\frac{\partial y}{\partial t} = e^{st}[s\sin(s+t) + \cos(s+t)].$$

Example 2.4.3. Let $f(u, v, w) = e^{uvw}$, where $u = x^2$, $v = x + 2y$, and $w = \frac{x}{y}$. Find $\frac{\partial f}{\partial x}$ and $\frac{\partial f}{\partial y}$.

Solution. This example is different from previous ones in several ways. First, we note that this is a function of three variables, and the intermediate variables are u, v, and w. Furthermore, u has one independent variable, while v and w both have two independent variables. With the help of the tree diagram shown in Figure 2.10(a), we find

$$\frac{\partial f}{\partial x} = \frac{\partial f}{\partial u}\frac{du}{dx} + \frac{\partial f}{\partial v}\frac{\partial v}{\partial x} + \frac{\partial f}{\partial w}\frac{\partial w}{\partial x}$$

$$= vwe^{uvw} \cdot 2x + uwe^{uvw} \cdot 1 + uve^{uvw} \cdot \frac{1}{y}$$

$$= e^{uvw}\left(2vwx + uw + \frac{uv}{y}\right)$$

$$= e^{x^3(x+2y)/y}\left(\frac{2x^2(x+2y) + x^3 + x^2(x+2y)}{y}\right) = e^{x^3(x+2y)/y}\left(\frac{4x^3}{y} + 6x^2\right).$$

Similarly,

$$\frac{\partial f}{\partial y} = \frac{\partial f}{\partial v}\frac{\partial v}{\partial y} + \frac{\partial f}{\partial w}\frac{\partial w}{\partial y}$$

$$= uwe^{uvw} \cdot 2 + uve^{uvw} \cdot \left(-\frac{x}{y^2}\right)$$

$$= e^{uvw}\left(\frac{2x^3}{y} - \frac{x^3(x+2y)}{y^2}\right)$$

$$= e^{x^3(x+2y)/y}\left(\frac{2x^3}{y} - \frac{x^3(x+2y)}{y^2}\right) = -\frac{x^4}{y^2}e^{x^3(x+2y)/y}.$$

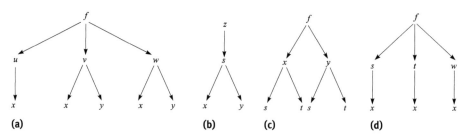

Figure 2.10: Tree diagrams for Examples 2.4.3, 2.4.4, 2.4.5, and 2.4.6.

2.4.3 Partial derivatives for abstract functions

In some theoretical analysis, we define some functions abstractly, for example, $u = f(x - y, \frac{y}{x})$, where we do not know the exact analytical equation for the function. With the help of the chain rule, we can still find their partial derivatives in term of some derivative notation.

Example 2.4.4. Let $z = \phi(x^2 + y^2)$. Find $\frac{\partial z}{\partial x}$.

Solution. Let $s = x^2 + y^2$. Then $z = \phi(s)$, and by the chain rule

$$\frac{\partial z}{\partial x} = \frac{dz}{ds}\frac{\partial s}{\partial x} = \phi'(u) \times 2x = 2x\phi'(x^2 + y^2).$$

Example 2.4.5. If $z = f(s^2 - t^2, st)$, find $\frac{\partial z}{\partial s}$ and $\frac{\partial z}{\partial t}$.

Solution. Let $x = s^2 - t^2$ and $y = st$. Then $z = f(x, y)$ and

$$\frac{\partial z}{\partial s} = \frac{\partial z}{\partial x}\frac{\partial x}{\partial s} + \frac{\partial z}{\partial y}\frac{\partial y}{\partial s} = f_x \cdot (2s) + f_y \cdot (t),$$

$$\frac{\partial z}{\partial t} = \frac{\partial z}{\partial x}\frac{\partial x}{\partial t} + \frac{\partial z}{\partial y}\frac{\partial y}{\partial t} = f_x \cdot (-2t) + f_y \cdot (s).$$

Example 2.4.6. If $u = f(x, x^2, x^3)$, find $\frac{du}{dx}$.

Solution. Let $s = x$, $t = x^2$, and $w = x^3$. Then

$$\frac{du}{dx} = \frac{\partial f}{\partial s}\frac{ds}{dx} + \frac{\partial f}{\partial t}\frac{dt}{dx} + \frac{\partial f}{\partial w}\frac{dw}{dx}$$
$$= \frac{\partial f}{\partial s} \cdot 1 + \frac{\partial f}{\partial t} \cdot 2x + \frac{\partial f}{\partial w} \cdot 3x^2.$$

Example 2.4.7. Let $u = f(x + y + z, xyz)$ and suppose that f has continuous second-order partials. Find $\frac{\partial u}{\partial z}$ and $\frac{\partial^2 u}{\partial x \partial z}$ in terms of the partial derivatives of f.

Solution. Let $s = x + y + z$ and $t = xyz$. Then $u = f(s, t)$ and

$$\frac{\partial u}{\partial z} = \frac{\partial f}{\partial s}\frac{\partial s}{\partial z} + \frac{\partial f}{\partial t}\frac{\partial t}{\partial z}$$
$$= f_1 + xyf_2,$$

where f_1 means the derivative of f with respect to the first variable, and f_2 means that with respect to the second variable. We have

$$\frac{\partial^2 u}{\partial x \partial z} = (f_1 + xyf_2)_x = (f_1)_x + yf_2 + xy(f_2)_x$$

$$= \frac{\partial f_1}{\partial s}\frac{\partial s}{\partial x} + \frac{\partial f_1}{\partial t}\frac{\partial t}{\partial x} + yf_2 + xy\left(\frac{\partial f_2}{\partial s}\frac{\partial s}{\partial x} + \frac{\partial f_2}{\partial t}\frac{\partial t}{\partial x}\right)$$
$$= f_{11} + f_{12}yz + yf_2 + xy(f_{21} + yzf_{22})$$
$$= f_{11} + yf_{12}(x+z) + yf_2 + xy^2 zf_{22}.$$

Note. Using the notation f_1, f_2, etc., helps to clarify some ambiguity in the notation $\frac{\partial f}{\partial x}$, which may mean f_1 (this is also f_u) or the partial derivative of the entire function with respect to x.

2.5 The Taylor expansion

For a function of one variable $y = f(x)$, which has an $(n+1)$th-order derivative at $x = a$, we have

$$f(x) = f(a) + f'(a)(x-a) + \frac{f''(a)}{2!}(x-a)^2 + \cdots + \frac{f^{(n)}(a)}{n!}(x-a)^n + \frac{f^{(n+1)}(\xi)}{(n+1)!}(x-a)^{n+1}$$

or

$$f(x) = f(a) + f'(a)\Delta x + \frac{f''(a)}{2!}(\Delta x)^2 + \cdots + \frac{f^{(n)}(a)}{n!}(\Delta x)^n + \frac{f^{(n+1)}(a+\theta\Delta x)}{(n+1)!}(\Delta x)^{n+1},$$

where ξ is between a and x, $0 < \theta < 1$, and

$$f(x) \approx f(a) + f'(x)(x-a) + \frac{f''(a)}{2!}(x-a)^2$$

is called the quadratic approximation. Similarly, for a function of two variables $z = f(x, y)$, which has $(n+1)$th-order continuous partials at (a, b), we define

$$\phi(t) = f(a + t\Delta x, b + t\Delta y).$$

So

$$\phi(t) = \phi(0) + \phi'(0)t + \frac{\phi''(0)}{2!}t^2 + \cdots + \frac{\phi^{(n)}(0)}{n!}t^2 + \frac{\phi^{(n+1)}(\theta t)}{(n+1)!}t^{n+1}.$$

Note that

$$\phi(t) = f(a + t\Delta x, b + t\Delta y), \quad \phi(1) = f(x, y), \quad \text{and} \quad \phi(0) = f(a, b).$$

Thus,

$$\phi'(t) = f_x(a + t\Delta x, b + t\Delta y)\Delta x + f_y(a + t\Delta x, b + t\Delta y)\Delta y \quad \text{and}$$
$$\phi'(0) = f_x(a, b)\Delta x + f_y(a, b)\Delta y.$$

Furthermore,

$$\phi''(t) = f_{xx}(a + t\Delta x, b + t\Delta y)(\Delta x)^2 + f_{xy}(a + t\Delta x, b + t\Delta y)\Delta x\Delta y$$
$$+ f_{yx}(a + t\Delta x, b + t\Delta y)\Delta x\Delta y + f_{yy}(a + t\Delta x, b + t\Delta y)(\Delta y)^2 \quad \text{and}$$
$$\phi''(0) = f_{xx}(a, b)(\Delta x)^2 + 2f_{xy}(a, b)\Delta x\Delta y + f_{yy}(a, b)(\Delta y)^2, \quad \text{since } f_{xy} = f_{yx}.$$

Letting $t = 1$, we have the quadratic approximation

$$f(x, y) \approx f(a, b) + f_x(a, b)\Delta x + f_y(a, b)\Delta y$$
$$+ \frac{1}{2}(f_{xx}(a, b)(\Delta x)^2 + 2f_{xy}(a, b)\Delta x\Delta y + f_{yy}(a, b)(\Delta y)^2).$$

Example 2.5.1. Use a linear and quadratic approximation to estimate $\dfrac{1}{\sqrt{10-(2.01)^2-2(0.98)^2}}$.

Solution. Let $z = f(x, y) = \dfrac{1}{\sqrt{10-x^2-2y^2}}$. The function has continuous partials at $(2, 1)$. The value of the function at $(2, 1)$ is $f(2, 1) = \frac{1}{2}$, and its first partials are

$$f_x = -\frac{1}{2}(10 - x^2 - 2y^2)^{-\frac{3}{2}}(-2x) = x(10 - x^2 - 2y^2)^{-\frac{3}{2}},$$

$$f_x(2, 1) = 2(10 - 2^2 - 2(1)^2)^{-\frac{3}{2}} = \frac{1}{4},$$

$$f_y = -\frac{1}{2}(10 - x^2 - 2y^2)^{-\frac{3}{2}}(-4y) = 2y(10 - x^2 - 2y^2)^{-\frac{3}{2}},$$

$$f_y(2, 1) = 2(1)(10 - 2^2 - 2(1)^2)^{-\frac{3}{2}} = \frac{1}{4}.$$

Thus, the linear approximation is

$$f(2.01, 0.98) \approx f(2, 1) + f_x(2, 1)\Delta x + f_y(2, 1)\Delta y$$
$$= \frac{1}{2} + \frac{1}{4}(0.01) + \frac{1}{4}(-0.02)$$
$$= 0.4975.$$

Now, we compute the second partial derivatives. We have

$$f_{xx} = (10 - x^2 - 2y^2)^{-\frac{3}{2}} + x\left(-\frac{3}{2}\right)(10 - x^2 - 2y^2)^{-\frac{5}{2}}(-2x)$$
$$= (10 - x^2 - 2y^2)^{-\frac{3}{2}} + 3x^2(10 - x^2 - 2y^2)^{-\frac{5}{2}},$$

$$f_{xx}(2, 1) = (10 - 2^2 - 2(1)^2)^{-\frac{3}{2}} + 3(2^2)(10 - 2^2 - 2(1)^2)^{-\frac{5}{2}} = \frac{1}{2},$$

$$f_{xy} = (x(10 - x^2 - 2y^2)^{-\frac{3}{2}})'_y$$
$$= x\left(-\frac{3}{2}\right)(10 - x^2 - 2y^2)^{-\frac{5}{2}}(-4y),$$

$$f_{xy}(2,1) = 2\left(-\frac{3}{2}\right)(10 - 2^2 - 2(1)^2)^{-\frac{5}{2}}(-4(1)) = \frac{3}{8},$$

$$f_{yy} = (2y(10 - x^2 - 2y^2)^{-\frac{3}{2}})'_y$$

$$= 2(10 - x^2 - 2y^2)^{-\frac{3}{2}} + 2y\left(-\frac{3}{2}\right)(10 - x^2 - 2y^2)^{-\frac{5}{2}}(-4y),$$

$$f_{yy}(2,1) = 2(10 - 2^2 - 2(1)^2)^{-\frac{3}{2}} + 2(1)\left(-\frac{3}{2}\right)(10 - 2^2 - 2(1)^2)^{-\frac{5}{2}}(-4(1))$$

$$= \frac{5}{8}.$$

Therefore, the quadratic approximation is

$$f(2.01, 0.98) \approx f(2,1) + f_x(2,1)\Delta x + f_y(2,1)\Delta y$$

$$+ \frac{1}{2}(f_{xx}(2,1)(\Delta x)^2 + 2f_{xy}(2,1)\Delta x \Delta y + f_{yy}(a,b)(\Delta y)^2)$$

$$= \frac{1}{2} + \frac{1}{4}(0.01) + \frac{1}{4}(-0.02)$$

$$+ \frac{1}{2}\left(\left(\frac{1}{2}\right)(0.01)^2 + 2\left(\frac{3}{8}\right)(0.01)(-0.02) + \left(\frac{5}{8}\right)(-0.02)^2\right)$$

$$= 0.49758.$$

Compared with the value of $\frac{1}{\sqrt{10-(2.01)^2-2(0.98)^2}}$ computed by a calculator, 0.49757, the quadratic approximation gives a better estimation.

If we use the notation

$$\left(\Delta x \frac{\partial}{\partial x} + \Delta y \frac{\partial}{\partial y}\right)^n f(x,y) = \sum_{k=0}^{n} \binom{n}{k}(\Delta x)^k (\Delta y)^{n-k} \frac{\partial^n f(x,y)}{\partial x^k \partial y^{n-k}},$$

then we can conclude with Taylor's theorem (or the Taylor expansion) for a function of two variables.

Theorem 2.5.1. *If $z = f(x,y)$ has continuous $(n+1)$th partial derivatives on some neighborhood D containing the point (a,b) and $(x,y) = (a + \Delta x, b + \Delta y)$ is a point in D, then*

$$f(x,y) = f(a,b) + \left(\Delta x \frac{\partial}{\partial x} + \Delta y \frac{\partial}{\partial y}\right)f(a,b) + \frac{1}{2!}\left(\Delta x \frac{\partial}{\partial x} + \Delta y \frac{\partial}{\partial y}\right)^2 f(a,b)$$

$$+ \cdots + \frac{1}{n!}\left(\Delta x \frac{\partial}{\partial x} + \Delta y \frac{\partial}{\partial y}\right)^n f(a,b)$$

$$+ \frac{1}{(n+1)!}\left(\Delta x \frac{\partial}{\partial x} + \Delta y \frac{\partial}{\partial y}\right)^{n+1} f(a + \theta \Delta x, b + \theta \Delta y)$$

for some $0 < \theta < 1$.

2.6 Implicit differentiation

We have shown some examples of explicitly defined functions where intermediate variables can be eliminated, allowing us to find partial derivatives of the function without using the chain rule. However, the chain rule can be extremely useful in finding partials for functions defined implicitly by one or more equations.

2.6.1 Functions implicitly defined by a single equation

Recall that an equation involving several variables can, in theory, be solved to give one variable, say, z, as a function of the other variables (with domain limitation), and in this case z is said to be implicitly defined as a function of the other variables by that equation. If the equation can actually be solved to give a formula for z, then z is defined explicitly by that formula. Implicit differentiation is a process for finding derivatives of implicitly defined functions.

The chain rules for functions of two variables can be used to find a formula to implicitly differentiate a function of one variable implicitly defined by an equation. We first consider an equation of the form $F(x, y) = 0$, which defines y implicitly as a differentiable function of x, for all x in some set D. This means that there exists some function $y = y(x)$ such that $F(x, y(x)) = 0$ for all $x \in D$ (D is the domain of y), but we may not have a formula for $y(x)$. In spite of the lack of a formula $y(x)$, we can find a formula for its derivative by the method of implicit differentiation, provided that F is differentiable. We develop the formula here for $\frac{dy}{dx}$ by differentiating both sides of the equation $F(x, y) = 0$ with respect to x (assuming y is a function of x) using the chain rule, i.e.,

$$\frac{\partial F}{\partial x} + \frac{\partial F}{\partial y}\frac{dy}{dx} = 0.$$

If $\partial F/\partial y \neq 0$, then we can solve for dy/dx, obtaining

$$\frac{dy}{dx} = -\frac{\frac{\partial F}{\partial x}}{\frac{\partial F}{\partial y}} = -\frac{F_x}{F_y}.$$

There is a theorem, called the *implicit function theorem*, which guarantees the existence of the derivative dy/dx under some conditions. It says that if F is defined on a region containing (a, b), where $F(a, b) = 0$, $F_y(a, b) \neq 0$, and F_x and F_y are both continuous on this region, then the equation $F(x, y) = 0$ defines a unique y as a function of x near a with $y(a) = b$, and the derivative dy/dx does exist and is equal to $-F_x/F_y$.

Example 2.6.1. Find y' if $x^3 + y^3 - 6xy + x^2 - 2y = 1$.

Solution. Let
$$F(x,y) = x^3 + y^3 - 6xy + x^2 - 2y - 1 = 0.$$
Then
$$F_x = 3x^2 - 6y + 2x \quad \text{and} \quad F_y = 3y^2 - 6x - 2.$$
Therefore,
$$y' = \frac{dy}{dx} = -\frac{F_x}{F_y} = -\frac{3x^2 - 6y + 2x}{3y^2 - 6x - 2}.$$

Now we consider a more complicated version, where z is defined implicitly as a function $z = z(x,y)$ by an equation of the form $F(x,y,z) = 0$. If this equation defines a differentiable function $z = f(x,y)$ for (x,y) in some region, and $F(x,y,z)$ is also differentiable, then we can use the chain rule to differentiate $F(x,y,z) = 0$ with respect to x (assuming that z is a function of the independent variables x and y) as follows:
$$\frac{\partial F}{\partial x} + \frac{\partial F}{\partial z}\frac{\partial z}{\partial x} = 0.$$
If $\partial F/\partial z \neq 0$, we solve for $\partial z/\partial x$, i. e.,
$$\frac{\partial z}{\partial x} = -\frac{\frac{\partial F}{\partial x}}{\frac{\partial F}{\partial z}} = -\frac{F_x}{F_z}.$$
In a similar manner, we have
$$\frac{\partial z}{\partial y} = -\frac{\frac{\partial F}{\partial y}}{\frac{\partial F}{\partial z}} = -\frac{F_y}{F_z}.$$

! **Example 2.6.2.** Use three methods to find $\frac{\partial z}{\partial x}$ and $\frac{\partial z}{\partial y}$ when the equation $x^3 + y^3 + z^3 + 6xyz = 7$ defines z implicitly as a function of x and y.

Solution. Method 1: Let $F(x,y,z) = x^3 + y^3 + z^3 + 6xyz - 7 = 0$. Then
$$\frac{\partial z}{\partial x} = -\frac{\frac{\partial F}{\partial x}}{\frac{\partial F}{\partial z}} = -\frac{3x^2 + 6yz}{3z^2 + 6xy} = -\frac{x^2 + 2yz}{z^2 + 2xy},$$
$$\frac{\partial z}{\partial y} = -\frac{\frac{\partial F}{\partial y}}{\frac{\partial F}{\partial z}} = -\frac{3y^2 + 6xz}{3z^2 + 6xy} = -\frac{y^2 + 2xz}{z^2 + 2xy}.$$

Method 2: We avoid the use of the formula by differentiating the equation with respect to x, using the one-variable chain rule and the one-variable implicit differentiation, assuming that z is a function of x, and treating y as a constant. Then we have
$$3x^2 + 3z^2\frac{\partial z}{\partial x} + 6yz + 6xy\frac{\partial z}{\partial x} = 0.$$

Solving this equation for $\partial z/\partial x$, we obtain

$$\frac{\partial z}{\partial x} = -\frac{x^2 + 2yz}{z^2 + 2xy}.$$

Similarly, implicit differentiation with respect to y gives the same formula as above for $\partial z/\partial y$.

Method 3: We use the differential operator "d." Then we obtain

$$d(x^3 + y^3 + z^3 + 6xyz) = d(7),$$
$$d(x^3) + d(y^3) + d(z^3) + d(6xyz) = d(7),$$
$$3x^2 dx + 3y^2 dy + 3z^2 dz + 6(xd(yz) + yzdx) = 0 \quad \text{(product rule)},$$
$$x^2 dx + y^2 dy + z^2 dz + 2(x(ydz + zdy) + yzdx) = 0,$$
$$(x^2 + 2yz)dx + (y^2 + 2xz)dy + (z^2 + 2xy)dz = 0.$$

Then

$$dz = -\frac{x^2 + 2yz}{z^2 + 2xy} dx - \frac{y^2 + 2xz}{z^2 + 2xy} dy.$$

This means $\frac{\partial z}{\partial x} = -\frac{x^2+2yz}{z^2+2xy}$ and $\frac{\partial z}{\partial y} = -\frac{y^2+2xz}{z^2+2xy}$.

2.6.2 Functions defined implicitly by systems of equations

In general, two equations will allow two variables to be defined implicitly as functions of the remaining variables, three equations will allow three variables to be defined implicitly as functions of the remaining variables, and so on.

We first consider the system of equations

$$\begin{cases} F(x, y, z) = 0, \\ G(x, y, z) = 0. \end{cases}$$

Assume that two functions $y(x)$ and $z(x)$ are implicitly defined as functions of the independent variable x. How do we find $\frac{dy}{dx}$ and $\frac{dz}{dx}$? We apply the chain rule to both equations simultaneously to obtain

$$\begin{cases} F_x + F_y \frac{dy}{dx} + F_z \frac{dz}{dx} = 0, \\ G_x + G_y \frac{dy}{dx} + G_z \frac{dz}{dx} = 0. \end{cases}$$

If the *Jacobian determinant* $\begin{vmatrix} F_y & F_z \\ G_y & G_z \end{vmatrix} = F_y G_z - F_z G_y \neq 0$, then, by using Cramer's rule, the determinant solution is

$$\frac{dy}{dx} = -\frac{\begin{vmatrix} F_x & F_z \\ G_x & G_z \end{vmatrix}}{\begin{vmatrix} F_y & F_z \\ G_y & G_z \end{vmatrix}} \quad \text{and} \quad \frac{dz}{dx} = -\frac{\begin{vmatrix} F_y & F_x \\ G_y & G_x \end{vmatrix}}{\begin{vmatrix} F_y & F_z \\ G_y & G_z \end{vmatrix}}. \tag{2.5}$$

Example 2.6.3. If $\begin{cases} x^2+y^2+z^2=4, \\ x^2-y+2z^2=2, \end{cases}$ find $\frac{dy}{dx}$ and $\frac{dz}{dx}$.

Solution. We could use equation (2.5), but instead we find the derivative with respect to x for each side of the equations using implicit differentiation to obtain

$$\begin{cases} 2x + 2y\frac{dy}{dx} + 2z\frac{dz}{dx} = 0, \\ 2x - \frac{dy}{dx} + 4z\frac{dz}{dx} = 0. \end{cases}$$

Now multiplying the first equation by 2 and subtracting it from the second equation, we have

$$\begin{cases} 2x + 2y\frac{dy}{dx} + 2z\frac{dz}{dx} = 0, \\ -2x - (1+4y)\frac{dy}{dx} = 0. \end{cases}$$

Solve for $\frac{dy}{dx}$ to obtain

$$\frac{dy}{dx} = \frac{-2x}{1+4y}.$$

Substituting $\frac{dy}{dx}$ back into the first equation, we have

$$\frac{dz}{dx} = -\frac{1}{2z}\left(2x + 2y\left(\frac{-2x}{1+4y}\right)\right) = -\frac{x+2xy}{z+4yz}.$$

Now we consider the situation where the two variables u and v are defined implicitly as functions $u = u(x,y)$ and $v = v(x,y)$ by two equations of the following form:

$$\begin{cases} F(x,y,u,v) = 0, \\ G(x,y,u,v) = 0. \end{cases}$$

We use the chain rule to differentiate the equations with respect to x (keeping y constant), and then we solve the two equations for $\frac{\partial u}{\partial x}$ and $\frac{\partial v}{\partial x}$. Then we have

$$\begin{cases} F_x + F_u\frac{\partial u}{\partial x} + F_v\frac{\partial v}{\partial x} = 0, \\ G_x + G_u\frac{\partial u}{\partial x} + G_v\frac{\partial v}{\partial x} = 0. \end{cases}$$

Note that these equations are linear in the variables $\frac{\partial u}{\partial x}$ and $\frac{\partial v}{\partial x}$ with coefficients F_x, F_u, F_v, G_x, G_u, and G_v. Consequently, we can solve for $\frac{\partial u}{\partial x}$ and $\frac{\partial v}{\partial x}$ using the usual methods for solving linear equations. By Cramer's rule, the determinant solution is

$$\frac{\partial u}{\partial x} = -\frac{\begin{vmatrix} F_x & F_y \\ G_x & G_y \end{vmatrix}}{\begin{vmatrix} F_u & F_v \\ G_u & G_v \end{vmatrix}} \quad \text{and} \quad \frac{\partial v}{\partial x} = -\frac{\begin{vmatrix} F_u & F_x \\ G_u & G_x \end{vmatrix}}{\begin{vmatrix} F_u & F_v \\ G_u & G_v \end{vmatrix}}, \tag{2.6}$$

given that the denominator is not 0. Similarly, we can find

$$\frac{\partial u}{\partial y} = -\frac{\begin{vmatrix} F_y & F_v \\ G_y & G_v \end{vmatrix}}{\begin{vmatrix} F_u & F_v \\ G_u & G_v \end{vmatrix}} \quad \text{and} \quad \frac{\partial v}{\partial y} = -\frac{\begin{vmatrix} F_u & F_y \\ G_u & G_y \end{vmatrix}}{\begin{vmatrix} F_u & F_v \\ G_u & G_v \end{vmatrix}} \qquad (2.7)$$

by differentiating both of the original equations with respect to y.

Example 2.6.4. Suppose two equations $xu - yv = 1$ and $yu + xv = 2$ implicitly define u and v as functions of x and y. Find $\frac{\partial u}{\partial x}, \frac{\partial u}{\partial y}, \frac{\partial v}{\partial x}$, and $\frac{\partial v}{\partial y}$.

Solution. Differentiating the system of equations with respect to x (assuming y is a constant), we have

$$u + x\frac{\partial u}{\partial x} - y\frac{\partial v}{\partial x} = 0 \quad \text{and} \quad y\frac{\partial u}{\partial x} + v + x\frac{\partial v}{\partial x} = 0.$$

Solving this system of linear equations (treating x, y, u, v as constants) for $\frac{\partial u}{\partial x}$ and $\frac{\partial v}{\partial x}$, we obtain

$$\frac{\partial u}{\partial x} = -\frac{xu + yv}{x^2 + y^2} \quad \text{and} \quad \frac{\partial v}{\partial x} = \frac{-xv + yu}{x^2 + y^2}.$$

We can also use equation (2.7). Since

$$F(x, y, u, v) = xu - yv - 1 = 0 \quad \text{and} \quad G(x, y, u, v) = yu + xv - 2 = 0,$$

we have

$$F_u = x, \quad F_v = -y, \quad F_y = -v, \quad G_u = y, \quad G_v = x, \quad \text{and} \quad G_y = u.$$

Then, by equation (2.7), we obtain

$$\frac{\partial u}{\partial y} = -\frac{\begin{vmatrix} F_y & F_v \\ G_y & G_v \end{vmatrix}}{\begin{vmatrix} F_u & F_v \\ G_u & G_v \end{vmatrix}} = -\frac{\begin{vmatrix} -v & -y \\ u & x \end{vmatrix}}{\begin{vmatrix} x & -y \\ y & x \end{vmatrix}} = -\frac{-xv + yu}{x^2 + y^2} = \frac{xv - yu}{x^2 + y^2}.$$

Similarly,

$$\frac{\partial v}{\partial y} = -\frac{\begin{vmatrix} F_u & F_y \\ G_u & G_y \end{vmatrix}}{\begin{vmatrix} F_u & F_v \\ G_u & G_v \end{vmatrix}} = -\frac{\begin{vmatrix} x & -v \\ y & u \end{vmatrix}}{\begin{vmatrix} x & -y \\ y & x \end{vmatrix}} = -\frac{xu + yv}{x^2 + y^2}.$$

Example 2.6.5. Assume x, y are independent variables and

$$\begin{cases} u = x^2 + y^2 + \cos v, \\ y \sin v + v \sin x = 0. \end{cases}$$

Find $\frac{\partial v}{\partial x}$ and $\frac{\partial u}{\partial y}$ at the point $(x, y, u, v) = (0, 1, 0, \pi)$.

Solution. We differentiate both equations with respect to x to get

$$\begin{cases} \frac{\partial u}{\partial x} = 2x - \sin v \frac{\partial v}{\partial x}, \\ y \cos v \frac{\partial v}{\partial x} + \frac{\partial v}{\partial x} \sin x + v \cos x = 0. \end{cases}$$

When $x = 0$, $y = 1$, $u = 0$, $v = \pi$, we have

$$\begin{cases} \frac{\partial u}{\partial x} = 0, \\ -\frac{\partial v}{\partial x} + \pi \cos 0 = 0. \end{cases}$$

Thus, $\frac{\partial v}{\partial x}\big|_{(0,1,0,\pi)} = \pi$. To compute $\frac{\partial u}{\partial y}$, we differentiate both equations with respect to y to get

$$\begin{cases} \frac{\partial u}{\partial y} = 2y - \sin v \frac{\partial v}{\partial y}, \\ \sin v + y \cos v \frac{\partial v}{\partial y} + \frac{\partial v}{\partial y} \sin x = 0. \end{cases}$$

When $x = 0$, $y = 1$, $u = 0$, $v = \pi$, the first equation gives $\frac{\partial u}{\partial y}\big|_{(0,1,0,\pi)} = 2$.

2.7 Tangent lines and tangent planes

2.7.1 Tangent lines and normal planes to a curve

We have already found tangent lines and normal planes for a curve C in space given by a vector-valued function $\mathbf{r}(t) = \langle x(t), y(t), z(t) \rangle$, where $x(t)$, $y(t)$, and $z(t)$ are differentiable functions of t. The line tangent to the curve at $t = t_0$ is

$$\mathbf{r}(t) = \mathbf{r}(t_0) + \mathbf{r}'(t_0)t,$$

where $\mathbf{r}'(t_0)$ is the tangent vector at $t = t_0$. The parametric equations and symmetric equations of the tangent line are, therefore,

$$\begin{cases} x = x_0 + x'(t_0)t, \\ y = y_0 + y'(t_0)t, \\ z = z_0 + z'(t_0)t, \end{cases} \text{and} \quad \frac{x - x_0}{x'(t_0)} = \frac{y - y_0}{y'(t_0)} = \frac{z - z_0}{z'(t_0)}, \text{ respectively.}$$

The normal plane at the same point is $x'(t_0)(x - x_0) + y'(t_0)(y - y_0) + z'(t_0)(z - z_0) = 0$.

Now we are able to find tangent lines and normal planes for a curve C that is implicitly defined by a system of equations of the form

$$C = \{(x, y, z) | F(x, y, z) = 0 \text{ and } G(x, y, z) = 0\}.$$

In general, these equations implicitly define two variables as functions of the third, say, $y = y(x)$ and $z = z(x)$. Thus, we can parameterize C with parameter x as follows:

$$x = x, \quad y = y(x), \quad \text{and} \quad z = z(x).$$

The implicit differentiation methods described previously allow us to compute $\frac{dy}{dx}$ and $\frac{dz}{dx}$ even though we do not have formulas for $y(x)$ and $z(x)$. Consequently, we are able to find the tangent vector $\langle 1, \frac{dy}{dx}, \frac{dz}{dx}\rangle$ and tangent line at any point on C.

Example 2.7.1. Find an equation of the tangent line and an equation of the normal plane at the point $(1, -2, 1)$ of the curve C defined implicitly by

$$C = \{(x,y,z) | x^2 + y^2 + z^2 = 6 \text{ and } x + y + z = 0\}.$$

Solution. If we regard x as the independent variable and as the parameter, then the system of equations implicitly defines two functions $y = y(x)$ and $z = z(x)$. Choosing the parameterizations $x = x$, $y = y(x)$, and $z = z(x)$, a tangent vector of the curve is $\langle 1, \frac{dy}{dx}, \frac{dz}{dx}\rangle$. In order to find the derivatives of these two functions, first implicitly differentiate the system of equations with respect to x, i.e.,

$$\begin{cases} 2x + 2y\frac{dy}{dx} + 2z\frac{dz}{dx} = 0, \\ 1 + \frac{dy}{dx} + \frac{dz}{dz} = 0. \end{cases}$$

Solving this system of linear equations for $\frac{dy}{dx}$ and $\frac{dz}{dx}$ gives

$$\frac{dy}{dx} = \frac{z-x}{y-z} \quad \text{and} \quad \frac{dz}{dx} = \frac{x-y}{y-z}.$$

Therefore, when $x = 1$, $y = -2$, and $z = 1$, we have $y'(1) = 0$ and $z'(1) = -1$, and the tangent vector $\langle 1, \frac{dy}{dx}, \frac{dz}{dx}\rangle$ is

$$\mathbf{v} = \langle 1, 0, -1 \rangle.$$

So, symmetric equations of the tangent line are

$$\frac{x-1}{1} = \frac{y+2}{0} = \frac{z-1}{-1}.$$

Parametric equations, with parameter t, of the tangent line are

$$x = 1 + t, \quad y = -2, \quad z = 1 - t.$$

An equation of the normal plane is

$$1 \cdot (x-1) + 0 \cdot (y+2) + (-1) \cdot (z-1) = 0.$$

This simplifies to the equation $x - z = 0$.

Note. If this example had asked for the tangent line at $(-2, 1, 1)$, then something different would have happened. If you try the parameterization

$$x = x, \quad y = y(x), \quad \text{and} \quad z = z(x),$$

then you will have the problem that

$$\begin{cases} 2x + 2y\frac{dy}{dx} + 2z\frac{dz}{dx} = 0, \\ 1 + \frac{dy}{dx} + \frac{dz}{dx} = 0 \end{cases}$$

has no solutions at $(-2, 1, 1)$ since the two equations would be inconsistent. This does not mean that there is no tangent there, but the tangent line is parallel to the yz-plane (perpendicular to the x-axis). To solve the problem, we try a different way to parameterize the curve, i. e.,

$$x = x(y), \quad y = y, \quad \text{and} \quad z = z(y).$$

Then, differentiating with respect to y, we have

$$\begin{cases} 2x\frac{dx}{dy} + 2y + 2z\frac{dz}{dy} = 0, \\ \frac{dx}{dy} + 1 + \frac{dz}{dy} = 0. \end{cases}$$

At $(-2, 1, 1)$ we obtain $\frac{dx}{dy} = 0$ and $\frac{dz}{dy} = -1$. Therefore, the tangent vector is $\mathbf{T} = \langle 0, 1, -1 \rangle$. So, the tangent line is

$$\frac{x+2}{0} = \frac{y-1}{1} = \frac{z-1}{-1}.$$

Figure 2.11 shows these graphs.

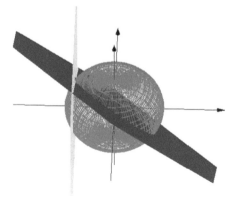

Figure 2.11: Tangent line to the intersection curve of a plane and a sphere.

2.7.2 Tangent planes and normal lines to a surface

Suppose S is a surface with equation $F(x,y,z) = 0$, and let $M(x_0, y_0, z_0)$ be a specific point on S. We show that there is a unique direction that is normal to (perpendicular to) the surface S at the point M, provided F is differentiable.

Consider any curve C that passes through the point M and lies entirely in the surface S, with parametric equations

$$x = x(t), \quad y = y(t), \quad \text{and} \quad z = z(t), (\alpha \le t \le \beta)$$

such that $t = t_0$ gives the point M. Since C lies on S, any point $(x(t), y(t), z(t))$ on C must satisfy the defining equation $F(x,y,z) = 0$ of S, so that

$$F(x(t), y(t), z(t)) = 0. \tag{2.8}$$

If $x(t)$, $y(t)$, and $z(t)$ are differentiable functions of t, and F is also differentiable, then we can use the chain rule to differentiate both sides of (2.8) as follows:

$$\frac{\partial F}{\partial x}\frac{dx}{dt} + \frac{\partial F}{\partial y}\frac{dy}{dt} + \frac{\partial F}{\partial z}\frac{dz}{dt} = 0.$$

Computing this at M where $t = t_0$ gives

$$F_x(M)x'(t_0) + F_y(M)y'(t_0) + F_z(M)z'(t_0) = 0, \tag{2.9}$$

and rewriting this as a dot product of two vectors shows

$$\langle F_x(M), F_y(M), F_z(M) \rangle \cdot \langle x'(t_0), y'(t_0), z'(t_0) \rangle = 0.$$

Denoting the vectors by

$$\mathbf{n} = \langle F_x(M), F_y(M), F_z(M) \rangle$$

and

$$\mathbf{v} = \langle x'(t_0), y'(t_0), z'(t_0) \rangle,$$

where $\mathbf{v}$ is the tangent vector to C at M, equation (2.9) can be written in terms of a dot product as

$$\mathbf{n} \cdot \mathbf{v} = 0. \tag{2.10}$$

This equation shows that $\mathbf{n}$ is perpendicular to the tangent vector $\mathbf{v}$ at M for any curve C on S that passes through M and satisfies the above differentiability conditions. Therefore, all the tangent lines of these curves at M must be coplanar as shown in Figure 2.12. Those tangent lines form a plane which we define as the *tangent plane* to the surface at M.

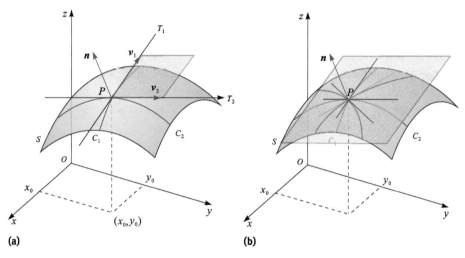

Figure 2.12: Tangent planes and normal line to a surface at a point.

Definition 2.7.1. Assume a surface in space has equation $F(x,y,z) = 0$, and $F(x,y,z)$ is differentiable at $M(x_0, y_0, z_0)$. Then the tangent plane to the surface at M is

$$F_x(M)(x - x_0) + F_y(M)(y - y_0) + F_z(M)(z - z_0) = 0.$$

The nonzero vector $\mathbf{n} = \langle F_x(M), F_y(M), F_z(M) \rangle$ is a *normal vector* of the tangent plane at M. The normal line at M is

$$\frac{x - x_0}{F_x(M)} = \frac{y - y_0}{F_y(M)} = \frac{z - z_0}{F_z(M)}.$$

Example 2.7.2. Find equations of the tangent plane and normal line to the ellipsoid

$$2x^2 + 4y^2 + z^2 = 10$$

at the point $(-1, 1, -2)$.

Solution. The equation of the ellipsoid can be written as

$$F(x, y, z) = 2x^2 + 4y^2 + z^2 - 10 = 0.$$

Therefore, we have

$$F_x(x,y,z) = 4x, \quad F_y(x,y,z) = 8y, \quad F_z(x,y,z) = 2z, \quad \text{and}$$
$$F_x(-1,1,-2) = -4, \quad F_y(-1,1,-2) = 8, \quad F_z(-1,1,-2) = -4.$$

The equation of the tangent plane at $(-1, 1, -2)$ is, therefore,

$$-4(x - (-1)) + 8(y - 1) + (-4)(z - (-2)) = 0,$$

which simplifies to $2y - x - z - 5 = 0$.

The symmetric equations of the normal line are
$$\frac{x+1}{-4} = \frac{y-1}{8} = \frac{z+2}{-4}.$$

Figure 2.13(a) shows the tangent plane and normal line at $(-1, 1, -2)$.

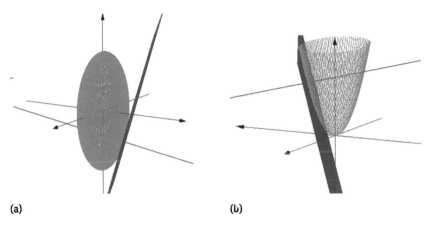

(a)   (b)

Figure 2.13: Tangent plane and normal line. Examples 2.7.2 and 2.7.3.

The equation of a surface is given explicitly by $z = f(x, y)$

In the special case in which the equation of a surface S is of the form $z = f(x, y)$, we can rewrite this equation in the original form as
$$F(x, y, z) = f(x, y) - z = 0.$$

If f is differentiable, then
$$F_x(x, y, z) = f_x(x, y), \quad F_y(x, y, z) = f_y(x, y), \quad \text{and} \quad F_z(x, y, z) = -1,$$

and a normal vector to the tangent plane is $\langle f_x, f_y, -1 \rangle$. Thus, an equation of the tangent plane to the surface at (x_0, y_0, z_0) becomes

$$f_x(x_0, y_0)(x - x_0) + f_y(x_0, y_0)(y - y_0) - (z - z_0) = 0, \tag{2.11}$$

or

$$z = z_0 + f_x(x_0, y_0)(x - x_0) + f_y(x_0, y_0)(y - y_0). \tag{2.12}$$

An equation of the normal line to the surface at (x_0, y_0, z_0) becomes

$$\frac{x - x_0}{f_x(x_0, y_0)} = \frac{y - y_0}{f_y(x_0, y_0)} = \frac{z - z_0}{-1}. \tag{2.13}$$

> **Example 2.7.3.** Find the tangent plane and normal line to the elliptic paraboloid $z = 2x^2 + y^2$ at the point $(1, 1, 3)$.

Solution. Let $f(x, y) = 2x^2 + y^2$. Then

$$f_x(x, y) = 4x, \quad f_y(x, y) = 2y,$$
$$f_x(1, 1) = 4, \quad f_y(1, 1) = 2.$$

This gives an equation of the tangent plane at $(1, 1, 3)$ as

$$4(x - 1) + 2(y - 1) - 1(z - 3) = 0,$$

or

$$4x + 2y - z = 3.$$

Symmetric equations of the normal line at this point are

$$\frac{x-1}{4} = \frac{y-1}{2} = \frac{z-3}{-1}.$$

Figure 2.13(b) shows the elliptic paraboloid and its tangent plane at $(1, 1, 3)$ that we found in this example.

Note. If the surface has a parameterization

$$\mathbf{r}(u, v) = \langle x(u, v), y(u, v), z(u, v) \rangle,$$

then $\mathbf{r}_u \times \mathbf{r}_v$ is a vector normal to the plane tangent to the surface. If the point P on the surface is $(u_0, v_0, \mathbf{r}(u, v_0))$, then $\mathbf{r}(u_0, v_0)$ is a curve on the surface passing through P, thus, $\mathbf{r}_u(u_0, v_0)$ is its tangent vector at P. Similarly, $\mathbf{r}_v(u_0, v_0)$ is a tangent vector of the curve $\mathbf{r}(u_0, v)$. Thus, a normal vector is obtained by taking the cross product of the two tangent vectors. If we choose the parameterization

$$\mathbf{r}(u, v) = \langle u, v, f(u, v) \rangle$$

for the surface $z = f(x, y)$, then a normal vector of its tangent plane is

$$\mathbf{r}_u \times \mathbf{r}_v = \langle 1, 0, f_u \rangle \times \langle 0, 1, f_v \rangle = \langle -f_u, -f_v, 1 \rangle,$$

which is the same as $\langle -f_x, -f_y, 1 \rangle$ found before.

Tangent plane approximation

Recall the total differential/linear approximation which we discussed in Section 2.3.3. Note that when $dx = \Delta x = x - x_0$, $dy = \Delta y = y - y_0$, and $\Delta z = z - z_0$, we have

$$\Delta z \approx dz = f_x(x_0, y_0)\Delta x + f_y(x_0, y_0)\Delta y,$$
$$z - z_0 \approx f_x(x_0, y_0)(x - x_0) + f_y(x_0, y_0)(y - y_0).$$

The linearization $z = z_0 + f_x(x_0, y_0)(x-x_0) + f_y(x_0, y_0)(y-y_0)$ is exactly an equation of the tangent plane at the point (x_0, y_0, z_0). The change Δz is approximated by the change dz in the corresponding tangent plane, as shown in Figure 2.14.

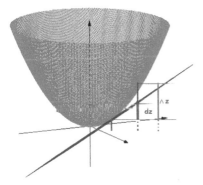

Figure 2.14: Tangent plane approximation.

2.8 Directional derivatives and gradient vectors

When you climb a mountain, at a certain point, you might ask, what is the slope of the surface? Your answer might be "it depends on which direction I go." This is the idea behind the directional derivative, i. e., to find the rate of change along some given direction. Recall that if $z = f(x, y)$, then the partial derivatives f_x and f_y represent the rates of change of z in the x- and y-direction, respectively (when the independent variables (x, y) change in the directions of the unit vectors **i** and **j**). How can we find the rate of change in other directions?

Suppose that we now wish to find the rate of change of z at the point $P(a, b)$ when the independent variables (x, y) change in the direction of an arbitrary unit vector $\mathbf{u} = \langle \cos \alpha, \cos \beta \rangle$. As shown in Figure 2.15, a point on the half-line L in the direction of **u** at a distance $l \geq 0$ from the point $P(a, b)$ is given by $(a + l \cos \alpha, b + l \cos \beta)$. The change of $f(x, y)$ in the direction **u** is

$$\Delta z = f(a + l \cos \alpha, b + l \sin \alpha) - f(a, b).$$

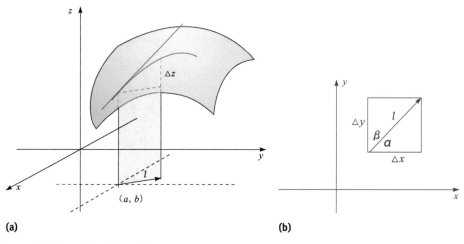

(a) (b)

Figure 2.15: Directional derivatives.

This involves one variable l, so we define the directional derivative along the direction **u** as follows.

Definition 2.8.1. Let $z = f(x,y)$ be a function of two variables and (a,b) be an interior point in its domain; $\mathbf{u} = \langle \cos\alpha, \sin\alpha \rangle$ is a unit vector. The directional derivative of z at the point (a,b) in the direction **u** is defined by

$$D_\mathbf{u} f(a,b) = \lim_{l \to 0^+} \frac{f(a+l\cos\alpha, b+l\sin\alpha) - f(a,b)}{l},$$

provided the limit exists.

Note. We also use notations such as $\frac{dz}{dl}$, $\frac{\partial z}{\partial l}$, $\frac{\partial f}{\partial l}$, or $\frac{\partial z}{\partial \rho}$ for directional derivatives.

Example 2.8.1. Find the directional derivative of $z = \sqrt{x^2 + y^2}$ at $(0,0)$ in the direction $\langle \frac{1}{\sqrt{2}}, \frac{1}{\sqrt{2}} \rangle$.

Solution. The direction $\mathbf{u} = \langle \frac{1}{\sqrt{2}}, \frac{1}{\sqrt{2}} \rangle$ is a unit vector already, so $\cos\alpha = \frac{1}{\sqrt{2}}$, $\cos\beta = \frac{1}{\sqrt{2}}$, and

$$D_\mathbf{u} f(0,0) = \lim_{l \to 0^+} \frac{f(a+l\cos\alpha, b+l\sin\alpha) - f(a,b)}{l}$$

$$= \lim_{l \to 0^+} \frac{\sqrt{(0 + l\frac{1}{\sqrt{2}})^2 + (0 + l\frac{1}{\sqrt{2}})^2} - \sqrt{0^2 + 0^2}}{l}$$

$$= \lim_{l \to 0^+} \frac{l}{l} = 1.$$

It is easy to see from the graph of this right circular cone, at the vertex, the slope along any direction is $\tan 45° = 1$.

Note. The cone has no partial derivative at $(0,0)$. So, this example shows that a function may have a directional directive at a point in some direction, even though it may not have partial derivatives at that point. This is because the directional derivative is defined as a one-sided limit!

Surprisingly, if a function is differentiable, its derivative in any direction exists, and we can find directional derivatives using its partials. This is shown in the following theorem.

Theorem 2.8.1. *If $z = f(x,y)$ is differentiable at $P_0(a,b)$, then the directional derivative of f exists at P_0 in the direction given by any unit vector $\mathbf{u} = \langle \cos \alpha, \cos \beta \rangle$ and*

$$D_\mathbf{u} f(a,b) = f_x(a,b) \cos \alpha + f_y(a,b) \cos \beta. \tag{2.14}$$

Proof. Since $f(x,y)$ is differentiable at the point $P_0(a,b)$, we have

$$\Delta z = \frac{\partial f}{\partial x} \Delta x + \frac{\partial f}{\partial y} \Delta y + o(\sqrt{\Delta x^2 + \Delta y^2}),$$

so

$$f(a + l\cos\alpha, b + l\sin\alpha) - f(a,b) = f_x(a,b) l \cos\alpha + f_y(a,b) l \sin\alpha$$
$$+ o(\sqrt{[l\cos\alpha]^2 + [l\sin\alpha]^2}).$$

It follows that

$$D_\mathbf{u} f(a,b) = \lim_{l \to 0} \frac{f(a + l\cos\alpha, b + l\sin\alpha) - f(a,b)}{l}$$
$$= \lim_{l \to 0} \frac{f_x(a,b) l \cos\alpha + f_y(a,b) l \sin\alpha + o(\sqrt{[l\cos\alpha]^2 + [l\sin\alpha]^2})}{l}$$
$$= \lim_{l \to 0} \frac{f_x(a,b) l \cos\alpha + f_y(a,b) l \sin\alpha + o(l)}{l}$$
$$= f_x(a,b) \cos\alpha + f_y(a,b) \sin\alpha. \qquad \square$$

Now we can rewrite the directional derivative in a dot product form, i. e.,

$$D_\mathbf{u} f(a,b) = f_x(a,b) \cos\alpha + f_y(a,b) \sin\alpha = \langle f_x(a,b), f_y(a,b) \rangle \cdot \langle \cos\alpha, \sin\alpha \rangle.$$

This is

$$D_\mathbf{u} f(a,b) = \langle f_x(a,b), f_y(a,b) \rangle \cdot \mathbf{u}. \tag{2.15}$$

Example 2.8.2. Find the directional derivative of $z = xe^{2y}$ at $P(1,0)$ in the direction from P to the point $Q(2,-1)$.

Solution. The unit vector in the direction of $\vec{PQ}$ is

$$\mathbf{u} = \frac{\langle 2-1, -1-0 \rangle}{\sqrt{(2-1)^2 + (-1-0)^2}} = \left\langle \frac{1}{\sqrt{2}}, \frac{-1}{\sqrt{2}} \right\rangle.$$

Since

$$\left.\frac{\partial z}{\partial x}\right|_{(1,0)} = \left.e^{2y}\right|_{(1,0)} = 1 \quad \text{and} \quad \left.\frac{\partial z}{\partial y}\right|_{(1,0)} = \left.2xe^{2y}\right|_{(1,0)} = 2,$$

by equation (2.15) the desired directional derivative is

$$D_{\mathbf{u}}f(a,b) = \langle 1, 2 \rangle \cdot \left\langle \frac{1}{\sqrt{2}}, -\frac{1}{\sqrt{2}} \right\rangle = 1 \cdot \frac{1}{\sqrt{2}} + 2 \cdot \left(-\frac{1}{\sqrt{2}}\right) = -\frac{\sqrt{2}}{2}.$$

Steepest ascent/descent and the gradient vector

Again, suppose you are somewhere on a mountain and want to know the direction of the steepest ascending path. The directional derivative can answer this question now. Assume $z = f(x, y)$ is differentiable at (a, b). The directional derivative in the direction $\mathbf{u}$ is

$$\begin{aligned} D_{\mathbf{u}}f &= \langle f_x(a,b), f_y(a,b) \rangle \cdot \mathbf{u} \\ &= |\langle f_x(a,b), f_y(a,b) \rangle| \|\mathbf{u}\| \cos\theta \\ &= \sqrt{f_x^2(a,b) + f_y^2(a,b)} \cos\theta. \end{aligned}$$

Therefore, the maximum directional derivative that f can obtain is $\sqrt{f_x^2 + f_y^2}$, and this happens when $\theta = 0$, that is, when the two vectors $\langle f_x(a,b), f_y(a,b) \rangle$ and $\mathbf{u}$ point in the same direction. The minimum directional derivative that f can obtain is $-\sqrt{f_x^2 + f_y^2}$, and this happens when $\theta = \pi$, that is, when the two vectors $\langle f_x(a,b), f_y(a,b) \rangle$ and $\mathbf{u}$ point in exactly opposite directions. Therefore, along the direction $\langle f_x(a,b), f_y(a,b) \rangle$, the function f obtains its greatest directional derivative; f attains its smallest directional derivative in the direction $-\langle f_x(a,b), f_y(a,b) \rangle$, as shown in Figure 2.16. We give the vector $\langle f_x(a,b), f_y(a,b) \rangle$ a special name.

Definition 2.8.2. If $z = f(x, y)$ is a differentiable function, then the gradient of f at the point (a, b) is the vector $\nabla f(a, b)$ defined by

$$\nabla f(a,b) = \langle f_x(a,b), f_y(a,b) \rangle = f_x(a,b)\mathbf{i} + f_y(a,b)\mathbf{j}.$$

The gradient vector at (x, y) is the vector function $\nabla f(x, y)$ defined by

$$\nabla f(x,y) = \langle f_x(x,y), f_y(x,y) \rangle = f_x(x,y)\mathbf{i} + f_y(x,y)\mathbf{j}.$$

Note. Sometimes $\nabla f(x, y)$ is also denoted by **grad**f.

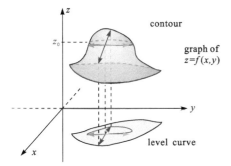

Figure 2.16: Gradient vectors, steepest ascent/descent.

By the above definition, we can write the directional derivative of f in the direction given by a unit vector **u** as

$$D_{\mathbf{u}}f(a,b) = \nabla f \cdot \mathbf{u}, \qquad (2.16)$$

and the steepest ascent/steepest slope of f at (x,y) is $|\nabla f|$, which occurs in the direction ∇f. The steepest descent of f at (x,y) is $-|\nabla f|$, which occurs in the direction of $-\nabla f$. In fact, the directional derivative of f at (a,b) in the direction **u** is the scalar projection of ∇f onto the vector **u**.

Gradient vectors and level curves

Recall that a *level curve* of $z = f(x,y)$ is a contour projected onto the xy-plane. Then, a level curve in the xy-plane has an equation $f(x,y) = k$, where k is some constant. Assume the equation $f(x,y) = k$ defines a function $y = y(x)$ implicitly at a point $P(a,b)$. Differentiating $f(x,y) = k$ at the point P gives

$$f_x + f_y \frac{dy}{dx} = 0,$$

$$f_x(a,b) + f_y(a,b)\frac{dy}{dx}\bigg|_{(a,b)} = 0 \quad \text{(computing at P)},$$

$$\langle f_x(a,b), f_y(a,b)\rangle \cdot \left\langle 1, \frac{dy}{dx}\bigg|_{(a,b)}\right\rangle = 0,$$

$$\nabla f(a,b) \cdot \left\langle \frac{dx}{dx}, \frac{dy}{dx}\right\rangle_{(a,b)} = 0.$$

Note that the tangent vector of the level curve written parametrically as $x = x$ and $y = y(x)$ is given by $\langle \frac{dx}{dx}, \frac{dy}{dx}\rangle$. The zero value for the dot product proves that the gradient vector of f is perpendicular to the level curve at P, as shown in Figure 2.17.

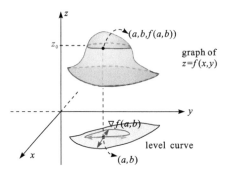
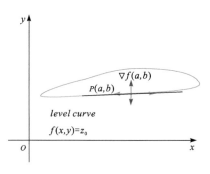

Figure 2.17: Gradient vectors and level curves.

Example 2.8.3. For the function $z = f(x,y) = x^2 - 3y^2$,
1. find the gradient of $f(x,y)$ at $(2,1)$.
2. find the derivative of f in the direction $\langle 3, -1 \rangle$ at $(2,1)$.
3. find an equation for the level curve C passing through the point $(2,1)$ in the xy-plane.
4. verify that the gradient vector of f and the tangent vector of C at $(2,1)$ are perpendicular to each other.

Solution.
1. The gradient of f at $(2,1)$ is

$$\nabla f(2,1) = \langle f_x(2,1), f_y(2,1) \rangle = \langle 2x, -6y \rangle|_{(2,1)} = \langle 4, -6 \rangle.$$

2. The directional derivative of f in the direction $\langle 3, -1 \rangle$ at $(2,1)$ is

$$D_{\mathbf{u}} f(2,1) = \nabla f \cdot \mathbf{u} = \langle 4, -6 \rangle \cdot \frac{\langle 3, -1 \rangle}{\sqrt{3^2 + (-1)^2}} = \frac{12 + 6}{\sqrt{10}} = \frac{18}{\sqrt{10}}.$$

3. When $x = 2$ and $y = 1$, $f(2,1) = 2^2 - 3(1)^2 = 1$. So, the level curve C has an equation

$$x^2 - 3y^2 = 1 \quad \text{and} \quad z = 0.$$

4. Applying implicit differentiation to $x^2 - 3y^2 = 1$, we have

$$2x - 6yy' = 0,$$
$$y' = \frac{x}{3y}.$$

So at the point $(2,1)$ on the level curve C, the slope is $y'(2) = \frac{2}{3}$, and the tangent line is

$$y - 1 = \frac{2}{3}(x - 2).$$

A tangent vector of C at $(2,1)$ is $\langle 3, 2 \rangle$, which is indeed perpendicular to the gradient $\nabla f(2,1) = \langle 4, -6 \rangle$ since

$$\langle 3, 2 \rangle \cdot \langle 4, -6 \rangle = 12 - 12 = 0.$$

2.8 Directional derivatives and gradient vectors — 119

Example 2.8.4. Suppose the temperature distribution on a plate at any point (x,y) satisfies

$$T(x,y) = 80 - 2x^2 - y^2 - x \; (°C).$$

An ant with bad luck unfortunately fell on the plate at $(1,1)$. Find the best escaping path for the ant.

Solution. Note that the temperature at $(1,1)$ is 76 °C! The strategy is to find the path along which the temperature decreases most rapidly. This is equivalent to finding the steepest descent path on the surface (the graph of the function $T(x,y)$) starting from the point $(1,1)$. Well, the direction of this path must be the opposite direction of ∇T. Assume the path is $y = y(x)$. Then $\langle dx, dy \rangle$, which is the tangent vector, must be parallel to $-\nabla T = \langle -T_x, -T_y \rangle$. So,

$$\langle dx, dy \rangle = \lambda \langle -T_x, -T_y \rangle \quad \text{for some } \lambda$$

or

$$\frac{dx}{T_x} = \frac{dy}{T_y}.$$

But $T_x = -4x - 1$ and $T_y = -2y$, so

$$\frac{1}{-4x-1} dx = \frac{1}{-2y} dy.$$

Integrating both sides gives

$$-\int \frac{1}{4x+1} dx = -\int \frac{1}{2y} dy.$$

This simplifies to $\frac{1}{4} \ln|4x+1| = \frac{1}{2} \ln|y| + C$, where C is an arbitrary constant. Then

$$\ln|4x+1| = 2\ln|y| + 4C,$$
$$|4x+1| = |y|^2 e^{4C},$$
$$4x+1 = y^2 D, \quad \text{where } D \text{ is also an arbitrary constant.}$$

Since $y(1) = 1$, this means $4 + 1 = 1^2 D$, so $D = 5$. Therefore, the escaping route is

$$4x + 1 = 5y^2.$$

This is certainly not the shortest path to the edge of the plate, but the path along which the temperature decreases most rapidly, as shown in Figure 2.18.

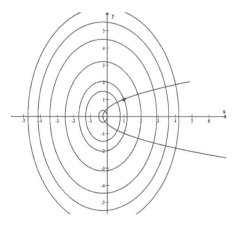 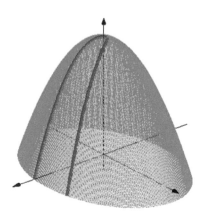

Figure 2.18: The escaping route.

Functions of more than two variables

The concept and notation of directional derivatives can be extended to functions of more than two variables. For example, for $u = f(x,y,z)$, the gradient vector is $\nabla f = \langle f_x, f_y, f_z \rangle$, and the derivative in the direction given by a unit vector **u** is $D_{\mathbf{u}}f(x,y,z) = \nabla f \cdot \mathbf{u}$.

For example, the directional derivative of $u = f(x,y,z)$ at the point (a,b,c) in the direction $\mathbf{u} = \langle \cos\alpha, \cos\beta, \cos\gamma \rangle$ is given by

$$D_{\mathbf{u}}f(a,b,c) = f_x(a,b,c)\cos\alpha + f_y(b,a,c)\cos\beta + f_z(a,b,c)\cos\gamma$$
$$= \langle f_x(a,b,c), f_y(a,b,c), f_z(a,b,c)\rangle \cdot \mathbf{u}$$
$$= \nabla f \cdot \mathbf{u}.$$

Example 2.8.5. Compute $\frac{\partial f}{\partial l}\big|_{(1,1,2)}$ in the direction with direction angles $\alpha = \pi/3$, $\beta = \pi/4$, and $\gamma = \pi/3$ when f is defined by

$$f(x,y,z) = xy + yz + zx.$$

Solution. The unit vector in the direction is

$$\mathbf{u} = \langle \cos\pi/3, \cos\pi/4, \cos\pi/3 \rangle = \left\langle \frac{1}{2}, \frac{\sqrt{2}}{2}, \frac{1}{2} \right\rangle.$$

Also,

$$f_x(1,1,2) = (y+z)\big|_{(1,1,2)} = 3,$$
$$f_y(1,1,2) = (x+z)\big|_{(1,1,2)} = 3,$$
$$f_z(1,1,2) = (y+x)\big|_{(1,1,2)} = 2,$$

so the directional derivative in the direction **u** is

$$\frac{\partial f(1,1,2)}{\partial l} = \langle 3,3,2\rangle \cdot \left\langle \frac{1}{2}, \frac{\sqrt{2}}{2}, \frac{1}{2} \right\rangle = 3\cdot\frac{1}{2} + 3\cdot\frac{\sqrt{2}}{2} + 2\cdot\frac{1}{2} = \frac{5}{2} + \frac{3}{2}\sqrt{2}.$$

Note. There is a similar idea to level curves for functions of three variables. If we set $f(x,y,z) = k$, for a constant k, then we get a *level surface* of the function $u = f(x,y,z)$. At any given point $P(a,b,c)$, for any curve passing through P that lies on the surface, we can show that the gradient vector ∇f is actually perpendicular to its tangent vector. (We provided a proof in the previous section.) Therefore, ∇f is perpendicular to the tangent plane to the level surface at P, as shown in Figure 2.19. Thus, the gradient vector $\nabla f = \langle f_x, f_y, f_z \rangle$ is a normal vector of the tangent plane to the level surface $f(x,y,z) = k$ through P.

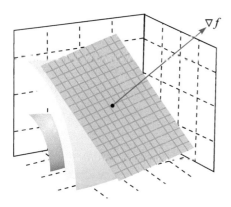

Figure 2.19: Level surfaces and gradient vectors for functions of three variables.

Now, we can use gradient vectors to find a tangent vector for a curve of intersection of two surfaces,

$$\begin{cases} f(x,y,z) = 0, \\ g(x,y,z) = 0 \end{cases}$$

at a given point P. Since a tangent vector of the curve at P is perpendicular to both ∇f and ∇g, one vector tangent to the curve at P is

$$\mathbf{v} = \nabla f \times \nabla g.$$

Using this idea, we consider Example 2.7.1 again. Since

$$f(x,y,z) = x^2 + y^2 + z^2 - 6 = 0,$$
$$g(x,y,z) = x + y + z = 0,$$

we have $\nabla f(1, -2, 1) = \langle 2x, 2y, 2z\rangle|_{(1,-2,1)} = \langle 2, -4, 2\rangle$ and $\nabla g(1, -2, 1) = \langle 1, 1, 1\rangle$. Thus,

$$\mathbf{v} = \begin{pmatrix} 2 \\ -4 \\ 2 \end{pmatrix} \times \begin{pmatrix} 1 \\ 1 \\ 1 \end{pmatrix} = \begin{pmatrix} -6 \\ 0 \\ 6 \end{pmatrix} = 6\begin{pmatrix} -1 \\ 0 \\ 1 \end{pmatrix}.$$

So any vector parallel to $\langle -1, 0, 1\rangle$ is a tangent vector to the curve at $(1, -2, 1)$. There is no need for any parameterization.

Note. Using vector notation, and knowledge in linear algebra, we now can write the Taylor series for a function of two variables in a more compact way. Let $\mathbf{x} = \langle x, y\rangle$ and $\mathbf{x}_0 = \langle a, b\rangle$. Then $f(x, y) = f(\mathbf{x})$, $f(a, b) = f(\mathbf{x}_0)$, and $\Delta \mathbf{x} = \mathbf{x} - \mathbf{x}_0 = \langle \Delta x, \Delta y\rangle$. Note that

$$\left(\Delta x \frac{\partial}{\partial x} + \Delta y \frac{\partial}{\partial y}\right) f(a, b) = \Delta x \frac{\partial f(\mathbf{x}_0)}{\partial x} + \Delta y \frac{\partial f(\mathbf{x}_0)}{\partial y}$$

$$= \langle f_x(\mathbf{x}_0), f_y(\mathbf{x}_0)\rangle \cdot \langle \Delta x, \Delta y\rangle = \nabla f(\mathbf{x}_0) \cdot \Delta \mathbf{x}.$$

Also, we have the **Hessian matrix** notation

$$H(\mathbf{x}_0) = \begin{vmatrix} f_{xx}(\mathbf{x}_0) & f_{xy}(\mathbf{x}_0) \\ f_{xy}(\mathbf{x}_0) & f_{yy}(\mathbf{x}_0) \end{vmatrix}.$$

Then, we can write

$$f(x) = f(\mathbf{x}_0) + \nabla f(\mathbf{x}_0) \cdot \Delta \mathbf{x} + \frac{1}{2!}\Delta \mathbf{x}^T H(x_0) \Delta \mathbf{x} + \cdots.$$

2.9 Maximum and minimum values

2.9.1 Extrema of functions of two variables

As shown in Figure 2.20, for a function of two variables, $z = f(x, y)$, there are also interesting features such as local or global extreme values, as we have seen in one-variable calculus. We first give the definition of local and global extrema.

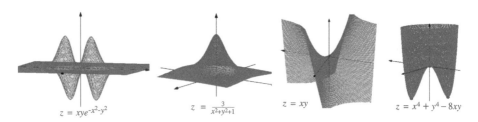

Figure 2.20: Local and global extreme values.

Definition 2.9.1. A function $z = f(x,y)$ has a domain $D \subset \mathbb{R}^2$. Then, $f(x,y)$ has
1. a **local/relative maximum** at an interior point $(a,b) \in D$ if $f(x,y) \leq f(a,b)$ for all (x,y) in D close to (a,b).
2. a **local/relative minimum** at an interior point $(a,b) \in D$ if $f(x,y) \geq f(a,b)$ for all (x,y) in D close to (a,b).
3. a **global/absolute maximum** at a point $(a,b) \in D$ if $f(x,y) \leq f(a,b)$ for all (x,y) in D.
4. a **global/absolute minimum** at a point $(a,b) \in D$ if $f(x,y) \geq f(a,b)$ for all (x,y) in D.

Figure 2.21 shows that a function whose graph is the upper hemisphere has a local maximum (also absolute maximum) above its center, and a function whose graph is the cone with vertex downwards has an absolute minimum at its vertex.

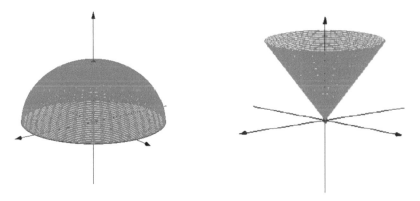

Figure 2.21: An upper sphere and an upper cone.

How can we identify these interesting points? As we saw in one-variable calculus, the answer is to use derivatives. In this section, we use partial derivatives to help locate maxima and minima of functions of two variables. We first consider the case that a differentiable function $z = f(x,y)$ has a local maximum point at (a,b). Then, the intersection curve

$$\begin{cases} z = f(x,y), \\ y = b \end{cases}$$

also has a local maximum at the same point (a,b). However, the curve $z = f(x,b)$ in the $y = b$ plane has just one variable. Therefore, the derivative with respect to x at $x = a$ must be 0. This means $f_x(a,b) = 0$. Similarly, we can obtain $f_y(a,b) = 0$, or, equivalently, $\nabla f(a,b) = 0$, as shown in Figure 2.22.

Note. Note that if $f_x(a,b) = 0$ and $f_y(a,b) = 0$, then $\nabla f(a,b) = \mathbf{0}$, and the tangent plane at (a,b) is $z = z_0$. This means the geometric interpretation of $\nabla f(a,b) = \mathbf{0}$ is that the graph of f has a horizontal tangent plane at the point (a,b).

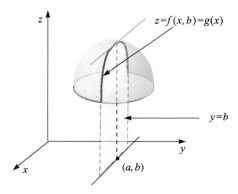

Figure 2.22: Candidates for local maximum or local minimum.

If a function has no partial derivatives at a point, it may still have extreme values there (similar to one-variable calculus, a function that is not differentiable may still have extreme values). For instance, the upper right circular cone $z = \sqrt{x^2 + y^2}$ has no partial derivatives at $(0, 0)$, but it does have a local minimum value 0 at $(0, 0)$. Therefore, candidates of extrema for any function are those points where $\nabla f = \mathbf{0}$ or ∇f does not exist. We give them a special name.

Definition 2.9.2 (Critical points). A point (a, b) is called a *critical point* of $z = f(x, y)$ if $\nabla f(a, b) = \mathbf{0}$ or if $\nabla f(a, b)$ does not exist.

These above arguments lead to the following theorem.

Theorem 2.9.1 (Candidate theorem). *If a function $z = f(x, y)$ has a local maximum/minimum at an interior point (a, b) in its domain, then $\nabla f(a, b) = \mathbf{0}$ or $\nabla f(a, b)$ does not exist.*

Theorem 2.9.1 shows that if f has a local maximum or minimum at (a, b), then (a, b) must be a critical point of f.

Example 2.9.1. Find all critical points for each of the following functions:

$$(a) \quad z = xy, \quad (b) \quad z = x^4 + y^4 - 8xy.$$

Solution. For (a), the function $z = xy$ is differentiable everywhere, so all critical points are those such that $\nabla f = \mathbf{0}$, i.e.,

$$\nabla f = \mathbf{0} \rightarrow f_x = 0 \quad \text{and} \quad f_y = 0,$$
$$f_x = y = 0 \quad \text{and} \quad f_y = x = 0.$$

So the only critical point is $(0, 0)$.

For (b), the function $z = x^4 + y^4 - 8xy$ is also differentiable everywhere, so all critical points are those such that $\nabla f = \mathbf{0}$.

$$\nabla f = \mathbf{0} \to f_x = 0 \quad \text{and} \quad f_y = 0,$$
$$f_x = 4x^3 - 8y = 0 \quad \text{and} \quad f_y = 4y^3 - 8x = 0.$$

Solve for x and y to obtain

$$x^3 = 2y \quad \text{and} \quad y^3 = 2x.$$

Thus, we have

$$x^9 = 8(2x),$$
$$x(x^8 - 16) = 0,$$
$$x(x^4 + 4)(x^2 + 2)(x - \sqrt{2})(x + \sqrt{2}) = 0.$$

Thus, $x = 0$, $x = \sqrt{2}$, or $x = -\sqrt{2}$, and all the critical points are

$$(0,0), (\sqrt{2}, \sqrt{2}), \text{ and } (-\sqrt{2}, -\sqrt{2}).$$

Graphs of the two functions are shown in Figure 2.20.

However, as in single-variable calculus, not all critical points give rise to maxima or minima. For instance, for the function $z = xy$, $\nabla f(0,0) = \mathbf{0}$, but the function value 0 at $(0,0)$ is neither a maximum nor a minimum because near $(0,0)$ there are points in the first and third quadrants of the xy-plane which make $z = xy$ positive and points in the second and fourth quadrants which make $z = xy$ negative. We also give a definition for this type of point.

Definition 2.9.3 (Saddle points). A critical point (a, b) of $z = f(x, y)$ at which $\nabla f(a, b) = \mathbf{0}$ but f does not have a local maximum or a local maximum is called a saddle point.

Note. In one-variable calculus, a saddle point is a point where the function has a horizontal tangent line, and nearby the point you can find places where the graph of the function is above the tangent line and other places where the graph is below the tangent line. The function $f(x) = x^3$ at $x = 0$ is a good example of a saddle point. Analogously, in two-variable calculus, a saddle point is a point where the function has a horizontal tangent plane, and nearby the point you can find places where the graph of the function is above the tangent plane and other places where the graph is below the tangent plane.

Theorem 2.9.2 (Second derivative test). *Assume all the second partial derivatives of f are continuous on a disk with center (a, b), and $f_x(a, b) = 0$ and $f_y(a, b) = 0$, so that (a, b) is a critical point of f. Let*

$$A = f_{xx}(a, b), \quad B = f_{xy}(a, b), \quad C = f_{yy}(a, b).$$

Then:
1. *If $AC - B^2 > 0$ and $A > 0$, then $f(a, b)$ is a local minimum.*
2. *If $AC - B^2 > 0$ and $A < 0$, then $f(a, b)$ is a local maximum.*
3. *If $AC - B^2 < 0$, then $f(a, b)$ is not a local minimum or maximum, so it is a saddle point.*

Note.
1. In case (3), where the point (a, b) is a saddle point of f, the graph $y = f(x, y)$ crosses its tangent plane at (a, b), that is, near the saddle point, part of the graph is above the tangent plane, and part of the graph is below the tangent plane.
2. If $AC - B^2 = 0$, the test fails to give any information. In this case, f could have a local maximum or local minimum at (a, b), or (a, b) could be a saddle point of f.
3. To help remember the formula for $AC - B^2$, we can write it in determinant form,

$$AC - B^2 = \begin{vmatrix} A & B \\ B & C \end{vmatrix} = \begin{vmatrix} f_{xx} & f_{xy} \\ f_{xy} & f_{yy} \end{vmatrix} = f_{xx}f_{yy} - (f_{xy})^2.$$

A proof of the second derivative test can be seen from the vector form of the Taylor expansion for a function of two variables. Assume $\mathbf{x} = \langle x, y \rangle$, $u = f(\mathbf{x})$, and $f(\mathbf{x})$ has continuous first and second partial derivatives at $\mathbf{x}_0 = (a, b)$. Then, for small $\Delta \mathbf{x}$,

$$f(\mathbf{x}_0 + \Delta \mathbf{x}) \approx f(\mathbf{x}_0) + \nabla f(\mathbf{x}_0) \cdot \Delta \mathbf{x}^T + \frac{1}{2!} \Delta \mathbf{x}^T H(\mathbf{x}_0) \Delta \mathbf{x},$$

where $\nabla f(\mathbf{x}_0) = \langle f_x(a, b), f_y(a, b) \rangle$, and $H(\mathbf{x}_0)$ is the Hessian matrix defined as

$$H(\mathbf{x}_0) = \begin{vmatrix} f_{xx} & f_{xy} \\ f_{yx} & f_{yy} \end{vmatrix}_{(a,b)}.$$

Since $\nabla f(\mathbf{x}_0) = \mathbf{0}$ at a candidate $\mathbf{x}_0$, $f(\mathbf{x}_0 + \Delta \mathbf{x}) > f(\mathbf{x}_0)$ if $H(\mathbf{x}_0)$ is positive definite, and $f(\mathbf{x}_0 + \Delta \mathbf{x}) < f(\mathbf{x}_0)$ if $H(\mathbf{x}_0)$ is negative definite.

Example 2.9.2. Locate and classify all the critical points for each of the following functions:

(a) $f(x, y) = xy$, (b) $f(x, y) = x^4 + y^4 - 8xy$, and (c) $z = xye^{-x^2-y^2}$.

Solution.
(a) We have found the critical point $(0, 0)$ for $z = xy$ in Example 2.9.1. Since $A = f_{xx} = 0$, $B = f_{xy} = 1$, and $C = f_{yy} = 0$, we have

$$AC - B^2 = 0 - 1 < 0.$$

So by the second derivative test, this is a saddle point.

(b) We calculate $A = f_{xx} = 12x^2$, $B = f_{xy} = -8$, and $C = f_{yy} = 12y^2$ and apply the second derivative test to critical points we have found in Example 2.9.1.
At $(0,0)$, $AC - B^2 = 144x^2y^2 - 64|_{(0,0)} = -64 < 0$, so it is a saddle point.
At $(\sqrt{2}, \sqrt{2})$, $AC - B^2 = 144x^2y^2 - 64|_{(\sqrt{2},\sqrt{2})} = 144(4) - 64 > 0$ and $A = 12(\sqrt{2})^2 > 0$, so it is a local minimum.
At $(-\sqrt{2}, -\sqrt{2})$, $AC - B^2 = 144x^2y^2 - 64|_{(-\sqrt{2},-\sqrt{2})} = 144(4) - 64 > 0$ and $A = 12(\sqrt{2})^2 > 0$, so it is a local minimum.

(c) Solving $\nabla f = \mathbf{0}$, we have
$$\begin{cases} f_x = ye^{-x^2-y^2} + xye^{-x^2-y^2}(-2x) = e^{-x^2-y^2}(y - 2x^2y) = 0, \\ f_y = xe^{-x^2-y^2} + xye^{-x^2-y^2}(-2y) = e^{-x^2-y^2}(x - 2xy^2) = 0. \end{cases}$$

Since $e^{-x^2-y^2} \neq 0$, these equations simplify to
$$\begin{cases} y - 2x^2y = 0, \\ x - 2xy^2 = 0. \end{cases}$$

This gives critical points $(0,0)$, $(\frac{1}{\sqrt{2}}, \frac{1}{\sqrt{2}})$, $(-\frac{1}{\sqrt{2}}, \frac{1}{\sqrt{2}})$, $(\frac{1}{\sqrt{2}}, \frac{1}{\sqrt{2}})$, and $(\frac{1}{\sqrt{2}}, -\frac{1}{\sqrt{2}})$. Then,

$$A = f_{xx} = e^{-x^2-y^2}(-2x)(y - 2x^2y) + e^{-x^2-y^2}(-4xy),$$
$$B = f_{xy} = e^{-x^2-y^2}(-2y)(y - 2x^2y) + e^{-x^2-y^2}(1 - 2x^2), \text{ and}$$
$$C = f_{yy} = e^{-x^2-y^2}(-2y)(x - 2xy^2) + e^{-x^2-y^2}(-4xy).$$

So at $(0,0)$, $AC - B^2 < 0$, and it is a saddle point. At the other points, note that $1 - 2x^2$ and $1 - 2y^2$ are 0. So, B is 0, and $AC - B^2 = 16e^{-2x^2-2y^2}x^2y^2 > 0$. Therefore, the function has local extrema at those points. When x and y have opposite signs, $A > 0$, and when x and y have the same sign, $A < 0$. Thus, the function has two local maxima at $(\frac{1}{\sqrt{2}}, \frac{1}{\sqrt{2}})$ and $(-\frac{1}{\sqrt{2}}, -\frac{1}{\sqrt{2}})$ and two local minima at $(-\frac{1}{\sqrt{2}}, \frac{1}{\sqrt{2}})$ and $(\frac{1}{\sqrt{2}}, -\frac{1}{\sqrt{2}})$. Graphs of the three functions are shown in Figure 2.20.

Global extreme values of functions defined on bounded closed regions

Similar to the closed interval test for global extreme values for a one-variable continuous function over a closed interval, for a continuous function of two variables $z = f(x,y)$ defined on a bounded closed set D, we also have the **closed region test**.
(1) Find the critical points of f in D.
(2) Find the critical points of f when f is considered as a function restricted to the boundary of D.
(3) Evaluate the function values at all the points found in (1) and (2). The greatest of these values is the absolute maximum value of f in D while the least of these values is the absolute minimum value of f in D.

Example 2.9.3. Find the global maximum and global minimum for the function

$$z = f(x,y) = x^2 + y^2 + 8x - 6y + 20, \quad \text{where } (x,y) \in D = \{(x,y) | x^2 + y^2 \leq 36\}.$$

Solution. We first find points such that $\nabla f = \mathbf{0}$. This means

$$f_x = 2x + 8 = 0,$$
$$f_y = 2y - 6 = 0.$$

So, $(-4, 3)$ is the critical point in D (if not in D, we will reject it). The function value at $(-4, 3)$ is

$$f(-4, 3) = (-4)^2 + (3)^2 + 8(-4) - 6(3) + 20 = -5.$$

Now we consider the function values on the boundary of D, $x^2 + y^2 = 36$. Using $x = 6\cos t$, $y = 6\sin t$ as the parameterization of the boundary gives

$$f = (6\cos t)^2 + (6\sin t)^2 + 8(6\cos t) - 6(6\sin t) + 20$$
$$= 48\cos t - 36\sin t + 56, \quad \text{for } 0 \leq t \leq 2\pi.$$

Of course, we can use one-variable calculus to find $\frac{df}{dt} = 0$ and then solve for t to find candidates for extreme values. However, if we note that $a\cos t + b\sin t = \sqrt{a^2 + b^2}\sin(t+\alpha)$ for some α, then the maximum value of the expression $a\cos t + b\sin t$ is $\sqrt{a^2 + b^2}$, and the minimum value of the expression $a\cos t + b\sin t$ is $-\sqrt{a^2 + b^2}$. Thus, we obtain the following maximum and minimum values of the function on the boundary:

$$f_{\text{maximum on the boundary}} = \sqrt{48^2 + 36^2} + 56 = 116,$$
$$f_{\text{minimum on the boundary}} = -\sqrt{48^2 + 36^2} + 56 = -4.$$

Comparing these function values, we conclude that

$$f_{\max} = 116 \quad \text{and} \quad f_{\min} = -5.$$

The graphs of this function and the cylinder are shown in Figure 2.23.

Note. In some practical circumstances an absolute maximum/minimum is known to exist and the domain set is not bounded or the boundary is obscure and/or consists of points of no practical interest. If only one critical point is found, then this critical point must be the desired absolute maximum/minimum.

Example 2.9.4. A rectangular container without a lid is to be made from 18 m² woodboard. Find the maximum volume of such a container.

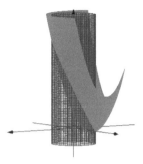

Figure 2.23: Global extreme values for functions defined over a closed region.

Solution. Let x = length, y = width, and z = height of the box (in meters). Then the volume of the box is given by

$$V = xyz.$$

Computing the area of the four sides and the bottom of the box, which must have a total area of 18 m², gives an extra equation (a constraint) linking x, y, and z, i.e.,

$$2xz + 2yz + xy = 18.$$

The domain satisfies $x \geq 0$, $y \geq 0$, and $z \geq 0$, but it is not bounded (for example, if $y \to 0$, then $2xz = 18$ and as $z \to 0$ it follows that $x \to \infty$). We solve the constraint for z to obtain

$$z = \frac{18 - xy}{2(x + y)}$$

and use this to express V as a function of just two variables x and y,

$$V = xy \frac{18 - xy}{2(x + y)} = \frac{18xy - x^2 y^2}{2(x + y)}.$$

We compute the partial derivatives, and we obtain

$$\frac{\partial V}{\partial x} = -\frac{y^2(x^2 + 2xy - 18)}{2(x + y)^2} \quad \text{and} \quad \frac{\partial V}{\partial y} = -\frac{x^2(y^2 + 2xy - 18)}{2(x + y)^2}.$$

Let $\partial V/\partial x = \partial V/\partial y = 0$. Note that $x \neq 0$ and $y \neq 0$. Hence, we must have

$$x^2 + 2xy - 18 = 0 \quad \text{and} \quad y^2 + 2xy - 18 = 0.$$

Subtracting these leads to $x^2 = y^2$ and so $x = y$ (note that x and y must both be positive). If we put $x = y$ in either equation, we obtain $3x^2 - 18 = 0$, which gives $x = \sqrt{6}$, $y = \sqrt{6}$, and $z = (18 - \sqrt{6} \cdot \sqrt{6})/[2(\sqrt{6} + \sqrt{6})] = \sqrt{6}/2$.

Of course, we can show that this indeed gives a local maximum of V by using the second derivative test. However, from the physical nature of this problem, we could simply argue that there must be an absolute maximum volume, and it has to occur at a critical point of V. Since there are no boundary values of interest and there is only one critical point, the function must take its absolute maximum at the only candidate.

2.9.2 Lagrange multipliers

In Example 2.9.3, we maximized the function $f(x, y)$ under the condition $x^2 + y^2 = 36$. We found the maximum by finding a parameterization of the boundary $\langle x(t), y(t) \rangle$ and then reducing $f(x, y)$ to $f(x(t), y(t))$, which is a one-variable function. In Example 2.9.4, we maximized a volume function $V = xyz$ subject to the constraint $2xz + 2yz + xy = 18$. We eliminated the constraint by replacing $z = (18 - xy)/(2x + 2y)$ in the objective function $V = xyz$. Thus, the problems were reduced to problems without constraints. However, this approach may be hard or even impossible in some cases. For example,

$$\text{maximize } f(x, y) = x^2 + 2y^2, \quad \text{where } x^2 - xy + y^2 = 1.$$

We need to consider the more general optimization problem

$$\max / \min z = f(x, y), \quad \text{subject to } g(x, y) = 0,$$

which is to find the maximum or minimum value of the *objective function* $z = f(x, y)$, subject to the constraint $g(x, y) = 0$. This type of problem is called a constrained maximum/minimum problem.

Sometimes we can convert a constrained maximum/minimum problem to a non-constrained one, as shown in Example 2.9.3 and Example 2.9.4, by expressing one variable in terms of other variables in the constraint condition. Now, the question is, how can we identify the candidates for constrained maximum/minimum if elimination of variables is hard or impossible? Note that if the curve $g(x, y) = 0$ in the xy-plane has a parameterization $\mathbf{r}(t) = \langle x(t), y(t) \rangle$, and at $t = t_0$ (this corresponds to some point (a, b) on the curve) there is a constrained maximum/minimum, then

$$z = f(x, y) = f(x(t), y(t))$$

also has a maximum/minimum at $t = t_0$, as shown in Figure 2.24.
Therefore,

$$\frac{dz}{dt} = 0 \rightarrow f_x \frac{dx}{dt} + f_y \frac{dy}{dt} = 0$$

$$\text{or } \langle f_x, f_y \rangle \cdot \left\langle \frac{dx}{dt}, \frac{dy}{dt} \right\rangle \bigg|_{t=t_0} = 0.$$

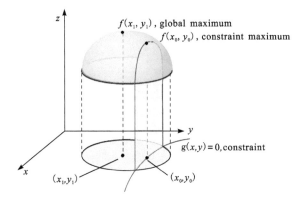

Figure 2.24: Constrained maximum/minimum.

This indicates that at the point $t = t_0$, ∇f is perpendicular to the tangent vector of the curve $\mathbf{r}(t)$. On the other hand the curve $\mathbf{r}(t) = \langle x(t), y(t) \rangle$ also satisfies $g(x(t), y(t)) = 0$. In a similar manner, we also have

$$g_x \frac{dx}{dt} + g_y \frac{dy}{dt} = 0 \quad \text{or} \quad \langle g_x, g_y \rangle \cdot \left\langle \frac{dx}{dt}, \frac{dy}{dt} \right\rangle = 0.$$

This means ∇g is also perpendicular to the tangent vector of the curve $\mathbf{r}(t)$ at t_0. Thus, at $t = t_0$, we must have $\nabla f \parallel \nabla g$. There must be some number λ such that $\nabla f = \lambda \nabla g$ at $t = t_0$. We conclude this by the following theorem.

Theorem 2.9.3. *Suppose both $f(x, y)$ and $g(x, y)$ are differentiable, and at some point (a, b), the optimization problem*

$$\max / \min z = f(x, y) \quad \text{subject to } g(x, y) = 0$$

has a constrained maximum or minimum. Then at the point (a, b), one must have the following conditions:

$$g(a, b) = 0 \quad \text{and} \quad \nabla f(a, b) = \lambda \nabla g(a, b) \quad \text{for some constant } \lambda.$$

Note. The constant λ in Theorem 2.9.3 is called the *Lagrange multiplier*. The theorem can be also stated in a form of a *Lagrange function* defined by $L(x, y, \lambda) = f(x, y) - \lambda g(x, y)$. The candidates for constrained maximum/minimum must then satisfy

$$\nabla L = \mathbf{0} \quad \text{or equivalently} \quad L_x = L_y = L_\lambda = 0.$$

The candidate theorem gives possible candidates for constrained maximum or minimum. When evaluating all function values at these candidate point, the greatest function value found is the desired constrained maximum, the least function value found is the constrained minimum.

Example 2.9.5. Find the constrained maximum and constrained minimum for the optimization problem

$$\max/\min f(x,y) = x + y, \text{ subject to } x^2 - xy + y^2 = 1.$$

Solution. Let $L(x, y, \lambda) = x + y - \lambda(x^2 - xy + y^2 - 1)$. Then any candidate must satisfy

$$L_x = 1 - 2\lambda x + \lambda y = 0,$$
$$L_y = 1 + \lambda x - 2\lambda y = 0,$$
$$L_\lambda = x^2 - xy + y^2 - 1 = 0.$$

From the first and second equations, we have

$$\frac{1}{2x - y} = \lambda = \frac{1}{2y - x}.$$

Then

$$2x - y = 2y - x, \quad \text{or} \quad x = y.$$

Substituting into the third equation, we obtain

$$x^2 - x(x) + x^2 - 1 = 0,$$

so we have $x = \pm 1$. Therefore, candidate points are $(1, 1)$ and $(-1, -1)$; $f(1, 1) = 2$ and $f(-1, -1) = -2$, so the constrained maximum is 2 at $(1, 1)$ and the constrained minimum is -2 at $(-1, -1)$. Figure 2.25(a) shows the graph of the plane $z = x + y$ and $z = x^2 - xy +$

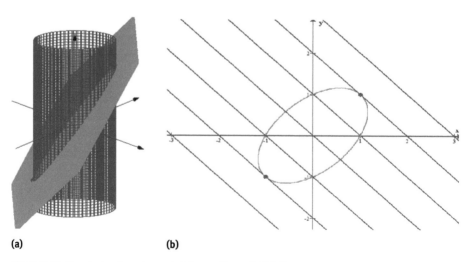

(a)   (b)

Figure 2.25: Constrained maximum/minimum, Example 2.9.5.

$y^2 - 1$. Figure 2.25(b) shows the level curves of the plane and the constraint. Note that the constraint candidates are exactly those where the constraint curve is tangent to a level curve.

Interpreting constrained max/min using level curves

In the previous example, the condition $g(x, y) = 0$, in fact, is a cylinder in space. The constrained maximum/minimum problem is to find the highest point and lowest point on the curve which is the intersection of the cylinder $g(x, y) = 0$ and the surface $z = f(x, y)$. There must be one contour of the surface $z = f(x, y)$ that touches the curve at its highest point (and, similarly, to its lowest point). When projecting the curve and the contour onto the xy-plane, we would expect that the two projection curves are tangent to each other at the projection of the highest point. This is indeed the case. If $f(a, b)$ is the constrained maximum (same argument for the minimum), then

$$\begin{cases} f(a, b) = f(x, y), \\ z = 0 \end{cases} \text{ is the level curve, and}$$

$$\begin{cases} g(x, y) = 0, \\ z = 0 \end{cases} \text{ is the constraint curve.}$$

Since, at (a, b), the tangent vectors of two curves are perpendicular to $\nabla f(a, b)$ and $\nabla g(a, b)$, respectively, it follows that $\nabla f(a, b)$ and $\nabla g(a, b)$ are parallel to each other at (a, b). Thus, the level curve $f(x, y) = f(a, b)$ and the curve $g(x, y) = 0$ are tangent to each other at (a, b)!

Figure 2.26(a) shows the graph of $z = x^2 + 2y^2$ and the constraint $x^2 + y^2 = 1$ (in the $z = 0$ plane). Figure 2.26(b) shows the level curves of $z = x^2 + 2y^2$. One can find candidates for the constrained maximum and minimum at those points where the level curve and constraint curve are tangent to each other.

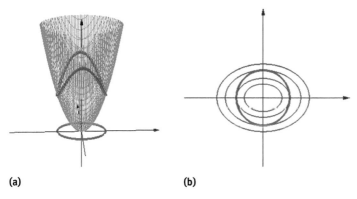

(a)   (b)

Figure 2.26: Level curves and constrained extrema.

Constrained extrema for functions of more than two variables

We have similar conclusions for functions of more than two variables. For instance, if the optimization problem

$$\max / \min u = f(x,y,z), \quad \text{subject to } g(x,y,z) = 0$$

has a constrained extremum at (a,b,c), then, if both f and g are differentiable, we must have

$$\nabla f(a,b,c) = \lambda \nabla g(a,b,c) \quad \text{and} \quad g(a,b,c) = 0.$$

Or we can define a new function, $L(x,y,x,\lambda)$, called the *Lagrangian*, with an extra variable λ called *Lagrange multiplier*, by

$$L(x,y,x,\lambda) = f(x,y,z) - \lambda g(x,y,z).$$

Then, we can find the constrained maximum/minimum for f by taking the following steps.

(1) Find all values of x, y, z, and λ such that the partial derivatives are zero (in other words, find the critical points of L), by solving

$$\begin{cases} L_x = f_x(x,y,z) - \lambda g_x(x,y,z) = 0, \\ L_y = f_y(x,y,z) - \lambda g_y(x,y,z) = 0, \\ L_z = f_z(x,y,z) - \lambda g_z(x,y,z) = 0, \\ L_\lambda = -g(x,y,z) = 0. \end{cases}$$

Note that the fourth equation is just the original constraint.

(2) Evaluate f at all solution points (x,y,z) found in step (1). The largest of these values is the constrained maximum value of f, the smallest is the constrained minimum value of f.

We illustrate the Lagrange multiplier method using the problem of a previous example.

Example 2.9.6. A rectangular container without a lid is to be made from 18 m² woodboard. Find the maximum volume of such a container.

Solution. As before, we wish to maximize

$$V = xyz$$

subject to the constraint

$$2xz + 2yz + xy = 18.$$

Using the method of Lagrange multipliers, $L(x, y, z, \lambda) = xyz - \lambda(2xz + 2yz + xy - 18)$, so the four partial derivatives give the equations

$$\begin{cases} L_x = yz - \lambda(2z + y) = 0, \\ L_y = xz - \lambda(2z + x) = 0, \\ L_z = xy - \lambda(2x + 2y) = 0, \\ L_\lambda = 2xz + 2yz + xy - 18 = 0. \end{cases}$$

There are no general rules for solving systems of nonlinear equations, and sometimes some ingenuity is required. In the present example, you might notice that if we multiply the first equation by x, the second equation by y, and the third equation by z, then we have

$$\begin{cases} xyz = \lambda(2xz + xy), \\ xyz = \lambda(2yz + xy), \\ xyz = \lambda(2xz + 2yz). \end{cases}$$

Subtracting the first two gives $0 = \lambda z(x - y)$, and we observe that z cannot be zero, and $\lambda \neq 0$ because $\lambda = 0$ would imply $yz = xz = xy = 0 \implies x = y = z = 0$. Therefore, we have $x = y$. Subtracting the first equation from the third equation then gives $0 = \lambda x(x - 2z) \implies x = 2z$. Hence, substituting $x = y = 2z$ into the constraint $2xz + 2yz + xy = 18$ and solving for z, which must be positive, gives

$$4z^2 + 4z^2 + 4z^2 = 18,$$
$$12z^2 = 18,$$
$$z = \sqrt{6}/2.$$

Hence, the only critical point is, as before, $x = \sqrt{6}$, $y = \sqrt{6}$, and $z = \sqrt{6}/2$, and this gives the maximum volume.

There is also a geometric interpretation for constrained extrema for differentiable functions of three variables. If $u = f(x, y, z)$ has a constrained extremum at (a, b, c) subject to $g(x, y, z) = 0$, then for any smooth curve $\mathbf{r}(t) = \langle x(t), y(t), z(t) \rangle$ passing through (a, b, c) and lying on the surface $g(x, y, z) = 0$, both ∇f and ∇g must be perpendicular to $\mathbf{r}'(t)$ at (a, b, c). Therefore, both ∇f and ∇g must be perpendicular (normal vectors) to the tangent plane to $g(x, y, z) = 0$ at (a, b, c). Thus, ∇f and ∇g are parallel to each other. So the level surface $f(a, b, c) = f(x, y, z)$ of $u = f(x, y, z)$ and the surface $g(x, y, z) = 0$ must have the same tangent plane at (a, b, c).

Constrained extrema for more than one constraint

We can also work with optimization problems with more than one constraint. For example, for the problem

$$\max / \min u = f(x, y, z), \quad \text{subject to } g(x, y, z) = 0 \quad \text{and} \quad h(x, y, z) = 0$$

we can define a Lagrange function as

$$L(x,y,z,\lambda,u) = f(x,y,z) - \lambda g(x,y,z) - uh(x,y,z).$$

By solving the equations

$$\nabla L = \mathbf{0} \rightarrow L_x = L_y = L_z = L_\lambda = L_u = 0,$$

we obtain candidates for constrained extreme values.

There is also a geometric interpretation for this case. Points that satisfy $g(x,y,z) = 0$ and $h(x,y,z) = 0$ must be on the intersection curve $\mathbf{r}(t)$ of the two surfaces. If $f(\mathbf{r}(t))$ has a local max/min at a candidate $t = t_0$, following an argument similar to the one before, we have $\nabla f \cdot \mathbf{r}'(t_0) = 0$, $\nabla g \cdot \mathbf{r}'(t_0) = 0$, and $\nabla h \cdot \mathbf{r}'(t_0) = 0$ at the candidate point. Therefore, ∇f, ∇g, and ∇h must be coplanar. This means

$$\nabla f = \lambda \nabla g + u \nabla h, \text{ for some constants } \lambda \text{ and } u. \tag{2.17}$$

Example 2.9.7. Find the shortest distance from the origin to the line of intersection of the two planes $y + 2z - 12 = 0$ and $x + y - 6 = 0$.

Solution. Suppose the point is (x,y,z) and the distance is $\sqrt{x^2 + y^2 + z^2}$. To minimize the distance, we minimize $x^2 + y^2 + z^2$. So

$$L(x,y,z,\lambda,u) = x^2 + y^2 + z^2 - \lambda(y + 2z - 12) - u(x + y - 6).$$

Then by computing $\nabla L = \mathbf{0}$, we have

$$2x - u = 0,$$
$$2y - \lambda - u = 0,$$
$$2z - 2\lambda = 0,$$
$$y + 2z - 12 = 0,$$
$$x + y - 6 = 0.$$

Solving the equations yields the only candidate $(2,4,4)$. Therefore, the shortest distance must be $\sqrt{2^2 + 4^2 + 4^2} = 6$.

2.10 Review

Main concepts discussed in this chapter are listed below.
1. Definitions of functions of more than one variable, such as $z = f(x,y)$ and $u = f(x,y,z)$.

2. Limits and continuity of functions of more than one variable.
3. Partial derivatives of $z = f(x, y)$ with respect to x or y:

$$f_x(a,b) = \lim_{\Delta x \to 0} \frac{f(a + \Delta x, b) - f(a, b)}{\Delta x}, \quad f_y(a,b) = \lim_{\Delta y \to 0} \frac{f(a, b + \Delta y) - f(a, b)}{\Delta y}.$$

4. Differentiability:

$$\Delta z = f_x(a,b)\Delta x + f_y(a,b)\Delta y + o(\sqrt{(\Delta x)^2 + (\Delta y)^2}).$$

5. The total differential:

$$dz = \frac{\partial f}{\partial x} dx + \frac{\partial f}{\partial y} dy.$$

6. The linear approximation:

$$\Delta z \approx f_x(a,b)\Delta x + f_y(a,b)\Delta y.$$

7. Clairaut theorem: if $\frac{\partial^2 f}{\partial x \partial y}$ and $\frac{\partial^2 f}{\partial y \partial x}$ are both continuous at (a, b), then

$$\frac{\partial^2 f}{\partial x \partial y} = \frac{\partial^2 f}{\partial y \partial x} \quad \text{at } (a, b).$$

8. The chain rules:

$$\text{if } z = z(x(t), y(t)), \text{ then } \frac{dz}{dt} = \frac{\partial z}{\partial x}\frac{dx}{dt} + \frac{\partial z}{\partial y}\frac{dy}{dt},$$

$$\text{if } z = z(x(s,t), y(s,t)), \text{ then } \frac{\partial z}{\partial s} = \frac{\partial z}{\partial x}\frac{\partial x}{\partial s} + \frac{\partial z}{\partial y}\frac{\partial y}{\partial s}.$$

9. The Taylor expansion:

$$f(\mathbf{x}) = f(\mathbf{x}_0) + \nabla f(\mathbf{x}_0) \cdot \Delta \mathbf{x} + \Delta \mathbf{x}^T H(\mathbf{x}_0) \Delta \mathbf{x} + \cdots.$$

10. Implicit differentiation: $F(x, y, z) = 0$,

$$\frac{\partial z}{\partial x} = -\frac{F_x}{F_z} \quad \text{and} \quad \frac{\partial z}{\partial y} = -\frac{F_y}{F_z}.$$

11. Implicit differentiation: $\begin{cases} F(x,y,u,v) = 0 \\ G(x,y,u,v) = 0 \end{cases}$ denote $\frac{\partial(f,g)}{\partial(x,y)} = \begin{vmatrix} f_x & f_y \\ g_x & g_y \end{vmatrix}$, then

$$\frac{\partial u}{\partial x} = -\frac{\frac{\partial(F,G)}{\partial(x,v)}}{\frac{\partial(F,G)}{\partial(u,v)}} \quad \text{and} \quad \frac{\partial v}{\partial x} = -\frac{\frac{\partial(F,G)}{\partial(u,x)}}{\frac{\partial(F,G)}{\partial(u,v)}}.$$

Similar results hold for $\frac{\partial u}{\partial y}$ and $\frac{\partial v}{\partial y}$.

12.
   (a) For a curve defined by $\begin{cases} f(x,y,z)=0, \\ g(x,y,z)=0, \end{cases}$ its tangent line at $P(x_0, y_0, z_0)$ is given by
   $$\mathbf{r} = \langle x_0, y_0, z_0 \rangle + t(\nabla f \times \nabla g)_P.$$
   (b) For a surface defined by $F(x, y, z) = 0$, its tangent plane at $P(x_0, y_0, z_0)$ is given by
   $$(\nabla F \cdot \Delta \mathbf{x})_P = 0.$$

13. The directional derivative of $z = f(x, y)$ in the direction given by unit vector $\mathbf{u}$ at point (a, b) is
    $$D_\mathbf{u} f(a, b) = \nabla f \cdot \mathbf{u}.$$

14. Candidates for local maxima/minima of the function f are points where $\nabla f = \mathbf{0}$ or ∇f does not exists.

15. If $A = f_{xx}$, $B = f_{xy} = f_{yx}$, and $C = f_{yy}$, then at point P where $\nabla f(P) = 0$,
    (a) if $AC - B^2 > 0$ and $A > 0$, there is a local minimum,
    (b) if $AC - B^2 > 0$ and $A < 0$, there is a local maximum,
    (c) if $AC - B^2 < 0$, there is a saddle point.

16. Candidates for maxima/minima of $z = f(x, y)$ subject to $g(x, y) = 0$ satisfy
    $$\nabla f = \lambda \nabla g \quad \text{and} \quad g(x, y) = 0 \quad \text{for some number } \lambda.$$

    Similar results hold for functions of three variables or with more than one restriction.

2.11 Exercises

2.11.1 Functions of two variables

1. Find the domain for each of the following functions and sketch it:
   (1) $z = \sqrt{1 - x^2} + \sqrt{y^2 - 1}$,
   (2) $z = \sqrt{x - \sqrt{y}}$,
   (3) $z = \ln(1 - x - y)$,
   (4) $z = \ln(y - x) + \dfrac{\sqrt{x}}{\sqrt{1-x^2-y^2}}$,
   (5) $u = \sqrt{R^2 - x^2 - y^2 - z^2} + \dfrac{1}{\sqrt{x^2+y^2+z^2-r^2}}$, $R > r$.

2. Find $f(x, y)$ if $f(x + y, xy) = x^3 + y^3$.

3. Find each of the following limits if it exists:
   (1) $\lim_{(x,y) \to (2, \frac{1}{2})} (2 + xy)^{\frac{1}{y+xy^2}}$,
   (2) $\lim_{\substack{x \to \infty \\ y \to \infty}} (x^2 + y^2) \sin \dfrac{3}{x^2+y^2}$,
   (3) $\lim_{(x,y) \to (0,0)} \dfrac{1-\cos(x^2+y^2)}{(x^2+y^2)e^{x^2y^2}}$,
   (4) $\lim_{(x,y) \to (2,0)} \dfrac{\sin(2xy)}{y}$,
   (5) $\lim_{(x,y) \to (1,0)} \dfrac{\ln(x+e^y)}{\sqrt{x^2+y^2}}$,
   (6) $\lim_{(x,y) \to (0,0)} \dfrac{xy \cos y}{3x^2+y^2}$,
   (7) $\lim_{(x,y) \to (0,0)} \dfrac{xy^2}{x^2+y^2+xy}$.

4. Show that $\lim_{(x,y)\to(0,0)} f(x,y)$ does not exist, where

$$f(x,y) = \begin{cases} \frac{x^2 y}{x^4+y^4}, & x^2+y^2 \neq 0, \\ 0, & x^2+y^2 = 0. \end{cases}$$

5. Determine the set of points at which each of the following functions is continuous:
   (1) $f(x,y) = \frac{\sin xy}{e^x - y^2}$,
   (2) $z = \frac{y^2+2x}{y^2-2x}$,
   (3) $u = \frac{1}{xyz}$,
   (4) $z = \ln(1 - x^2 - y^2)$.

2.11.2 Partial derivatives and differentiability

1. Find the first partial derivatives for each of the following functions:
   (1) $u = xy + \frac{y}{x}$,
   (2) $u = \frac{x}{\sqrt{x^2+y^2}}$,
   (3) $u = x \sin(x - 2y)$,
   (4) $u = x \arctan(xe^y)$,
   (5) $u = x^y$,
   (6) $u = (\frac{x}{y})^z$,
   (7) $u = z^{xy}$,
   (8) $u = \tan \frac{x}{y}$.

2. Find the indicated partial derivatives for each of the following functions:
   (1) for $f(x,y) = \ln(x + \ln y)$, $f_x(1,e)$ and $f_y(1,e)$,
   (2) for $f(x,y) = \sin \frac{x}{y} \cos \frac{y}{x}$, $f_x(2,\pi)$.

3. Find the indicated second- or third-order partial derivatives for each of the following functions:
   (1) for $u = x \ln(x + y)$, find $\frac{\partial^2 u}{\partial x^2}$, $\frac{\partial^2 u}{\partial y^2}$, and $\frac{\partial^2 u}{\partial x \partial y}$,
   (2) for $u = x^3 \sin y + y^3 \sin x$, find $\frac{\partial^3 u}{\partial x^2 \partial y}$.

4. If $u = \ln \sqrt{(x-a)^2 + (y-b)^2}$, where a and b are constants, show that

$$\frac{\partial^2 u}{\partial x^2} + \frac{\partial^2 u}{\partial y^2} = 0.$$

5. If $f(x,y) = \begin{cases} xy\frac{x^2-y^2}{x^2+y^2} & \text{when } x^2+y^2 \neq 0, \\ 0 & \text{when } x^2+y^2 = 0, \end{cases}$ show that $f_{xy}(0,0) \neq f_{yx}(0,0)$.

6. If $f(x,y) = \begin{cases} \frac{xy}{x^2+y^2} & \text{when } x^2+y^2 \neq 0, \\ 0 & \text{when } x^2+y^2 = 0, \end{cases}$ show that both $f_x(0,0)$ and $f_y(0,0)$ exist but f is not differentiable at $(0,0)$.

7. Show that $f(x,y) = \begin{cases} (x^2+y^2)\sin(\frac{1}{x^2+y^2}) & \text{when } x^2+y^2 \neq 0, \\ 0 & \text{when } x^2+y^2 = 0 \end{cases}$ is differentiable at $(0,0)$, but neither f_x nor f_y is continuous at $(0,0)$.

8. Let f be the function

$$f(x,y) = \begin{cases} \frac{x^2 y^2}{(x^2+y^2)^{\frac{3}{2}}} & \text{if } (x,y) \neq (0,0), \\ 0 & \text{if } (x,y) = (0,0). \end{cases}$$

(a) Find the limit $\lim_{(x,y)\to(0,0)} f(x,y)$ or show that it does not exist.
(b) Is the function continuous at $(0,0)$?
(c) Is the function differentiable at $(0,0)$?

9. Find the total differential for each of the following functions:
   (1) $z = x^3 \ln(y^2)$, (2) $z = \arctan \frac{xy}{1-xy}$, (3) $u = \sqrt{x^2 + y^2 + z^2}$.

10. Explain why the following functions are differentiable at the given point. Then find the linearization $L(x,y)$ of the function at that point, and use it to approximate the given number.
    (1) $f(x,y) = x^y$ at $(1,1)$, $f(0.97, 1.06)$,
    (2) $f(x,y,z) = \sqrt{x^2 + y^2 + z^2}$ at $(3,2,6)$, $\sqrt{3.02^2 + 1.97^2 + 5.99^2}$.

2.11.3 Chain rules and implicit differentiation

1. Find the given partial derivative for each given explicitly or implicitly defined function.
   (1) $u = \ln(e^x + e^y)$, $y = x^3$. Find $\frac{du}{dx}$.
   (2) $z = \sin(x^2 y)x$. Find $\frac{\partial z}{\partial x}$.
   (3) $u = x^2 y - xy^2 + z$, where $x = t\cos(s)$, $y = t\sin(t)$, and $z = t + s$. Find $\frac{\partial u}{\partial t}$ and $\frac{\partial u}{\partial s}$.
   (4) $u = f(x^2 - y^2, e^{xy})$. Find $\frac{\partial u}{\partial x}$ and $\frac{\partial u}{\partial y}$.
   (5) $u = f(x, xy, xyz)$. Find $\frac{\partial u}{\partial x}$, $\frac{\partial u}{\partial y}$, and $\frac{\partial u}{\partial z}$.
   (6) $u = f(x, \frac{x}{y})$. Find $\frac{\partial^2 u}{\partial x^2}$, $\frac{\partial^2 u}{\partial y^2}$, and $\frac{\partial^2 u}{\partial x \partial y}$.
   (7) $e^z = xyz$. Find $\frac{\partial z}{\partial x}$.
   (8) $yz = \ln(x + z)$. Find $\frac{\partial z}{\partial x}$.
   (9) $u = f(x^2, y, xy)$. Find $\frac{\partial u}{\partial x}$ and $\frac{\partial u}{\partial y}$.

2. If $z = f(x,y)$, where $x = r\cos\theta$ and $y = r\sin\theta$, show that
$$\frac{\partial^2 z}{\partial x^2} + \frac{\partial^2 z}{\partial y^2} = \frac{\partial^2 z}{\partial r^2} + \frac{1}{r^2}\frac{\partial^2 z}{\partial \theta^2} + \frac{1}{r}\frac{\partial z}{\partial r}.$$

3. If $u = \phi(x^2 + y^2)$, show that $y\frac{\partial u}{\partial x} - x\frac{\partial u}{\partial y} = 0$.

4. If $x = x(y,z)$, $y = y(x,z)$, and $z = z(x,y)$ are three functions implicitly defined by the equation $F(x,y,z) = 0$, show that the partial derivatives of these functions satisfy
$$\frac{\partial x}{\partial y}\frac{\partial y}{\partial z}\frac{\partial z}{\partial x} = -1.$$

5. If $\begin{cases} z = x^2+y^2, \\ x+y+z=1, \end{cases}$ find $\frac{dz}{dx}$ and $\frac{dy}{dx}$.

6. Assume $y = f(x,t)$ is differentiable, and an equation $F(x,y,t) = 0$ defines $t = t(x,y)$ implicitly as a function of x and y. If f and F have continuous partials, show that
$$\frac{dy}{dx} = \frac{f_x F_t - f_t F_x}{F_t + f_t F_y}.$$

7. **(Derivative under integrals)** The famous Leibniz theorem says that if $f(x,t)$ is a function such that $f(x,t)$ and its partial derivative $f_x(x,t)$ are both continuous in

some region of the xt-plane, with $a(x) \le t \le b(x)$ for two differentiable functions $a(x)$ and $b(x)$, then

$$\frac{d}{dx}\int_{a(x)}^{b(x)} f(x,t)dt = f(x,b(x))\frac{d}{dx}b(x) - f(x,a(x))\frac{d}{dx}a(x) + \int_{a(x)}^{b(x)} \frac{\partial f(x,t)}{\partial x}dt.$$

When $f(x,t) = f(t)$, a one-variable function in t, then this is proved by the fundamental theorem of calculus, part I. When $a(x) = a$ and $b(x) = b$ are two constants, then the theorem becomes

$$\frac{d}{dx}\int_a^b f(x,t)dt = \int_a^b f_x(x,t)dt.$$

This means that the derivative operator can pass through the integral sign. This is essentially the interchange of two limits (can you see why?).
(a) Prove the Leibniz theorem for the special case where $a(x) = a$ and $b(x) = b$.
(b) By considering the function $\phi(t) = \int_0^1 \frac{\ln(1+tx)}{1+x^2}dx$ or otherwise, evaluate $\int_0^1 \frac{\ln(1+x)}{1+x^2}dx$.
(c) Find the integral $\int_0^{\frac{\pi}{2}} \ln(\cos^2 x + a^2 \sin^2 x)dx$, $a > 0$.

2.11.4 Tangent lines/planes, directional derivatives

1. Find equations of (a) the tangent line and (b) the normal plane to each of the following curves at the specified point:
   (1) $x = t^2$, $y = 1-t$, $z = t^3$, $(1,0,1)$, (2) $\mathbf{r}(t) = \langle \frac{\sin^2 t}{2}, \frac{t+\sin t \cdot \cos t}{2}, \sin t \rangle$, $t = \frac{\pi}{4}$,
   (3) $\begin{cases} x^2+y^2+z^2=6, \\ x+y+z=0, \end{cases}$ $(1,-2,1)$, (4) $\begin{cases} (x-1)^2+y^2=1, \\ x^2+y^2+z^2=4, \end{cases}$ $(1,1,\sqrt{2})$.
2. Find equations of (a) the tangent plane and (b) the normal line to the following surfaces at the given point:
   (1) $z = 2x^2 + 4y^2$, $(2,1,12)$, (2) $x^2 = 2z$, $(2,0,2)$,
   (3) $\cos \pi x + x^2 y + e^{xz} - yz + 4 = 0$, $(0,1,6)$.
3. Find the directional derivative of the following functions at the given point in the direction of the vector $\mathbf{v}$:
   (1) $f(x,y) = x^2 - y^2$ at the point $(1,1)$, given $\mathbf{v} = \langle 1, \sqrt{3} \rangle$,
   (2) $f(x,y,z) = \frac{x}{y+z}$ at the point $(4,1,1)$, given $\mathbf{v} = \langle 1,2,-1 \rangle$.
4. Find all points at which the direction of fastest change of the function $f(x,y) = x^2 + y^2 - 2x - 4y$ is $\vec{i} + \vec{j}$.
5. Find the maximum rate of change of f at the given points and the direction in which it occurs:
   (1) $f(x,y) = x^2 y + e^{xy} \sin y$, $(1,0)$, (2) $f(x,y,z) = xy^2 z$, $(1,-1,2)$,
   (3) $f(x,y,z) = \ln(x^2 + y^2 - 1) + y + 6z$, $(1,1,0)$.

6. A vector tangent to the curve $\begin{cases} f(x,y,z)=0, \\ g(x,y,z)=0 \end{cases}$ at $P(x,y,z)$ is $\nabla f \times \nabla g$. Use this to find an equation for the tangent line and normal plane to the following curves at the indicated point:

(1) $\begin{cases} x^2 + y^2 + z^2 - 3y = 0, \\ 2x + y - z - 2 = 0, \end{cases}$ $(1,1,1)$, (2) $\begin{cases} e^x - z + xy = 0, \\ x^2 - y^2 + 2z^3 = 1, \end{cases}$ $(0,1,1)$.

2.11.5 Maximum/minimum problems

1. Find and classify all the critical points for each of the following functions:
   (1) $f(x,y) = x^2 - xy + y^2 + 9x - 6y + 20$, (2) $f(x,y) = 4(x-y) - x^2 - y^2$,
   (3) $f(x,y) = 3x - x^3 - 2y^2 + y^4$.
2. If the function $z = z(x,y)$ is implicitly defined by the equation
$$x^2 - 6xy + 10y^2 - 2yz - z^2 + 18 = 0,$$
   find all critical points of z and local maximum and local minimum values of z.
3. Find the absolute maximum and minimum values of $f(x,y)$ on the set D:
   (1) $f(x,y) = 2 - 4x - 5y$, where D is the closed triangular region with vertices $(0,0)$, $(2,0)$, and $(0,3)$,
   (2) $f(x,y) = xy^2$, where $D = \{(x,y) | x \geq 0, y \geq 0, x^2 + y^2 \leq 3\}$,
   (3) $f(x,y) = 24xy - 8x^3 - 6y^2$ on the rectangular region D: $0 \leq x \leq 1$ and $0 \leq y \leq 2$.
4. Find three positive numbers x, y, and z whose sum is 100 and whose product is a maximum.
5. Find all points on the ellipse $x^2 + 4y^2 = 4$ that are closest to the line $2x + 3y - 6 = 0$.
6. Find the dimensions of the closed rectangular box with least total surface area if the volume is given by V m^3.
7. Find the maximum value of $f(x,y,z) = xyz$ on the line of intersection of the two planes $x + y + z = 40$ and $x + y = z$.
8. **(Least square method)** Suppose the two variables x and y are related linearly by the equation $y = kx + b$ for some constants k and b. However, in practice, observed pairs of data (x_1, y_1), (x_2, y_2), (x_3, y_3) ... (x_n, y_n) that should satisfy this equation usually do not lie exactly on a straight line. So scientists want to find the constant k and b such that the line $y = kx + b$ best "fits" these points.
Let $d_i = y_i - (kx_i + b)$ be the vertical deviation of the point (x_i, y_i) from the line $y = kx + b$. The *least square method* determines k and b by minimizing $\sum_{i=1}^{n} d_i^2$ (the sum of the squares of these deviations). Show that the "*best fit line of y on x*" is given by
$$y = \frac{n\sum_{i=1}^{n} x_i y_i - \sum_{i=1}^{n} x_i \sum_{i=1}^{n} y_i}{n\sum_{i=1}^{n} x_i^2 - (\sum_{i=1}^{n} x_i)^2} x + \frac{\sum_{i=1}^{n} x_i^2 \sum_{i=1}^{n} y_i - \sum_{i=1}^{n} x_i y_i \sum_{i=1}^{n} x_i}{n\sum_{i=1}^{n} x_i^2 - (\sum_{i=1}^{n} x_i)^2}.$$

9. For functions of more than two variables, there are similar tests for local extreme values using the Hessian matrix $H(\mathbf{x})$. For example, for $u = f(x,y,z)$, its Hessian matrix is

$$H(\mathbf{x}) = \begin{vmatrix} \dfrac{\partial^2 f}{\partial x^2} & \dfrac{\partial^2 f}{\partial x \partial y} & \dfrac{\partial^2 f}{\partial x \partial z} \\ \dfrac{\partial^2 f}{\partial y \partial x} & \dfrac{\partial^2 f}{\partial y^2} & \dfrac{\partial^2 f}{\partial y \partial z} \\ \dfrac{\partial^2 f}{\partial z \partial x} & \dfrac{\partial^2 f}{\partial z \partial y} & \dfrac{\partial^2 f}{\partial z^2} \end{vmatrix}.$$

If all its second derivatives are continuous, then $H(\mathbf{x})$ is a symmetric matrix. If $H(\mathbf{x})$ is positive definite at $\mathbf{x}_0$, then f attains an isolated local minimum at $\mathbf{x}_0$. If the Hessian is negative definite at $\mathbf{x}_0$, then f attains an isolated local maximum at $\mathbf{x}_0$. If the Hessian has both positive and negative eigenvalues at $\mathbf{x}_0$, then $\mathbf{x}_0$ is a saddle point for f. Otherwise the test is inconclusive.

Use a suitable Hessian matrix to classify the critical points of the function $f(x,y,z) = x^2 + y^2 - x + z^2$.

3 Multiple integrals

In this chapter, we extend the idea of the definite integral of a function of one variable to an analogous concept of a function of two or three variables, called a multiple integral. Multiple integrals are used in a number of applications, including computing volumes, surface areas, and masses of two- or three-dimensional objects.

3.1 Definition and properties

We start with some interesting questions. It rained very hard in a city area last night. How much rain was received by the city? How much water is there in a certain water reservoir? If a rectangular plate is not uniform, and at each point (x, y) the density is given by a density function, say, $f(x, y)$, then what is the total mass of this plate? Well, as in one-variable calculus, to answer these questions, we need to identify the "elements" which we add up to form a Riemann sum and then take a limit to have an integral. We first consider an ideal reservoir whose surface is a rectangle and whose base is a smooth surface. If we turn it upside down, then finding the amount of water in the reservoir is the same as finding the volume of a solid in space.

Volumes of solids

Now suppose that f is a two-variable function with a rectangular domain D satisfying $f(x, y) \geq 0$ for all $(x, y) \in D$, a rectangular region in the xy-plane. Hence, the graph of f is a surface S above the xy-plane with equation $z = f(x, y)$, and the projection of S onto the xy-plane is the domain set D. The solid (three-dimensional) region Ω that lies above D in the xy-plane and under the graph of f is

$$\Omega = \{(x, y, z) \in \mathbb{R}^3 \mid 0 \leq z \leq f(x, y),\ (x, y) \in D\}.$$

Our initial goal is to find a method for computing the volume of Ω, and this provides a motivation for multiple integrals.

The first step is to subdivide the region D into n small closed subregions $\Delta\sigma_1, \Delta\sigma_2, \ldots, \Delta\sigma_n$, as illustrated in Figure 3.1, where the subregions are created by drawing lines parallel to the x- and y-axes. The value of n is left unspecified because eventually we will use a limiting process in which $n \to \infty$.

Arbitrarily choose a point (ξ_i, η_i) in each $\Delta\sigma_i$ for $i = 1, 2, \ldots, n$. We approximate the part of Ω that lies above each $\Delta\sigma_i$ by a thin rectangular box (or "column") with base $\Delta\sigma_i$ and height $f(\xi_i, \eta_i)$, as shown in Figure 3.1. The volume ΔV_i of this column is approximately the height of the column, $f(\xi_i, \eta_i)$, multiplied by the base area $\Delta\sigma_i$ of the base region, $\Delta\sigma_i$ (we are using $\Delta\sigma_i$ both as the name of the subregion and as the area of this subregion), so we have

$$\Delta V_i \approx f(\xi_i, \eta_i)\Delta\sigma_i.$$

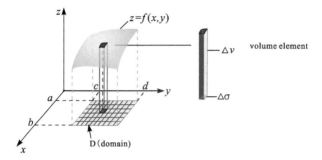

Figure 3.1: Double integral, volume of a solid in space.

If we form the sum of these approximations over all subregions (a *Riemann sum*), we get an approximation to the total volume, V, of the three-dimensional region Ω, i. e.,

$$V \approx \sum_{i=1}^{n} f(\xi_i, \eta_i) \Delta \sigma_i.$$

If the limit of this sum exists when $n \to \infty$ and the maximum $\Delta\sigma_i$ approaches zero, then we define this limit to be the volume of Ω, i. e.,

$$V = \lim_{\max |\Delta\sigma_i| \to 0,\, n \to \infty} \sum_{i=1}^{n} f(\xi_i, \eta_i) \Delta \sigma_i.$$

Note. This limit must be the same value no matter how the subregions $\Delta\sigma_i$ are made and no matter where the point (ξ_i, η_i) is chosen in each subregion $\Delta\sigma_i$. If the limit is taken only as $n \to \infty$ (the number of subregions approaches ∞), it would then still be possible for some subregions $\Delta\sigma_i$ to stay quite large. To avoid this problem, the limit must also be taken so that the largest subregion approaches zero both in area and in physical dimensions. We write $|\Delta\sigma_i|$ to indicate the greatest dimension of the subregion $\Delta\sigma_i$ and then write $\max |\Delta\sigma_i| \to 0$ to indicate that this greatest dimension must approach zero for all the subregions. This limit seems very complicated and hard to compute but, surprisingly, it can be shown to exist whenever the function f is continuous and the domain D is of a suitable form. We will show later that it can be computed using two one-variable integrals (iterated integrals) that you have studied previously.

Mass of a lamina

We investigate a second quite different problem of computing the mass of a lamina (thin plate), and surprisingly it can be found by exactly the same process as we did for finding the volume of a solid. Suppose a rectangular lamina (thin plate) is represented by region D of the xy-plane. Suppose further that the density (mass per unit area) of the lamina at a point corresponding to (x, y) in D is given by $\mu(x, y)$, where $\mu(x, y)$ is a continuous function on D, as shown in Figure 3.2. We now derive a way to compute the

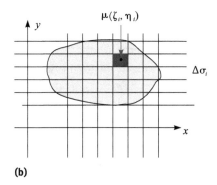

(a) (b)

Figure 3.2: Double integral, mass of a lamina.

total mass M of the lamina using methods similar to the volume computation above. We divide D into n small closed subregions $\Delta\sigma_1, \Delta\sigma_2, \ldots, \Delta\sigma_n$ by drawing "nets," and arbitrarily choose a point (ξ_i, η_i) in each $\Delta\sigma_i$. If $\Delta\sigma_i$ is very small, so that the density does not change much over $\Delta\sigma_i$, then the mass of the part of the lamina represented by $\Delta\sigma_i$ is approximately the density at (ξ_i, η_i) multiplied by the area, i.e., $\mu(\xi_i, \eta_i)\Delta\sigma_i$ (we are using $\Delta\sigma_i$ both as the name of the subregion and as the area of this subregion). If we add all such mass approximations, we get an approximation to the total mass, i.e.,

$$M \approx \sum_{i=1}^{n} \mu(\xi_i, \eta_i)\Delta\sigma_i.$$

If the limit of this sum exists as $n \to \infty$ and $\max|\Delta\sigma_i| \to 0$, and it is independent of choices of subdivisions of D and sample point (ξ_i, η_i) in each $\Delta\sigma_i$, then we define this limit to be the *mass of the lamina*, written

$$M = \lim_{\max|\Delta\sigma_i| \to 0,\ n \to \infty} \sum_{i=1}^{n} \mu(\xi_i, \eta_i)\Delta\sigma_i.$$

Note this is exactly the same type of limit as that used before to compute the volume of a three-dimensional region. We will see later that many other applied problems can be reduced to computing a limit exactly of this type.

We now link this limiting process to a more general definition of the double integral of a function $f(x, y)$ over a general region D in the xy-plane.

Definition 3.1.1 (Double integrals). The **double integral** of f over the region D is defined to be the following limit:

$$\iint_D f(x, y)\, d\sigma = \lim_{\max|\Delta\sigma_i| \to 0,\ n \to \infty} \sum_i f(\xi_i, \eta_i)\Delta\sigma_i,$$

where $\Delta\sigma_1, \Delta\sigma_2, \ldots, \Delta\sigma_n$ are n closed subregions which are a partition of the region D ($\Delta\sigma_i$ also denotes the area of $\Delta\sigma_i$) and (ξ_i, η_i) is an arbitrarily chosen point in $\Delta\sigma_i$, assuming that this limit exists. The limit must have the same value for any choice of subdivision and the choice of sample points (ξ_i, η_i). The double integral is also often written as $\iint_D f(x,y)dA$.

Note. The motivation for the double integral assumed that $f(x,y) \geq 0$ for $(x,y) \in D$, but the definition here does not assume that f is nonnegative. When f takes both positive and negative values on D, then the double integral $\iint_D f(x,y)d\sigma$, if it exists, is equal to the volume of the part of the solid that lies above the xy-plane minus the volume of the part of the solid that lies below the xy-plane.

Properties of double integrals

We list the following properties of double integrals, which can be proved using the definition of a double integral, in a manner similar to the proofs for functions of a single variable (we assume that all integrals exist):

(1) $\iint_D 1 d\sigma = A(D)$, the area of D;
(2) $\iint_D f(x,y) + g(x,y) d\sigma = \iint_D f(x,y)d\sigma + \iint_D g(x,y)d\sigma$;
(3) $\iint_D kf(x,y)d\sigma = k\iint_D f(x,y)d\sigma$, where k is a constant ((2) and (3) are the linearity property);
(4) if $f(x,y) \geq g(x,y)$ for all $(x,y) \in D$, then $\iint_D f(x,y)d\sigma \geq \iint_D g(x,y)d\sigma$;
(5) if $D = D_1 \cup D_2$, where D_1 and D_2 do not overlap except perhaps on their boundaries, then

$$\iint_D f(x,y)d\sigma = \iint_{D_1} f(x,y)d\sigma + \iint_{D_2} f(x,y)d\sigma;$$

(6) if $m \leq f(x,y) \leq M$ for all $(x,y) \in D$ and $A(D)$ denotes the area of D, then

$$m \cdot A(D) \leq \iint_D f(x,y)d\sigma \leq M \cdot A(D).$$

(7) (mean value theorem) if $f(x,y)$ is continuous on the closed region D and $A(D)$ is the area of D, then there exists $(\xi, \eta) \in D$ such that

$$\iint_D f(x,y)d\sigma = f(\xi, \eta)A(D).$$

Example 3.1.1. If $D = \{(x,y) | x^2 + y^2 \leq 4\}$, evaluate the integral

$$\iint_D \sqrt{x^2 + y^2} d\sigma.$$

Solution. Evaluating this double integral directly from the definition as a limit is hard. However, because $\sqrt{x^2 + y^2} \geq 0$, we can find the integral by interpreting it as a volume of a solid Ω. The surface $z = \sqrt{x^2 + y^2}$ is a cone with vertex downwards at the origin and axis along the z-axis and with height 2. Therefore, the given double integral represents the volume of the solid below this cone and above the disk D. The volume of Ω is the volume of a cylinder with base D and height 2 minus the volume of a cone with the same base and height. Thus,

$$\iint_D \sqrt{x^2 + y^2} d\sigma = \pi 2^2 \times 2 - \frac{1}{3}\pi 2^2 \times 2 = \frac{16\pi}{3}.$$

Example 3.1.2. Use the properties of double integrals to estimate the integral $\iint_D e^{\sin x \cos y} d\sigma$, where D is the disk in the xy-plane with radius 2 and center the origin.

Solution. Because $-1 \leq \sin x \leq 1$ and $-1 \leq \cos y \leq 1$, we have $-1 \leq \sin x \cos y \leq 1$ and, therefore,

$$e^{-1} \leq e^{\sin x \cos y} \leq e^1 = e.$$

Let $m = e^{-1} = 1/e$ and $M = e$. By using property (6) and noting that the area of D is given by $A(D) = \pi(2)^2 = 4\pi$, we obtain

$$\frac{4\pi}{e} \leq \iint_D e^{\sin x \cos y} d\sigma \leq 4\pi e.$$

Symmetry in double integrals

Sometimes, we can take advantage of the symmetry properties in the integrand or the region of integration, as shown in the following example.

Example 3.1.3. Find the integral $\iint_D (x^3(1+y^2) + 5) d\sigma$ for $D = \{-1 \leq x \leq 1, -1 \leq y \leq 1\}$.

Solution. The region of integration D is a square with center the origin. By the linearity property,

$$\iint_D (x^3(1+y^2) + 5) d\sigma = \iint_D x^3(1+y^2) d\sigma + \iint_D 5 d\sigma.$$

In $\iint_D x^3(1+y^2) d\sigma$, the integrand is an odd function with respect to x. This means that half of the integrand is positive and the other half is negative over D. That is, half of the graph of $f(x, y) = x^3(1+y^2)$ is above the xy-plane, and the other half is below it, and the two halves are symmetric. This double integral is equal to 0. Since $\iint_D 5 d\sigma = 5 \iint_D 1 d\sigma$, the double integral is equal to 5 times the area of region of integration. Thus,

$$\iint_D (x^3(1+y^2) + 5) d\sigma = 0 + 5 \times (2)^2 = 20.$$

Figure 3.3 shows the graph of the function $f(x, y) = x^3(1+y^2)$ on the region D.

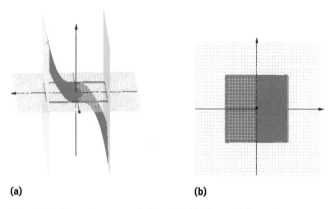

(a) (b)

Figure 3.3: Symmetry example for a double integral, Example 3.1.3.

3.2 Double integrals in rectangular coordinates

It is very difficult to evaluate a double integral using its definition as a limit. However, in this section we show how to express a double integral as iterated integrals that can be evaluated by calculating two single-variable integrals.

We first consider the volume problem where $f(x,y) \geq 0$ for all (x,y) in the rectangular domain

$$D = \{(x,y) \mid a \leq x \leq b, \text{and } c \leq y \leq d\}.$$

If we divide D horizontally and vertically into nm subregions, we note that the area element $\Delta\sigma$ is $\Delta x \Delta y$. If we let x be fixed, say, $x = x_i^* \in [x_{i-1}, x_i] \subset [a,b]$, then

$$\lim_{\max |\Delta y_j| \to 0} \sum_{j=1}^{m} f(x_i^*, y_j^*) \Delta y_j = \int_c^d f(x_i^*, y) dy$$

gives the area of the region that lies inside the solid, as shown in Figure 3.4.

If we multiply this area by a tiny thickness Δx_i, this would give us a volume element

$$\Delta V_i \approx \int_c^d f(x_i^*, y) dy \Delta x_i.$$

Taking a limit of a Riemann sum will give the volume

$$V = \lim_{\max |\Delta x_i| \to 0} \sum_{i=1}^{n} \int_c^d f(x_i^*, y) dy \Delta x_i.$$

3.2 Double integrals in rectangular coordinates — 151

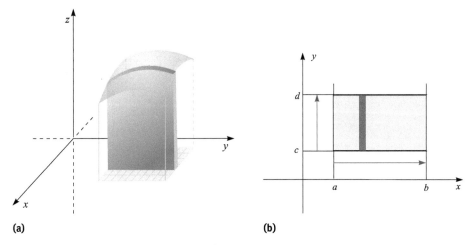

Figure 3.4: Iterated integrals over a rectangular region.

This is one-variable integration with an integrand the function $A(x) = \int_c^d f(x,y)dy$. So if $f(x,y)$ is continuous, we have

$$V = \int_a^b A(x)dx = \int_a^b \left(\int_c^d f(x,y)dy \right) dx.$$

It is sometimes convenient to write this as

$$V = \int_a^b dx \int_c^d f(x,y)dy.$$

In a similar manner, we can also integrate with respect to x first. This gives

$$V = \int_c^d A(y)dy = \int_c^d \left(\int_a^b f(x,y)dx \right) dy.$$

Note.
1. We can also interpret the two integrals as in the mass of lamina model. The inner integral $\int_c^d f(x,y)dy$ gives the mass of a vertical rod. Then we sum/integrate those masses of rods to get the total mass of the lamina. This is illustrated in Figure 3.4(b).
2. As we defined the differentials dx, dy, and dz, we can define $d\sigma = dxdy$ or $d\sigma = dydz$ in rectangular coordinates. We will see this will help a lot in algebraic manipulations.

Recall that when $f(x,y) \geq 0$, the volume V is exactly represented by the double integral $\iint_D f(x,y)d\sigma$. The method discussed above also works even if $f(x,y)$ takes positive and negative values over a region D. We summarize these arguments in the following theorem.

Theorem 3.2.1 (Fubini's theorem: rectangular region). *If $f(x, y)$ is continuous on a rectangular region*

$$D = \{(x, y) | a \le x \le b, c \le y \le d\},$$

then the double integral

$$\iint_D f(x,y)\,d\sigma = \iint_D f(x,y)\,dxdy = \int_c^d dy \int_a^b f(x,y)\,dx,$$

and also

$$\iint_D f(x,y)\,d\sigma = \iint_D f(x,y)\,dydx = \int_a^b dx \int_c^d f(x,y)\,dy.$$

Example 3.2.1. Find the double integrals

(a) $\displaystyle\iint_D x + y^2\,d\sigma,$ where $D = \{(x,y) | 1 \le x \le 5, -1 \le y \le 4\}$,

(b) $\displaystyle\iint_D x^2 y \sin y^2\,d\sigma,$ where $D = \{(x,y) | 0 \le x \le 6, 0 \le y \le \sqrt{\pi}\}$.

Solution.
(a) By Fubini's theorem

$$\iint_D x + y^2\,d\sigma = \int_1^5 dx \int_{-1}^4 (x + y^2)\,dy$$

$$= \int_1^5 \left(xy + \frac{y^3}{3}\right)_{-1}^4 dx$$

$$= \int_1^5 \left(5x + \frac{4^3}{3} - \left(-\frac{1}{3}\right)\right) dx = \frac{440}{3}.$$

(b) Note that the integrand has the form $g(x)f(y)$. Thus,

$$\iint_D x^2 y \sin y\,d\sigma = \int_0^6 dx \int_0^{\sqrt{\pi}} (x^2 y \sin y^2)\,dy$$

$$= \int_0^6 x^2\,dx \int_0^{\sqrt{\pi}} y \sin y^2\,dy$$

$$= \frac{x^3}{3}\Big|_0^6 \cdot \frac{-1}{2}\cos y^2\Big|_0^{\sqrt{\pi}}$$

$$= 72 \cdot \frac{-1}{2}(\cos \pi - \cos 0) = 72.$$

Note that we have evaluated the two definite integrals separately. That is,

$$\int_a^b dx \int_c^d f(x)g(y)dydx = \int_a^b f(x)dx \int_c^d g(y)dy.$$

Now we consider more general regions of integration. If D is the region between the graphs of two continuous functions of x (this type of region is called a *type I region*, as shown in Figure 3.5(a)), that is, there is some interval $a \leq x \leq b$ and functions of one variable $y = \phi_1(x)$ and $y = \phi_2(x)$ such that

$$D = \{(x,y) \mid a \leq x \leq b,\ \phi_1(x) \leq y \leq \phi_2(x)\},$$

then, for each x in $[a, b]$, the range for y now depends on x with lower bound $\phi_1(x)$ and upper bound $\phi_2(x)$. Therefore, if we keep x constant, we can also interpret $A(x) = \int_{\phi_1(x)}^{\phi_2(x)} f(x,y)dy$ as the area of a cross-section of the solid. Thus,

$$V = \int_a^b A(x)dx.$$

Therefore, we can still write the volume V and double integral of $f(x,y)$ as two one-variable integrals (*iterated integrals*), i. e.,

$$V = \iint_D f(x,y)d\sigma = \int_a^b A(x)dx = \int_a^b \left[\int_{\phi_1(x)}^{\phi_2(x)} f(x,y)dy\right]dx.$$

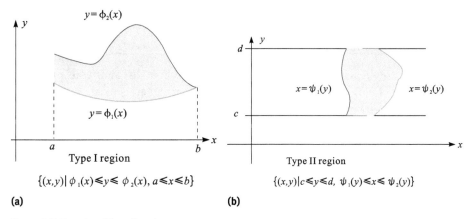

(a) Type I region
$\{(x,y) \mid \phi_1(x) \leq y \leq \phi_2(x),\ a \leq x \leq b\}$

(b) Type II region
$\{(x,y) \mid c \leq y \leq d,\ \psi_1(y) \leq x \leq \psi_2(y)\}$

Figure 3.5: Type I and type II regions.

Similarly, D could be a *type II region* (see Figure 3.5(b)) bounded by two continuous functions in the xy-plane, $x = \psi_1(y)$ and $x = \psi_2(y)$ for some interval of y values $c \le y \le d$,

$$D = \{(x,y) \mid \psi_1(y) \le x \le \psi_2(y),\ c \le y \le d\}.$$

A similar derivation to that used above for type I regions shows that

$$\iint_D f(x,y)d\sigma = \int_c^d \left[\int_{\psi_1(y)}^{\psi_2(y)} f(x,y)dx\right]dy = \int_c^d dy \int_{\psi_1(y)}^{\psi_2(y)} f(x,y)dx.$$

The above results on iterated integrals are true even if $f(x,y)$ is not nonnegative. The formal statement is given in Fubini's theorem. Rigorous proofs of Fubini's theorem can be found in more theoretical calculus textbooks.

Theorem 3.2.2 (Fubini's theorem: general region). *If $z = f(x,y)$ is continuous on its domain D, a **type I region***

$$D = \{(x,y) \mid a \le x \le b,\ \phi_1(x) \le y \le \phi_2(x)\},$$

then the double integral can be evaluated by the iterated integrals

$$\iint_D f(x,y)d\sigma = \int_a^b dx \int_{\phi_1(x)}^{\phi_2(x)} f(x,y)dy.$$

*If $z = f(x,y)$ is continuous on its domain D, a **type II region***

$$D = \{(x,y) \mid \psi_1(y) \le x \le \psi_2(y),\ c \le y \le d\},$$

then the double integral can be evaluated by the iterated integrals

$$\iint_D f(x,y)d\sigma = \int_c^d dy \int_{\psi_1(y)}^{\psi_2(y)} f(x,y)dx.$$

Note. The theorem is also true if f is bounded on D and is discontinuous only on a finite number of smooth curves, provided the iterated integrals exist. However, the proof of this fact is beyond the scope of this book.

Example 3.2.2. Evaluate $\iint_D (x+2y)d\sigma$, where D is the region bounded by straight lines $y = 2$ and $y = x$ and the hyperbola $xy = 1$.

Solution. The hyperbola intersects the two lines at two points $(\frac{1}{2}, 2)$ and $(1,1)$ and the two lines intersect at the point $(2,2)$, as shown in Figure 3.6.

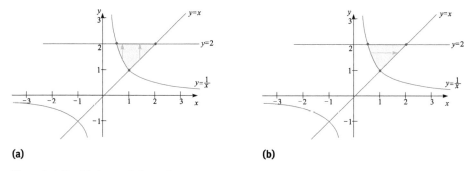

Figure 3.6: Double integral, Example 3.2.2.

We note that the region D is both a type I region and a type II region, but the description of D as a type I region is more complicated since the lower boundary consists of two parts. Therefore, it is better to express D as a type II region bounded on the left by $x = \frac{1}{y}$ and on the right by $x = y$, so

$$D = \left\{(x,y) \mid \frac{1}{y} \le x \le y,\ 1 \le y \le 2\right\}.$$

We compute the double integral (recall that the inner iterated integral is evaluated as though y is a constant – like a partial derivative with respect to x)

$$\iint_D (x+2y)\,d\sigma = \int_1^2 \int_{\frac{1}{y}}^{y} (x+2y)\,dx\,dy = \int_1^2 \left[\frac{x^2}{2} + 2yx\right]_{x=\frac{1}{y}}^{x=y} dy$$

$$= \int_1^2 \left(\frac{y^2}{2} + 2y^2 - \frac{1}{2y^2} - 2\right) dy = \frac{43}{12}.$$

If we had expressed D as a type I region, then we would evaluate it in two parts, D_1 for $\frac{1}{2} \le x \le 1$, bounded above by $y = 2$ and below by $y = \frac{1}{x}$, and D_2 for $1 \le x \le 2$, bounded above by $y = 2$ and below by $y = x$. Hence,

$$\iint_D x+2y\,d\sigma = \iint_{D_1} (x+2y)\,d\sigma + \iint_{D_2} (x+2y)\,d\sigma$$

$$= \int_{\frac{1}{2}}^{1} \int_{\frac{1}{x}}^{2} (x+2y)\,dy\,dx + \int_1^2 \int_x^2 (x+2y)\,dy\,dx$$

$$= \int_{\frac{1}{2}}^{1} \left(2x - \frac{1}{x^2} + 3\right) dx + \int_1^2 (-2x^2 + 2x + 4)\,dx$$

$$= \frac{5}{4} + \frac{7}{3} = \frac{43}{12}.$$

This clearly involves more work than the first method.

Example 3.2.3. Evaluate $\iint_D xy\,d\sigma$, where D is the region bounded by the line $y = x$ and the parabola $y^2 = 2x + 8$.

Solution. The region is shown in Figure 3.7, and it lies between $x = \frac{y^2}{2} - 4$ and $x = y$. So,

$$D = \left\{(x,y)\mid -2 \le y \le 4,\ \frac{y^2}{2} - 4 \le x \le y\right\}.$$

Again, D is both a type I and a type II region, but we prefer to express D as a type II region because it is less complicated. The double integral becomes

$$\iint_D xy\,d\sigma = \int_{-2}^{4}\left[\int_{\frac{y^2}{2}-4}^{y} xy\,dx\right] dy = \int_{-2}^{4}\left[\frac{x^2 y}{2}\right]_{\frac{y^2}{2}-4}^{y} dy = \frac{1}{2}\int_{-2}^{4}\left(y^3 - \left(\frac{y^2}{2} - 4\right)^2 y\right) dy$$

$$= \frac{1}{2}\int_{-2}^{4}\left(-\frac{1}{4}y^5 + 5y^3 - 16y\right)dy = 18.$$

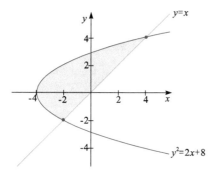

Figure 3.7: Double integral, Example 3.2.3.

Change the order of integration

Example 3.2.4. Evaluate the iterated integral $\int_0^1 \int_x^{\sqrt{x}} \frac{\sin y}{y}\,dy\,dx$.

Solution. Note that evaluating $\int \frac{\sin y}{y}\,dy$ is impossible as it is not an elementary function. So, we cannot compute the integral as it stands. However, if we change the order of integration, then it may be possible to evaluate the iterated integral. We first express

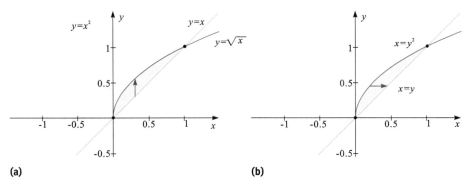

(a) **(b)**

Figure 3.8: Change the order of integration.

the given iterated integral as a double integral, i.e.,

$$\int_0^1 \int_x^{\sqrt{x}} \frac{\sin y}{y} dy dx = \iint_D \frac{\sin y}{y} d\sigma,$$

where D is the type I region shown in Figure 3.8(a) between the curves $y = x$ and $y = \sqrt{x}$,

$$D = \{(x,y)|0 \le x \le 1,\ x \le y \le \sqrt{x}\}.$$

From Figure 3.8(b), we can see that there is an alternative description of D as a type II region between the curves $x = y^2$ and $x = y$, i.e.,

$$D = \{(x,y)|0 \le y \le 1,\ y^2 \le x \le y\}.$$

This enables us to express the double integral as an iterated integral in a different order,

$$\iint_D \frac{\sin y}{y} dx dy = \int_0^1 dy \int_{y^2}^{y} \frac{\sin y}{y} dx = \int_0^1 \left[x \frac{\sin y}{y} \right]_{x=y^2}^{x=y} dy$$

$$= \int_0^1 (\sin y - y \sin y) dy = 1 - \sin 1.$$

3.3 Double integral in polar coordinates

If we are going to integrate the double integral

$$\iint_D x^2 + y^2 d\sigma \quad \text{for } D = \{(x,y)|x^2 + y^2 \le a^2\},$$

we will have to evaluate the integral

$$\int_{-a}^{a} dx \int_{-\sqrt{a^2-x^2}}^{\sqrt{a^2-x^2}} x^2 + y^2 \, dy.$$

This is certainly not fun. However, if we describe D in polar coordinates, then we will have $D'_{r\theta} = \{(r,\theta)| 0 \leq r \leq a, 0 \leq \theta \leq 2\pi\}$. This is a rectangular region in the $r\theta$-plane on which the integration might be easier. In rectangular coordinates, we see the area element $d\sigma = dxdy$. What would this be in polar coordinates? Recall that we found the area element by dividing the region into many subregions using lines that are parallel to the x- or y-axis. So an area element is represented by $\Delta x \Delta y$ in rectangular coordinates. Similarly, we can draw circles all centered at the origin with different radii and half-lines with initial point the origin but different angles from the positive x-axis. This produces Δr and $\Delta \theta$. Looking at an area element as shown in Figure 3.9, we approximate it as the difference of areas of two sectors. So, the area approximation is

$$\Delta \sigma \approx \frac{1}{2}\left(r^* + \frac{\Delta r}{2}\right)^2 \Delta \theta - \frac{1}{2}\left(r^* - \frac{\Delta r}{2}\right)^2 \Delta \theta = r^* \Delta r \Delta \theta.$$

This means that we can consider the limit of the Riemann sum

$$\lim_{\max|\Delta \sigma_i| \to 0} \sum_{i=1}^{m} f(r_i^*, \theta_i^*) \Delta \sigma_i = \lim_{\max|\Delta r, \Delta \theta| \to 0} \sum_{i=1}^{m} f(r_i^*, \theta_i^*) r_i^* \Delta r_i \Delta \theta_i.$$

Then, if the limit exists independent of the way of subdividing the region and the choice of sample points, we can define

$$\iint_D f(x,y) d\sigma = \iint_{D'_{r\theta}} f(r\cos\theta, r\sin\theta) r \, dr \, d\theta.$$

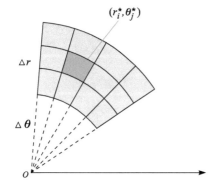

Figure 3.9: Double integral in polar coordinates.

In particular, by taking $f(x, y) = 1$, we can see that the area of the region D bounded by $\theta = \alpha$, $\theta = \beta$, and $r = r(\theta)$ is given by

$$A(D) = \iint_D 1 d\sigma = \int_\alpha^\beta \int_0^{r(\theta)} r dr d\theta = \int_\alpha^\beta \left[\frac{r^2}{2}\right]_0^{r(\theta)} d\theta = \frac{1}{2} \int_\alpha^\beta [r(\theta)]^2 d\theta. \quad (3.1)$$

Example 3.3.1. Find the area of the disk $D = \{(x, y)|x^2 + y^2 \leq R^2\}$.

Solution. In polar coordinates, the circle has equation $r = R$ and D is

$$\{(r, \theta)|0 \leq r \leq R,\ 0 \leq \theta \leq 2\pi\}.$$

The area is given by

$$A(D) = \frac{1}{2} \int_\alpha^\beta r^2(\theta) d\theta = \frac{1}{2} \int_0^{2\pi} R^2 d\theta = 2\pi \times \frac{1}{2} R^2 = \pi R^2.$$

Example 3.3.2. Find the double integral $\iint_D x^2 + y^2 d\sigma$ for $D = \{(x, y)|x^2 + y^2 \leq a^2\}$.

Solution. Using polar coordinates, D becomes $D'_{r\theta} = \{(r, \theta)|0 \leq r \leq a, 0 \leq \theta \leq 2\pi\}$. So,

$$\iint_D x^2 + y^2 d\sigma = \iint_{D'_{r\theta}} ((r\cos\theta)^2 + (r\sin\theta)^2) r dr d\theta$$

$$= \iint_D r^3 dr d\theta = \int_0^{2\pi} d\theta \int_0^a r^3 dr = \frac{\pi a^4}{2}.$$

Example 3.3.3. Find the volume of the solid bounded by the plane $z = 0$ and the paraboloid $z = 1 - x^2 - y^2$, using polar coordinates.

Solution. Let $z = 0$ in the equation of the paraboloid. We get $x^2 + y^2 = 1$. Thus, the solid lies under the paraboloid and above the circular disk $D: x^2 + y^2 \leq 1$ in the xy-plane. In polar coordinates, D is described by $0 \leq r \leq 1$ and $0 \leq \theta \leq 2\pi$. Since $1 - x^2 - y^2 = 1 - r^2$, the volume is, therefore, given by

$$V = \iint_D (1 - x^2 - y^2) d\sigma = \int_0^{2\pi} \int_0^1 (1 - r^2) r dr d\theta$$

$$= \int_0^{2\pi} d\theta \int_0^1 (r - r^3) dr = \int_0^{2\pi} \left[\frac{r^2}{2} - \frac{r^4}{4}\right]_0^1 d\theta$$

$$= \frac{1}{4} \int_0^{2\pi} d\theta = \frac{\pi}{2}.$$

Note. If we had used rectangular coordinates instead of polar coordinates, then we would have to evaluate

$$V = \iint_D (1 - x^2 - y^2) d\sigma = \int_{-1}^{1} \int_{-\sqrt{1-x^2}}^{\sqrt{1-x^2}} (1 - x^2 - y^2) dy dx.$$

This integral can be evaluated using trigonometric substitution and using trigonometric identities, but it is quite complicated.

! **Example 3.3.4.** Find the volume of the solid that lies under the sphere $x^2 + y^2 + z^2 \leq 4$, above the xy-plane, and inside the cylinder $x^2 + y^2 = 2x$.

Solution. The solid lies above the disk D whose boundary circle, $x^2 + y^2 = 2x$ (center $(1, 0)$, radius 1), is determined by the cylinder. In polar coordinates, we have $x^2 + y^2 = r^2$ and $x = r\cos\theta$. Then, the boundary circle becomes $r^2 = 2r\cos\theta \implies r = 2\cos\theta$ for $-\frac{\pi}{2} \leq \theta \leq \frac{\pi}{2}$. Thus, the disk D is given by

$$D = \{(r, \theta) | -\pi/2 \leq \theta \leq \pi/2, \ 0 \leq r \leq 2\cos\theta\}.$$

Hence, the volume is

$$V = \iint_D \sqrt{(4 - x^2 - y^2)} d\sigma = \int_{-\pi/2}^{\pi/2} \int_0^{2\cos\theta} \sqrt{4 - r^2} \, r \, dr \, d\theta$$

$$= -\frac{1}{2} \int_{-\pi/2}^{\pi/2} \frac{2}{3} (4 - r^2)^{\frac{3}{2}} \Big|_0^{2\cos\theta} d\theta = -\frac{1}{3} \int_{-\pi/2}^{\pi/2} ((4 - 4\cos^2\theta)^{\frac{3}{2}} - 4^{\frac{3}{2}}) d\theta$$

$$= -\frac{1}{3} \int_{-\pi/2}^{\pi/2} (8(\sin^2\theta)^{\frac{3}{2}} - 8) d\theta = -\frac{2}{3} \int_0^{\pi/2} (8(\sin^2\theta)^{\frac{3}{2}} - 8) d\theta$$

$$= -\frac{16}{3} \int_0^{\pi/2} (\sin^3\theta - 1) d\theta = -\frac{16}{3} \int_0^{\pi/2} \left(\frac{1}{4} (3\sin\theta - \sin 3\theta) - 1 \right) d\theta$$

$$= -\frac{16}{3} \left(\frac{1}{12} \cos 3\theta - \frac{3}{4} \cos\theta - \theta \right) \Big|_0^{\frac{\pi}{2}} = \frac{16}{3} \left(\frac{\pi}{2} - \frac{2}{3} \right).$$

Note. In the above, note that $-\frac{1}{3} \int_{-\pi/2}^{\pi/2} (8(\sin^2\theta)^{\frac{3}{2}} - 8) d\theta \neq -\frac{8}{3} \int_{-\pi/2}^{\pi/2} (\sin^3\theta - 1) d\theta$. That is because $(\sin^2\theta)^{\frac{3}{2}}$ must use the positive square root and so $(\sin^2\theta)^{\frac{3}{2}} = |\sin\theta|^3$. Also, $\sin 3\theta = 3\sin\theta - 4\sin^3\theta$ is an identity.

3.4 Change of variables formula for double integrals

In one-dimensional calculus we often use a change of variables (a substitution) to modify an integral of a continuous function f. If we let $x = g(t)$, where g is a one-to-one continuous function, then the one-dimensional change of variables theorem is

$$\int_a^b f(x)dx = \int_c^d f(g(t))\frac{dx}{dt}dt = \int_c^d f(g(t))g'(t)dt,$$

where $c = g^{-1}(a)$ and $d = g^{-1}(b)$.

Change of variables in double integrals is analogous to this but more complicated. Intuitively, when one has a one-to-one transformation $T : (u,v) \to (x,y)$ with $x = x(u,v)$ and $y = y(u,v)$, the change of variables in a double integral $\iint_D f(x,y)d\sigma$ becomes an integration over the region D'_{uv} with the integrand $f(x(u,v),y(u,v))$. What happens to the area element $d\sigma = dxdy$ in terms of $dudv$? We demonstrate the idea as follows. Let $T(u,v) = \langle x(u,v), y(u,v) \rangle = x(u,v)\mathbf{i} + y(u,v)\mathbf{j}$.

As shown in Figure 3.10, the area of the image is approximated by the parallelogram whose area is the modulus of a cross product, i. e.,

$$|(T(u+\Delta u,v) - T(u,v)) \times (T(u,v+\Delta v) - T(u,v))| \approx |T_u(u,v)\Delta u \times T_v(u,v)\Delta v|$$
$$= |T_u(u,v) \times T_v(u,v)|\Delta u \Delta v.$$

Note that $T_u(u,v) = \langle \frac{\partial x}{\partial u}, \frac{\partial y}{\partial u} \rangle$ and $T_v(u,v) = \langle \frac{\partial x}{\partial v}, \frac{\partial y}{\partial v} \rangle$. Then

$$|T_u(u,v) \times T_v(u,v)|\Delta u \Delta v = \left\| \begin{matrix} \frac{\partial x}{\partial u} & \frac{\partial y}{\partial u} \\ \frac{\partial x}{\partial v} & \frac{\partial y}{\partial v} \end{matrix} \right\| \Delta u \Delta v.$$

Therefore, we have

$$\Delta x \Delta y \approx \left| \frac{\partial(x,y)}{\partial(u,v)} \right| \Delta u \Delta v.$$

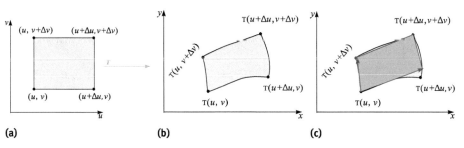

Figure 3.10: Change of variables in a double integral, area element, transformation.

We call the determinant $\begin{vmatrix} \frac{\partial x}{\partial u} & \frac{\partial y}{\partial u} \\ \frac{\partial x}{\partial v} & \frac{\partial y}{\partial v} \end{vmatrix}$ a *Jacobian determinant* and denote it by J or $\frac{\partial(x,y)}{\partial(u,v)}$. The Jacobian determinant is a magnification (or reduction) factor. That is, it relates the area $dxdy$ of a small region in the xy-plane to the area of the corresponding region $dudv$ in the uv-plane.

Note that

$$\begin{vmatrix} \frac{\partial x}{\partial u} & \frac{\partial y}{\partial u} \\ \frac{\partial x}{\partial v} & \frac{\partial y}{\partial v} \end{vmatrix} = \begin{vmatrix} \frac{\partial x}{\partial u} & \frac{\partial x}{\partial v} \\ \frac{\partial y}{\partial u} & \frac{\partial y}{\partial v} \end{vmatrix},$$

so we use either of them and denote them by $\frac{\partial(x,y)}{\partial(u,v)}$.

Theorem 3.4.1 (Change of variables in a double integral). *Let $f(x,y)$ be a continuous function on a bounded and closed region $D \in \mathbb{R}^2$, and let the functions $x = x(u,v)$ and $y = y(u,v)$ be a continuously differentiable ($x(u,v)$ and $y(u,v)$ both have continuous first-order partial derivatives) transformation (mapping) from a region D' onto the region D. If the transformation is one-to-one, and $\frac{\partial(x,y)}{\partial(u,v)} \neq 0$, for all $(u,v) \in D'$, then*

$$\iint_D f(x,y)\,dxdy = \iint_{D'} f(x(u,v), y(u,v)) \left| \frac{\partial(x,y)}{\partial(u,v)} \right| dudv. \tag{3.2}$$

Example 3.4.1. Evaluate the integral

$$\iint_D \frac{e^{(x-y)}}{(x+y)}\,dxdy,$$

where D is the rectangular region bounded by $x+y=\frac{1}{2}, x+y=1, x-y=-\frac{1}{2}, x-y=\frac{1}{2}$.

Solution. We simplify the problem by using the transformation $T^{-1}: u = x - y$ and $v = x + y$ from the xy-plane to the uv-plane. In order to use the change of variables theorem, we solve these for x and y to find the transformation T from the uv-plane to the xy-plane, i.e.,

$$x = \frac{1}{2}(u+v) \quad \text{and} \quad y = \frac{1}{2}(v-u).$$

The Jacobian of T is

$$\frac{\partial(x,y)}{\partial(u,v)} = \begin{vmatrix} \frac{\partial x}{\partial u} & \frac{\partial x}{\partial v} \\ \frac{\partial y}{\partial u} & \frac{\partial y}{\partial v} \end{vmatrix} = \begin{vmatrix} \frac{1}{2} & \frac{1}{2} \\ -\frac{1}{2} & \frac{1}{2} \end{vmatrix} = \frac{1}{2}.$$

To find the region D' corresponding to D, we find the transformation of each of the boundary lines of D (see Figure 3.11),

$$x+y=\frac{1}{2} \rightarrow v=\frac{1}{2}, \quad x+y=1 \rightarrow v=1,$$
$$x-y=-\frac{1}{2} \rightarrow u=-\frac{1}{2}, \quad x-y=\frac{1}{2} \rightarrow u=\frac{1}{2}.$$

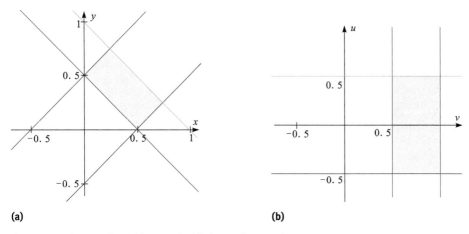

Figure 3.11: Change of variables in a double integral, Example 3.4.1.

Thus, the region D', shown in Figure 3.11(b), is defined by

$$D' = \left\{(u,v) \mid -\frac{1}{2} \le u \le \frac{1}{2}, \frac{1}{2} \le v \le 1\right\}.$$

Hence, the change of variables formula gives

$$\iint_D \frac{e^{(x-y)}}{x+y} dxdy = \iint_{D'} \frac{e^u}{v} \begin{vmatrix} \frac{1}{2} & \frac{1}{2} \\ -\frac{1}{2} & \frac{1}{2} \end{vmatrix} dudv - \frac{1}{2} \iint_{D'} \frac{e^u}{v} dudv$$

$$= \frac{1}{2} \int_{\frac{1}{2}}^{1} \frac{1}{v} dv \int_{-\frac{1}{2}}^{\frac{1}{2}} e^u du = \frac{\ln 2}{2}\left(\sqrt{e} - \frac{1}{\sqrt{e}}\right).$$

The Jacobian determinant has a nice property for a one-to-one transformation $T: (u,v) \to (x,y)$, $T^{-1}: (x,y) \to (u,v)$, i.e.,

$$\frac{\partial(x,y)}{\partial(u,v)} \frac{\partial(u,v)}{\partial(x,y)} = 1.$$

This is similar to the one-variable case where if $y = f(x)$ is a differentiable one-to-one function with inverse $x = \psi(y)$, then $\frac{dy}{dx}\frac{dx}{dy} = 1$.

Example 3.4.2. Find the area of the region bounded by the four curves

$$y = ax^2, \quad y = bx^2, \quad xy = c, \quad \text{and} \quad xy = d,$$

where a, b, c, and d are four constants satisfying $0 < a < b$ and $0 < c < d$.

Solution. The area is given by $\iint_D 1 d\sigma$, which is hard to compute. Instead, we use the transformation

$$u = \frac{y}{x^2} \text{ and } v = xy, \quad \text{thus } a < u < b \text{ and } c < v < d.$$

To compute $\frac{\partial(x,y)}{\partial(u,v)}$, we write

$$\frac{\partial(x,y)}{\partial(u,v)} = \frac{1}{\left|\frac{\partial(u,v)}{\partial(x,y)}\right|} = \frac{1}{\begin{vmatrix} \frac{\partial u}{\partial x} & \frac{\partial u}{\partial y} \\ \frac{\partial v}{\partial x} & \frac{\partial v}{\partial y} \end{vmatrix}} = \frac{1}{\begin{vmatrix} -2\frac{y}{x^3} & \frac{1}{x^2} \\ y & x \end{vmatrix}}$$

$$= \frac{1}{-2\frac{y}{x^2} - \frac{y}{x^2}} = \frac{-1}{3\frac{y}{x^2}} = \frac{-1}{3u}.$$

Therefore,

$$\iint_D 1 d\sigma = \iint_{D'} 1 \cdot \left|\frac{\partial(x,y)}{\partial(u,v)}\right| dudv = \iint_{D'} 1 \cdot \left|-\frac{1}{3u}\right| dudv$$

$$= \iint_{D'} \frac{1}{3u} dudv = \frac{1}{3}\int_a^b \frac{1}{u} du \int_c^d dv = \frac{d-c}{3} \ln(b-a).$$

Figure 3.12 shows the graphs.

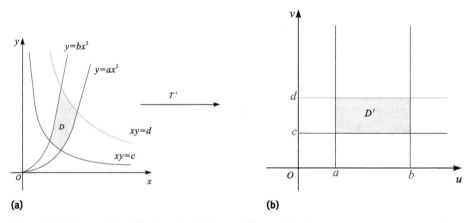

Figure 3.12: Change of variables in a double integral, Example 3.4.2.

The transformation from x- and y-coordinates to polar coordinates (with the same origin and with the initial line of the polar coordinates along the x-axis) is given by

$$\begin{cases} x = r\cos\theta, \\ y = r\sin\theta. \end{cases}$$

Hence,

$$\left|\frac{\partial(x,y)}{\partial(r,\theta)}\right| = \left\|\begin{array}{cc} x_r & x_\theta \\ y_r & y_\theta \end{array}\right\| = \left\|\begin{array}{cc} \cos\theta & -r\sin\theta \\ \sin\theta & r\cos\theta \end{array}\right\| = |r| = r.$$

Applying the change of variables theorem to this transformation gives

$$\iint_D f(x,y)dxdy = \iint_{D'} f(r\cos\theta, r\sin\theta)rdrd\theta.$$

This agrees with what we have done before for double integrals in polar coordinates.

3.5 Triple integrals

3.5.1 Triple integrals in rectangular coordinates

If we have a solid box which is not a uniform one, that is, at each point (x,y,z) inside the box the density is a continuous function $f(x,y,z)$, then how do you find its total mass?

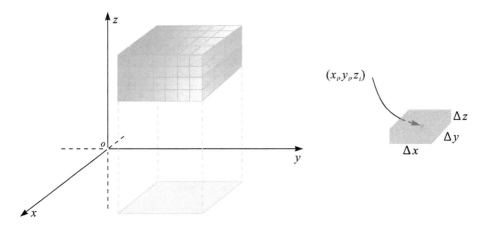

Figure 3.13: Triple integrals: mass of a box.

We now follow a process very similar to that used for double integrals. For a bounded function of three variables, $f(x,y,z)$ defined on a closed bounded region (a solid) $\Omega \subset \mathbb{R}^3$, we construct a *Riemann sum* as

$$R_n = \sum_{i=1}^n f(x_i, y_i, z_i)\Delta V_i,$$

where ΔV_i is the volume of a subregion of Ω, (x_i, y_i, z_i) is an arbitrarily chosen point in this subregion, and the nonoverlapping subregions for $i = 1, 2, 3, \ldots, n$ cover all of Ω. If the limit of R_n exists as $n \to \infty$ and as the size of the largest subregion approaches zero, then we say that f is *integrable* on Ω and denote this as

$$\iiint_\Omega f(x,y,z)dV = \lim_{\max |\Delta V_i| \to 0, n \to \infty} \sum_{i=1}^n f(x_i, y_i, z_i)\Delta V_i,$$

and $\iiint_\Omega f(x,y,z)dV$ is called the *triple integral of f over the region* Ω. This means that the limit must exist and have the same value no matter how the subregions are created and how (x_i, y_i, z_i) are chosen. It can be shown that the limit always exist when $f(x,y,z)$ is continuous on Ω provided Ω satisfies some fairly mild condition.

When $f(x, y, z) = 1$ for all $(x, y, z) \in \Omega$, then the triple integral gives the volume $V(\Omega)$ of the region Ω, so

$$V(\Omega) = \iiint_\Omega 1 \cdot dV. \tag{3.3}$$

If the density function of a solid Ω is $\rho(x, y, z)$ mass/unit volume at any point (x, y, z) of Ω, then the mass M of the solid Ω is

$$M = \iiint_\Omega \rho(x, y, z)dV. \tag{3.4}$$

Triple integrals also have properties such as linearity and additivity, as double integrals do.

Example 3.5.1. Evaluate the triple integral $\iiint_\Omega (x \cos(yz^2) + 5)dV$, where

$$\Omega = \{(x, y, z) | -1 \le x \le 1, -2 \le y \le 2, -3 \le z \le 3\}.$$

Solution. Note that $x \cos(yz^2)$ is an odd function with respect to x, while Ω is symmetric about x. Therefore,

$$\iiint_\Omega (x \cos(yz^2) + 5)dV = \iiint_\Omega x \cos(yz^2)dV + 5\iiint_\Omega 1dV \quad \text{(linearity property)}$$

$$= 0 + 5\iiint_\Omega 1dV \quad \text{(symmetry property)}$$

$$= 5 \times \text{volume of } \Omega = 5 \times 2 \times 4 \times 6 = 240.$$

In general, how do you evaluate a triple integral? First of all, we consider the mass model where Ω is a rectangular box given by

$$\Omega = \{(x, y, z) | a_1 \le x \le a_2, b_1 \le y \le b_2, c_1 \le z \le c_2\}.$$

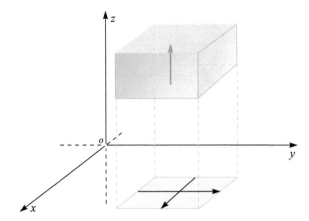

Figure 3.14: Triple integrals: iterated integrals, rectangular base.

The projection of the region Ω onto the xy-plane is a rectangular region on the xy-plane $D = \{(x,y)|a_1 \leq x \leq a_2 \text{ and } b_1 \leq y \leq b_2\}$ as shown in Figure 3.14. If we let x and y be fixed, then the integral $\int_{c_1}^{c_2} f(x,y,z)dz$ gives the mass of a rod. If we add the masses of all such rods, then the total mass is given by

$$\iint_D \left(\int_{c_1}^{c_2} f(x,y,z)dz \right) d\sigma.$$

This means

$$\iiint_\Omega f(x,y,z)dV = \iint_D \left(\int_{c_1}^{c_2} f(x,y,z)dz \right) d\sigma.$$

We can then evaluate a triple integral by finding a definite integral followed by evaluating a double integral. Recalling what we did for double integrals, this eventually leads to the iterated integrals

$$\iiint_\Omega f(x,y,z)dV = \int_{a_1}^{a_2} dx \int_{b_1}^{b_2} dy \int_{c_1}^{c_2} f(x,y,z)dz.$$

We can write the above equation as

$$\iiint_\Omega f(x,y,z)dV = \int_{a_1}^{a_2} \int_{b_1}^{b_2} \int_{c_1}^{c_2} f(x,y,z)dzdydx, \quad \text{or}$$

$$\iiint_\Omega f(x,y,z)dV = \iiint_\Omega f(x,y,z)dzdydx,$$

where $dV = dzdydx$ is interpreted as the volume element in rectangular coordinates. Of course, we can write $dV = dxdydz$, $dV = dydxdz$, etc., but these mean using different orders of the iterated integrals.

Example 3.5.2. A box with dimensions $2 \times 4 \times 8$ with height 8 has a density ρ at each of point in the box. The density ρ is proportional to the product of the distance from the point to the bottom of the box and the distance from the point to the top surface of the box. The proportionality constant is 3. Find the total mass of the box.

Solution. Set up a coordinate system with the left-most corner as the origin, as shown in Figure 3.15. Then the density $\rho(x,y,z) = 3z(8-z)$. The total mass is given by

$$\iiint_\Omega \rho(x,y,z)dV = \iiint_\Omega 3z(8-z)dV = \int_0^2 \int_0^4 \int_0^8 3z(8-z)dzdydx$$

$$= 3\int_0^2 dx \int_0^4 dy \int_0^8 z(8-z)dz$$

$$= 3 \times 2 \times 4 \times \int_0^8 (8z - z^2)dz = 2048 \text{ units}.$$

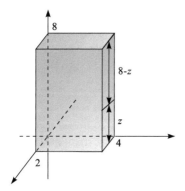

Figure 3.15: Triple integrals: Example 3.5.2.

Now we consider the case of a so-called type I region where the region Ω is enclosed by two smooth surfaces $z = z_1(x,y)$ and $z = z_2(x,y)$, as shown in Figure 3.16. If the projection of the region onto the xy-plane is D, then in a way similar to the double integral, we have

$$\iiint_\Omega f(x,y,z)dV = \iint_D \left(\int_{z_1(x,y)}^{z_2(x,y)} f(x,y,z)dz \right) d\sigma.$$

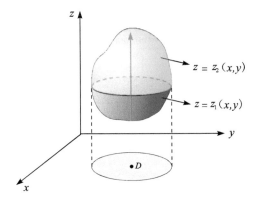

Figure 3.16: Triple integrals: general region.

Similarly, we can evaluate a triple integral on a type II or type III region. We summarize these results in the following definition and Fubini's theorem.

Definition 3.5.1. A region Ω is of **type I** if, for each $(x,y) \in D$ (a region of the xy-plane), all points in Ω for all z-values lie between two surfaces $z_1 = z_1(x,y)$ and $z_1 = z_2(x,y)$, that is,

$$z_1(x,y) \leq z \leq z_2(x,y) \quad \text{and} \quad (x,y) \in D.$$

Then $(x,y,z) \in \Omega$ and all points of Ω are of this type.

In other words, D is the projection of the region Ω onto the xy-plane and Ω is the set of points

$$\Omega = \{(x,y,z) \in \mathbb{R}^3 | (x,y) \in D, z_1(x,y) \leq z \leq z_2(x,y)\}.$$

Similarly, a **type II** or **type III** region is defined to lie between two functions with domain D in the xz- or yz-plane, respectively.

Theorem 3.5.1 (Iterated integral theorem). *Assume that $f(x,y,z)$ is continuous on a type I region Ω of the form*

$$\Omega = \{(x,y,z) \in \mathbb{R}^3 | (x,y) \in D_{xy}, z_1(x,y) \leq z \leq z_2(x,y)\}.$$

Then f is integrable on Ω and can be evaluated as a single-variable integration with respect to z (x and y are held constant) followed by a double integral over the region D in the xy-plane as

$$\iiint_\Omega f(x,y,z)\,dV = \iint_{D_{xy}} \left(\int_{z_1(x,y)}^{z_2(x,y)} f(x,y,z)\,dz \right) dxdy.$$

Similarly, for a type II region $\Omega = \{(x,y,z) \in \mathbb{R}^3 | (x,z) \in D_{xz}, y_1(x,z) \leq y \leq y_2(x,z)\}$ we have

$$\iiint_\Omega f(x,y,z)\,dV = \iint_{D_{xz}} \left(\int_{y_1(x,z)}^{y_2(x,z)} f(x,y,z)\,dy \right) dxdz,$$

where D_{xz} is the projection of Ω onto the xz-plane. For a type III region $\Omega = \{(x,y,z) \in \mathbb{R}^3 | (y,z) \in D_{yz}, x_1(y,z) \le x \le x_2(y,z)\}$ we have, when D_{yz} is the projection of Ω onto the yz-plane,

$$\iiint_\Omega f(x,y,z)dV = \iint_{D_{yz}} \left(\int_{x_1(y,z)}^{x_2(y,z)} f(x,y,z)dx \right) dydz.$$

Example 3.5.3. Evaluate $\iiint_\Omega x dV$, where Ω is the solid tetrahedron bounded by the four planes $x = 0$, $y = 0$, and $z = 0$, and $x + y + z = 1$, using the method described above.

Solution. It is always helpful if we draw two diagrams: one is the solid region Ω, and the other is its projection D onto an appropriate coordinate planes when evaluating a triple integral. The diagrams for this example are shown in Figure 3.17.

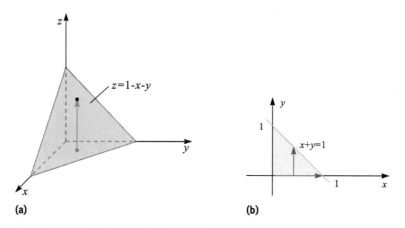

Figure 3.17: Triple integrals: Example 3.5.3.

The lower boundary of the tetrahedron is the plane $z = 0$, and the upper boundary in the z-direction is the plane $z = 1 - x - y$. Note that the planes $x + y + z = 1$ and $z = 0$ intersect in the line $x + y = 1$ in the xy-plane. Thus, the projection of Ω onto the xy-plane is the triangular region (see Figure 3.17(b)) bounded by the x-axis, the y-axis, and $x + y = 1$.

We can treat Ω as a type I region

$$\Omega = \{(x,y,z) | (x,y) \in D, \ 0 \le z \le 1 - x - y\}, \quad \text{where}$$
$$D = \{(x,y) | 0 \le y \le 1 - x, \ 0 \le x \le 1\}.$$

Then this enables us to evaluate the integral as follows:

$$\iiint_\Omega x dV = \iint_D \left(\int_0^{1-x-y} x dz \right) dA = \iint_D [xz]_{z=0}^{z=1-x-y} dydx$$

$$= \iint_D x(1-x-y)dydx = \int_0^1 \int_0^{1-x} (x - x^2 - xy)dydx$$

$$= \int_0^1 \left[xy - x^2y - \frac{xy^2}{2} \right]_{y=0}^{y=1-x} dx$$

$$= \int_0^1 (x-x^2)(1-x) - \frac{x}{2}(1-x)^2 dx = \frac{1}{24}.$$

Sometimes, we may also evaluate the triple integral by first evaluating a double integral and then evaluating a one-variable integral. For example, if D_z is the projection (onto the xy-plane) of the cross-section (D_z) of Ω by a horizontal plane with distance z units from the xy-plane, and all cross-sections of Ω satisfy $c_1 \leq z \leq c_2$, then Ω is defined by

$$\Omega = \{(x,y,z) | c_1 \leq z \leq c_2 \text{ and } (x,y) \in D_z\}.$$

In this case, we have

$$\iiint_\Omega f(x,y,z)dV = \int_{c_1}^{c_2} dz \iint_{D_z} f(x,y,z)dxdy.$$

In less precise language you can think of the double integral $\iint_{D_z} f(x,y,z)dxdz$ as the mass of the lamina (D_z) when the density per unit volume is $f(x,y,z)$, and then the integration with respect to z computes the mass of the solid Ω by adding the masses of all of the laminae. This is illustrated in Figure 3.18.

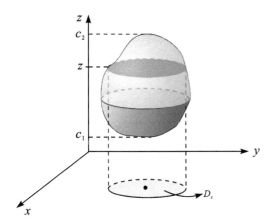

Figure 3.18: Triple integrals: a double integral first.

Similarly, if D_x is the projection (onto the yz-plane) of the cross-section (D_x) parallel to the yz-plane at distance x with $a_1 \le x \le a_2$, and D_y is the projection (onto the xz-plane) of the cross-section (D_y) parallel to the xz-plane at distance y with $b_1 \le y \le b_2$, then we also have

$$\iiint_\Omega f(x,y,z)dV = \int_{a_1}^{a_2} dx \iint_{D_x} f(x,y,z)d\sigma,$$

$$\iiint_\Omega f(x,y,z)dV = \int_{b_1}^{b_2} dy \iint_{D_y} f(x,y,z)d\sigma.$$

Example 3.5.4. Find $\iiint_\Omega x\,dV$, where Ω is the same region in Example 3.5.3, by evaluating a double integral first.

Solution. Evaluate the integral $\iiint_\Omega x\,dV$ by first evaluating a double integral over

$$D_z = \{(x,y) | 0 \le y \le 1-x-z,\ 0 \le x \le 1-z\}.$$

This is a triangular cross-section of Ω at height z from the xy-plane. Its projection D_{xy} onto the xy-plane is bounded by $x+y = 1-z$, $x = 0$, and $y = 0$, as shown in Figure 3.19. We have

$$\iiint_\Omega z\,dV = \int_0^1 dz \iint_{D_z} x\,d\sigma = \int_0^1 dz \int_0^{1-z} \int_0^{1-x-z} x\,dy\,dx$$

$$= \int_0^1 \int_0^{1-z} x(1-x-z)\,dx\,dz = \int_0^1 \left(-\frac{1}{6}(z-1)^3\right)dz = \frac{1}{24}.$$

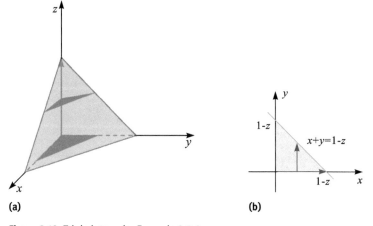

Figure 3.19: Triple integrals: Example 3.5.4.

Example 3.5.5. Attempt to evaluate the following integral by first considering E to be a type I region, then a type II region, and then a third method using a double integral as the inner integral:

$$\iiint_E \sqrt{x^2 + z^2}\, dV,$$

where E is the region bounded by the paraboloid $y = x^2 + z^2$ and the plane $y = 1$.

Solution. We first sketch the region of E, as shown in Figure 3.20.

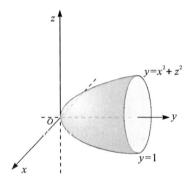

Figure 3.20: Triple integrals: Example 3.5.5.

Method 1: We regard the solid E as a type I region, as shown in Figure 3.21(a). The projection of D_z onto the xy-plane is the parabolic region $x^2 \le y \le 1$. In order to find the upper and lower bounding z-value functions, solve $y = x^2 + z^2$ to obtain $z = \pm\sqrt{y - x^2}$. Hence, the lower boundary surface of E is $z = -\sqrt{y - x^2}$ and the upper surface is $z = \sqrt{y - x^2}$. Therefore, the description of E as a type I region is

$$E = \{(x, y, z) \mid -1 \le x \le 1,\ x^2 \le y \le 1,\ -\sqrt{y - x^2} \le z \le \sqrt{y - x^2}\},$$

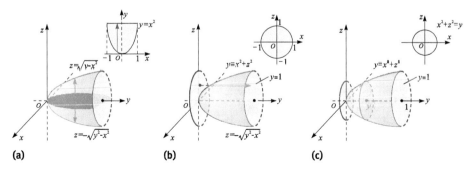

(a)   (b)   (c)

Figure 3.21: Triple integrals: Example 3.5.5.

so we obtain

$$\iiint_E \sqrt{x^2+z^2}\,dV = \int_{-1}^{1}\int_{x^2}^{1}\left[\int_{-\sqrt{y-x^2}}^{\sqrt{y-x^2}} \sqrt{x^2+z^2}\,dz\right] dy\,dx.$$

Although this expression is correct, it is extremely difficult to evaluate. So we try a second method.

Method 2: Let us regard E as a type II region. As such, the projection of E onto the xz-plane is the disk D_{xz}, $x^2 + z^2 \le 1$. Then the left boundary of E is the paraboloid $y = x^2 + z^2$ and the right boundary is the plane $y = 1$ (see Figure 3.21(b)), so we have

$$\iiint_E \sqrt{x^2+z^2}\,dV = \iint_{D_{xz}}\left[\int_{x^2+z^2}^{1} \sqrt{x^2+z^2}\,dy\right]d\sigma$$

$$= \iint_{D_{xz}} \left([y\sqrt{x^2+z^2}]_{x^2+z^2}^{1}\right) d\sigma$$

$$= \iint_{D_{xz}} (1 - x^2 - z^2)\sqrt{x^2+z^2}\,d\sigma.$$

Since the domain D_{xz} is a circular disk, it is easier to convert this to polar coordinates in the xz-plane, using the substitution $x = r\cos\theta$, $z = r\sin\theta$; D_{xz} is now given by $0 \le \theta \le 2\pi$ and $0 \le r \le 1$, which gives

$$\iiint_E \sqrt{x^2+z^2}\,dV = \iint_{D_{xz}} (1-x^2-z^2)\sqrt{x^2+z^2}\,d\sigma$$

$$= \int_0^{2\pi}\int_0^1 (1-r^2) r \cdot r\,dr\,d\theta = \int_0^{2\pi} d\theta \int_0^1 (r^2 - r^4)\,dr$$

$$= 2\pi\left[\frac{r^3}{3} - \frac{r^5}{5}\right]_0^1 = \frac{4\pi}{15}.$$

Method 3: Now we consider computing a double integral first in the xz-plane. The cross-section of E by the vertical plane passing through $(0, y, 0)$ and perpendicular to the y-axis is the circular disk $D_y : x^2 + z^2 \le y$ with center at $(0, y, 0)$ and radius $\sqrt{y}$ (see Figure 3.21(c)). The triple integral becomes

$$\iiint_E \sqrt{x^2+z^2}\,dV = \int_0^1 dy \iint_{D_y} \sqrt{x^2+z^2}\,d\sigma.$$

Converting to polar coordinate in the xz-plane ($x = r\cos\theta$, $z = r\sin\theta$) and computing the double integral first over the region D_y given by $r \le \sqrt{y}$ and $0 \le \theta \le 2\pi$ gives

$$\iint_{D_y} \sqrt{x^2 + z^2}\, dA = \int_0^{2\pi} d\theta \int_0^{\sqrt{y}} r \cdot r\, dr = \frac{2\pi y^{\frac{3}{2}}}{3}.$$

Hence,

$$\iiint_E \sqrt{x^2 + z^2}\, dV = \int_0^1 dy \iint_{D_y} \sqrt{x^2 + z^2}\, d\sigma = \int_0^1 \frac{2\pi y^{\frac{3}{2}}}{3}\, dy = \frac{4\pi}{15}.$$

3.5.2 Cylindrical and spherical coordinates

Cylindrical coordinates

The cylindrical coordinates of a point P are (r, θ, z), as shown in Figure 3.22(a).

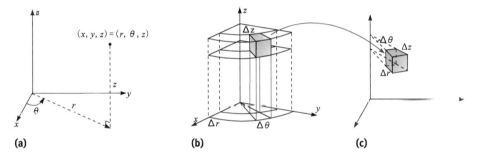

Figure 3.22: Triple integrals: cylindrical coordinates.

The coordinates r and θ are the polar coordinates of the projection of P onto the xy-plane, and z is the directed distance from the xy-plane to P (the usual z-coordinate).

In cylindrical coordinates, the surfaces analogous to coordinate planes in Cartesian coordinates are as follows. If k, l, and m are constants:

$r = k$ is a cylinder whose axis is the z-axis,
$\theta = l$ is a half-plane whose edge is the z-axis and its angle with the xz-plane is $\theta = l$,
$z = m$ is a horizontal plane with distance m from the xy-plane.

The equations relating Cartesian coordinates and cylindrical coordinates of a point are

$$x = r\cos\theta, \quad y = r\sin\theta, \quad \text{and} \quad z = z, \tag{3.5}$$

where $0 \le r < +\infty$, $0 \le \theta \le 2\pi$, and $-\infty < z < \infty$. The volume element in cylindrical coordinates is $r\, dr\, d\theta\, dz$ (see Figure 3.22(b)).

Clearly, a transformation from Cartesian coordinates to cylindrical coordinates could also be defined with the polar coordinates in the yz-plane or in the xz-plane.

Integration is often easier in cylindrical coordinates when the region of integration has a cylindrical (not necessarily circular) form with axis parallel to the x-, y-, or z-axis.

Example 3.5.6. A solid E lies within the cylinder $x^2 + y^2 = 1$ below the plane $z = 2$ and above the paraboloid $z = 1 - x^2 - y^2$. The density (mass per unit volume) at any point (x, y, z) is $\rho(x, y, z) = z\sqrt{x^2 + y^2}$. Find the mass of E.

Solution. In cylindrical coordinates, the cylinder has equation $r = 1$, the upper boundary is unchanged ($z = 2$), and the paraboloid has equation $z = 1 - r^2$. So, the region in r-, θ-, and z-coordinates is written

$$E' = \{(r, \theta, z) | 0 \leq \theta \leq 2\pi,\ 0 \leq r \leq 1,\ 1 - r^2 \leq z \leq 2\}.$$

The density function in cylindrical coordinates is $\rho(x, y, z) = zr$, and, therefore, the mass M of E is

$$M = \iiint_E z\sqrt{x^2 + y^2}\,dV = \iiint_{E'} (zr) r\,dz\,dr\,d\theta = \int_0^{2\pi}\int_0^1 \int_{1-r^2}^2 (rz) r\,dz\,dr\,d\theta$$

$$= \int_0^{2\pi} d\theta \int_0^1 r^2\,dr \int_{1-r^2}^2 z\,dz$$

$$= 2\pi \int_0^1 r^2 \left[\frac{1}{2}z^2\right]_{1-r^2}^2 dr$$

$$= \pi \int_0^1 r^2 (4 - (1 - r^2)^2)\,dr = \frac{44\pi}{35}.$$

The solid E is shown in Figure 3.23.

Example 3.5.7. Evaluate $\int_{-1}^{1}\int_{-\sqrt{1-x^2}}^{\sqrt{1-x^2}}\int_{\sqrt{x^2+y^2}}^{1} z\,dz\,dy\,dx$.

Solution. This iterated integral is a triple integral of $f(x, y, z) = z$ over the solid region Ω, i.e.,

$$\Omega = \{(x, y, z) | -1 \leq x \leq 1,\ -\sqrt{1 - x^2} \leq y \leq \sqrt{1 - x^2},\ \sqrt{x^2 + y^2} \leq z \leq 1\}.$$

The projection of Ω onto the xy-plane is the disk $x^2 + y^2 \leq 1$, the lower surface of Ω is the cone $z = \sqrt{x^2 + y^2}$, and the upper surface is the plane $z = 1$. This region has a much

Figure 3.23: Triple integrals: cylindrical coordinates, Example 3.5.6.

simpler description in cylindrical coordinates, i. e.,

$$\Omega = \{(r, \theta, z) | 0 \le \theta \le 2\pi,\ 0 \le r \le 1,\ r \le z \le 1\}.$$

Hence, converting the triple integral to cylindrical coordinates gives

$$\int_{-1}^{1} \int_{-\sqrt{1-x^2}}^{\sqrt{1-x^2}} \int_{\sqrt{x^2+y^2}}^{1} z\,dz\,dy\,dx = \iiint_{\Omega} z\,dV = \int_{0}^{2\pi} \int_{0}^{1} \int_{r}^{1} z\,r\,dz\,dr\,d\theta$$

$$= \int_{0}^{2\pi} d\theta \int_{0}^{1} r\,dr \int_{r}^{1} z\,dz = \pi \int_{0}^{1} r(1-r^2)\,dr = \frac{\pi}{4}.$$

Spherical coordinates

The spherical coordinates (ρ, θ, ϕ) of a point P are usually defined as in Figure 3.24(a). The coordinate $\rho = |OP|$ is the distance from the origin to P, θ is the same angle as in cylindrical coordinates, and ϕ is the angle between the positive z-axis and the line segment $\overline{OP}$. Thus, all points in space have unique spherical coordinates (ρ, θ, ϕ), provided ρ, θ, ϕ are restricted by $\rho \ge 0$, $0 \le \theta \le 2\pi$, and $0 \le \phi \le \pi$. The spherical coordinate system is especially useful in problems where the formula of the function being integrated contains the quantity $x^2 + y^2 + z^2$ or where the domain has a spherical nature with center at the origin. In spherical coordinates, the surfaces analogous to the coordinate planes in Cartesian coordinates are as follows. If k, l, and m are any constants:

$\rho = k$ is a sphere with center the origin and radius k,
$\theta = l$ is a half-plane whose edge is the z-axis and angle with the xz-plane is $\theta = l$,
$\phi = m$ is a half-cone making an angle ϕ with the positive z-axis.

The equations relating spherical and rectangular coordinates of a point are

$$x = \rho \sin \phi \cos \theta, \quad y = \rho \sin \phi \sin \theta, \quad \text{and} \quad z = \rho \cos \phi. \tag{3.6}$$

The volume element in spherical coordinates is $\rho^2 \sin \phi\,d\rho\,d\theta\,d\phi$, as shown in Figure 3.24(b).

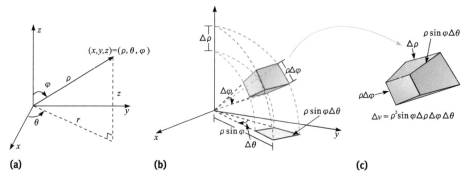

Figure 3.24: Triple integrals: spherical coordinates.

Example 3.5.8. Evaluate $\iiint_E e^{(x^2+y^2+z^2)^{3/2}}\, dV$, where E is the top half of the unit ball

$$E = \{(x,y,z) | x^2 + y^2 + z^2 \leq 1 \text{ and } z \geq 0\}.$$

Solution. Since the boundary of E is part of a sphere, it is wise to try spherical coordinates. The region corresponding to E in spherical coordinates is

$$E' = \left\{(\rho,\theta,\phi) | 0 \leq \rho \leq 1,\ 0 \leq \theta \leq 2\pi,\ 0 \leq \phi \leq \frac{\pi}{2}\right\}.$$

In addition, spherical coordinates give $x^2+y^2+z^2 = \rho^2$ and this simplifies the integrand. Thus,

$$\iiint_E e^{(x^2+y^2+z^2)^{3/2}}\, dV = \iiint_{E'} e^{(\rho^2)^{3/2}} \rho^2 \sin\phi\, d\rho d\theta d\phi$$

$$= \int_0^{2\pi}\int_0^{\pi/2}\int_0^1 e^{(\rho^2)^{3/2}} \rho^2 \sin\phi\, d\rho d\phi d\theta$$

$$= \int_0^{2\pi} d\theta \int_0^{\pi/2} \sin\phi\, d\phi \int_0^1 e^{\rho^3} \rho^2\, d\rho$$

$$= 2\pi \cdot (-\cos\phi)\Big|_0^{\pi/2} \cdot \left(\frac{1}{3}e^{\rho^3}\right)\Big|_0^1$$

$$= \frac{2}{3}\pi(e-1).$$

Example 3.5.9. Find the volume of the solid Ω enclosed by the sphere $x^2 + y^2 + (z-a)^2 = a^2$ and inside the half-cone $z = \sqrt{3x^2 + 3y^2}$.

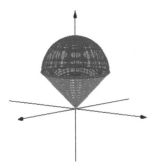

Figure 3.25: Triple integrals: spherical coordinates, Example 3.5.9.

Solution. The solid is shown in Figure 3.25. In spherical coordinates, the boundary surfaces become, after simplification,

$$\rho = 2a\cos\phi \quad \text{and} \quad \phi = \frac{\pi}{6}.$$

So, the solid Ω is defined by the region $\Omega': 0 \le \rho \le 2a\cos\phi,\ 0 \le \theta \le 2\pi,\ 0 \le \phi \le \frac{\pi}{6}$ in spherical coordinates. Therefore,

$$V = \iiint_\Omega dxdydz = \iiint_{\Omega'} \rho^2 \sin\phi\, d\rho d\phi d\theta = \int_0^{2\pi} d\theta \int_0^{\frac{\pi}{6}} \sin\phi\, d\phi \int_0^{2a\cos\phi} \rho^2\, d\rho$$

$$= 2\pi \int_0^{\frac{\pi}{6}} \sin\phi \cdot \left[\frac{\rho^3}{3}\right]_0^{2a\cos\phi} d\phi = \frac{16\pi a^3}{3}\int_0^{\frac{\pi}{6}} (\cos^3\phi\sin\phi)\, d\phi$$

$$= \frac{16\pi a^3}{3}\left(-\frac{1}{4}\cos^4\phi\right)_0^{\frac{\pi}{6}} = \frac{7}{12}\pi a^3 \text{ cubic units.}$$

3.6 Change of variables in triple integrals

The change of variables formula for triple integrals is similar to that for double integrals. Let T be a continuously differentiable one-to-one transformation that maps a region Ω' in uvw-space to a region Ω in the xyz-space with equations

$$x = x(u,v,w),\quad y = y(u,v,w),\quad \text{and}\quad z = z(u,v,w).$$

The **Jacobian** of T is the following 3×3 determinant:

$$\frac{\partial(x,y,z)}{\partial(u,v,w)} = \begin{vmatrix} \frac{\partial x}{\partial u} & \frac{\partial x}{\partial v} & \frac{\partial x}{\partial w} \\ \frac{\partial y}{\partial u} & \frac{\partial y}{\partial v} & \frac{\partial y}{\partial w} \\ \frac{\partial z}{\partial u} & \frac{\partial z}{\partial v} & \frac{\partial z}{\partial w} \end{vmatrix}.$$

We have the following change of variables formula for triple integrals:

$$\iiint_\Omega f(x,y,z)\,dV = \iiint_{\Omega'} f(x(u,v,w),\, y(u,v,w),\, z(u,v,w)) \left|\frac{\partial(x,y,z)}{\partial(u,v,w)}\right| du\,dv\,dw. \tag{3.7}$$

Example 3.6.1. Evaluate $\iiint_\Omega |xy|\,dV$, where Ω is the solid bounded by the ellipsoid $\frac{x^2}{a^2} + \frac{y^2}{b^2} + \frac{z^2}{c^2} = 1$.

Solution. We use the substitution $x = au$, $y = bv$, $z = cw$ and compute its Jacobian

$$\frac{\partial(x,y,z)}{\partial(u,v,w)} = \begin{vmatrix} \frac{\partial x}{\partial u} & \frac{\partial x}{\partial v} & \frac{\partial x}{\partial w} \\ \frac{\partial y}{\partial u} & \frac{\partial y}{\partial v} & \frac{\partial y}{\partial w} \\ \frac{\partial z}{\partial u} & \frac{\partial z}{\partial v} & \frac{\partial z}{\partial w} \end{vmatrix} = \begin{vmatrix} a & 0 & 0 \\ 0 & b & 0 \\ 0 & 0 & c \end{vmatrix} = abc.$$

Then we note that we have $u^2 + v^2 + w^2 = 1$, so

$$\iiint_\Omega |xy|\,dV = \iiint_{\Omega'_{uvw}} (|au \cdot bv|abc)\,du\,dv\,dw$$

$$= a^2 b^2 c \iiint_{u^2+v^2+w^2 \le 1} (|uv|)\,du\,dv\,dw$$

$$= a^2 b^2 c \times 8 \iiint_{u^2+v^2+w^2 \le 1,\, u\ge 0,\, v\ge 0,\, w\ge 0} uv\,du\,dv\,dw$$

$$= 8a^2 b^2 c \int_0^{\frac{\pi}{2}} d\theta \int_0^{\frac{\pi}{2}} d\phi \int_0^1 \rho \sin\phi \cos\theta \cdot \rho \sin\phi \sin\theta \cdot \rho^2 \sin\phi\, d\rho$$

$$= 8a^2 b^2 c \int_0^{\frac{\pi}{2}} \cos\theta \sin\theta\, d\theta \int_0^{\frac{\pi}{2}} \sin^3\phi\, d\phi \int_0^1 \rho^4\, d\rho$$

$$= \frac{8a^2 b^2 c}{15}.$$

Cylindrical and spherical coordinates are special transformations in a triple integral. The Jacobian of the transformation from Cartesian to cylindrical coordinates is

$$\frac{\partial(x,y,z)}{\partial(r,\theta,z)} = \begin{vmatrix} \frac{\partial x}{\partial r} & \frac{\partial x}{\partial \theta} & \frac{\partial x}{\partial z} \\ \frac{\partial y}{\partial r} & \frac{\partial y}{\partial \theta} & \frac{\partial y}{\partial z} \\ \frac{\partial z}{\partial r} & \frac{\partial z}{\partial \theta} & \frac{\partial z}{\partial z} \end{vmatrix} = \begin{vmatrix} \cos\theta & -r\sin\theta & 0 \\ \sin\theta & r\cos\theta & 0 \\ 0 & 0 & 1 \end{vmatrix} = r.$$

Hence, the absolute value of this Jacobian (used in the change of variables formula) is

$$|r| = r, \quad \text{since } r \ge 0.$$

Therefore, the change of variables in a triple integral from a region Ω in Cartesian to a region Ω' in cylindrical coordinates is

$$\iiint_\Omega f(x,y,z)\,dV = \iiint_{\Omega'} f(r\cos\theta,\, r\sin\theta,\, z)\, r\,dr\,d\theta\,dz,$$

thus giving us the formula for triple integration in cylindrical coordinates.

We now can compute the Jacobian of the transformation from Cartesian coordinates to spherical coordinates as follows:

$$\frac{\partial(x,y,z)}{\partial(\rho,\theta,\phi)} = \begin{vmatrix} \frac{\partial x}{\partial \rho} & \frac{\partial x}{\partial \theta} & \frac{\partial x}{\partial \phi} \\ \frac{\partial y}{\partial \rho} & \frac{\partial y}{\partial \theta} & \frac{\partial y}{\partial \phi} \\ \frac{\partial z}{\partial \rho} & \frac{\partial z}{\partial \theta} & \frac{\partial z}{\partial \phi} \end{vmatrix} = \begin{vmatrix} \sin\phi\cos\theta & -\rho\sin\phi\sin\theta & \rho\cos\phi\cos\theta \\ \sin\phi\sin\theta & \rho\sin\phi\cos\theta & \rho\cos\phi\sin\theta \\ \cos\phi & 0 & -\rho\sin\phi \end{vmatrix}$$
$$= -\rho^2 \sin\phi.$$

Since $0 \le \phi \le \pi$, we have $\sin\phi \ge 0$, and, therefore, the absolute value of the Jacobian (used in the change of variables formula) is

$$\left|\frac{\partial(x,y,z)}{\partial(\rho,\theta,\phi)}\right| = |-\rho^2 \sin\phi| = \rho^2 \sin\phi.$$

So, the change of variables from Cartesian to spherical makes the following changes in a triple integral:

$$\iiint_\Omega f(x,y,z)\,dV = \iiint_{\Omega'} f(\rho\sin\phi\cos\theta, \rho\sin\phi\sin\theta, \rho\cos\phi)\rho^2 \sin\phi\, d\rho\,d\theta\,d\phi.$$

3.7 Other applications of multiple integrals

3.7.1 Surface area

Parameterized surfaces

When using a graphing calculator to sketch a sphere, you may notice that the calculator does not do a good job in sketching functions such as $z = \sqrt{1 - x^2 - y^2}$. However, if you use the parametric form for the same surface $x = a\sin u \cos v$, $y = a\sin u \sin v$, and $z = a\cos u$, the graphing calculator does a much better job. In fact, a parameterization of a surface can be written in a form of a vector-valued function

$$\mathbf{r}(u,v) = x(u,v)\mathbf{i} + y(u,v)\mathbf{j} + z(u,v)\mathbf{k} \quad \text{or} \quad \mathbf{r}(u,v) = \langle x(u,v), y(u,v), z(u,v)\rangle,$$

where u and v are two independent variables (parameters).

> **Example 3.7.1.** Find parametric descriptions for the following surfaces:
> $$(1) x^2 + y^2 = a^2, \quad (2) z = a\sqrt{x^2+y^2}, \quad (3) z = x^2 + 2y^2.$$

Solution. There are many ways to parameterize a surface.
(1) One way to parameterize the cylinder is to set $z = v$, $x = a\cos u$, and $y = a\sin u$. We can also write it as a vector-valued function,

$$\mathbf{r}(u,v) = \langle a\cos u, a\sin u, v\rangle, \quad 0 \le u \le 2\pi, \; -\infty < v < \infty.$$

Note that $\mathbf{r}(u,v) = \langle a\cos\frac{u}{2}, a\sin\frac{u}{2}, v^3\rangle$ is also a parametric description of the same cylinder.

(2) A parametric description of the circular cone is

$$\mathbf{r}(u,v) = \left\langle \frac{v}{a}\cos u, \frac{v}{a}\sin u, v\right\rangle, \; 0 \le u \le 2\pi, v \ge 0.$$

(3) One parametric description is

$$\mathbf{r}(u,v) = \langle u, v, u^2 + 2v^2\rangle, \quad -\infty < u, v < \infty.$$

Also

$$\mathbf{r}(u,v) = \left\langle \sqrt{v}\cos u, \sqrt{\frac{v}{2}}\sin u, v\right\rangle, \quad 0 \le u \le 2\pi, v \ge 0$$

is a parametric description.

Surface area

We now apply double integrals to the problem of computing the area of a surface S defined by $\mathbf{r}(u,v) = \langle x(u,v), y(u,v), z(u,v)\rangle$, where $x = x(u,v), y = y(u,v),$ and $z = z(u,v)$ all have continuous partial derivatives at $(u,v) \in D$. A special parametric description where the surface has an explicit equation $z = f(x,y)$, and a parametric description for this surface is

$$\mathbf{r}(u,v) = \langle u, v, f(u,v)\rangle \quad \text{or} \quad \mathbf{r}(x,y) = \langle x, y, f(x,y)\rangle.$$

To find the "surface area element dS," we can use the tangent plane approximation, as shown in Figure 3.26, where the area element on the tangent plane is given by

$$|\mathbf{r}_u \Delta u \times \mathbf{r}_v \Delta v| = |\mathbf{r}_u \times \mathbf{r}_v|\Delta u \Delta v.$$

Therefore, adding up these elements, we will have

$$\text{surface area } S = \iint_{D_{uv}} |\mathbf{r}_u \times \mathbf{r}_v| du dv. \tag{3.8}$$

Figure 3.27 shows more general cases.

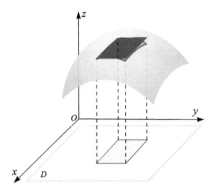

Figure 3.26: Surface area: tangent plane approximation.

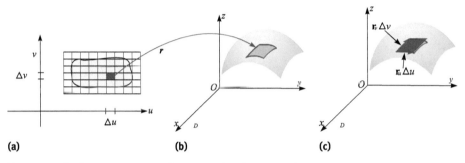

(a)   **(b)**   **(c)**

Figure 3.27: Surface area: tangent plane approximation, general case.

Example 3.7.2. Find the surface area of a ball with radius R.

Solution. The ball can be described by

$$\mathbf{r}(u,v) = \langle R\sin u \cos v, R\sin u \sin v, R\cos u\rangle, \quad 0 \le v \le 2\pi, \ 0 \le u \le \pi,$$

where u and v are actually ϕ and θ in spherical coordinates. Since

$$\mathbf{r}_u \times \mathbf{r}_v = \begin{vmatrix} \mathbf{i} & \mathbf{j} & \mathbf{k} \\ x_u & y_u & z_u \\ x_v & y_v & z_v \end{vmatrix} = \begin{vmatrix} \mathbf{i} & \mathbf{j} & \mathbf{k} \\ R\cos u \cos v & R\cos u \sin v & -R\sin u \\ -R\sin u \sin v & R\sin u \cos v & 0 \end{vmatrix}$$
$$= -R^2 \sin^2 u \cos v \,\mathbf{i} + R^2 \sin^2 u \sin v \,\mathbf{j} - R^2 \sin u \cos u\,\mathbf{k},$$

it follows that

$$|\mathbf{r}_u \times \mathbf{r}_v| = \sqrt{(-R^2 \sin^2 u \cos v)^2 + (R^2 \sin^2 u \sin v)^2 + (-R^2 \sin u \cos u)^2}$$
$$= R^2 \sin u.$$

Thus,

$$\text{surface area } S = \iint_{D_{uv}} |\mathbf{r}_u \times \mathbf{r}_v| du dv = \iint_{D_{uv}} R^2 \sin u \, du dv$$

$$= \int_0^{2\pi} dv \int_0^{\pi} R^2 \sin u \, du = 4\pi R^2.$$

Surface area when the surface is given by an explicit equation $z = f(x, y)$

In this case, we let $x = x$, $y = y$, and $z = z(x, y)$. Then $\mathbf{r}(x, y) = \langle x, y, z(x, y) \rangle$ is a parametric description, and $\mathbf{r}_x = \langle 1, 0, z_x \rangle$, $\mathbf{r}_y = \langle 1, 0, z_y \rangle$, and

$$\mathbf{r}_x \times \mathbf{r}_y = \begin{vmatrix} \mathbf{i} & \mathbf{j} & \mathbf{k} \\ 1 & 0 & z_x \\ 0 & 1 & z_y \end{vmatrix} = -z_x \mathbf{i} - z_y \mathbf{j} + \mathbf{k},$$

so, in rectangular coordinates,

$$\text{surface area } S = \iint_{D_{xy}} |\mathbf{r}_x \times \mathbf{r}_y| d\sigma = \iint_{D_{xy}} \sqrt{1 + z_x^2 + z_y^2} d\sigma. \tag{3.9}$$

We use this formula to compute the surface area of a ball with radius R again in the following example.

Example 3.7.3. Find the surface area of the sphere with radius R.

Solution. By symmetry, we compute the area of the top half of the sphere and double this. To make the calculation easier, assume that the center of the sphere is at the origin so that the equation of the top half of the sphere is

$$z = f(x, y) = \sqrt{R^2 - x^2 - y^2},$$

so that

$$z_x = \frac{\partial z}{\partial x} = \frac{-x}{\sqrt{R^2 - x^2 - y^2}} \quad \text{and} \quad z_y = \frac{\partial z}{\partial y} = \frac{-y}{\sqrt{R^2 - x^2 - y^2}}.$$

The projection D_{xy} of this surface onto the xy-plane is

$$x^2 + y^2 \le R^2 \quad \text{and} \quad z = 0.$$

Applying the surface area formula, equation (3.9), we obtain

$$S = 2 \iint_{D_{xy}} \sqrt{z_x^2 + z_y^2 + 1} d\sigma = 2 \iint_{D_{xy}} \sqrt{\left(\frac{-x}{\sqrt{R^2 - x^2 - y^2}}\right)^2 + \left(\frac{-y}{\sqrt{R^2 - x^2 - y^2}}\right)^2 + 1} d\sigma$$

$$= 2 \iint_{D_{xy}} \frac{R}{\sqrt{R^2 - x^2 - y^2}} d\sigma.$$

The integrand (the function being integrated) is unbounded on the region $D_{xy} : x^2 + y^2 \leq R^2$, so we consider $D'_{xy} : x^2 + y^2 \leq b^2$, $0 < b < R$ and then let $b \to R$. Converting to cylindrical coordinates, we obtain

$$S = \lim_{b \to R} 2 \int_0^{2\pi} d\theta \int_0^b \frac{R}{\sqrt{R^2 - r^2}} r \, dr = \lim_{b \to R} 4\pi R \int_0^b \frac{r}{\sqrt{R^2 - r^2}} dr$$

$$= \lim_{b \to R} 4\pi R[-\sqrt{R^2 - r^2}]_0^b = 4\pi R \lim_{b \to R} (R - \sqrt{R^2 - b^2}) = 4\pi R^2.$$

Example 3.7.4. Find the area of the part of the paraboloid $z = x^2 + y^2$ that lies under the plane $z = 2$. !

Solution. The plane $z = 2$ intersects the paraboloid in the circle $x^2 + y^2 = 2$, $z = 2$. Therefore, the given surface lies above the disk $D: x^2 + y^2 \leq 2$ with center the origin and radius $\sqrt{2}$.

So, the required area A is given by

$$A = \iint_D \sqrt{1 + \left(\frac{\partial z}{\partial x}\right)^2 + \left(\frac{\partial z}{\partial y}\right)^2} dxdy = \iint_D \sqrt{1 + (2x)^2 + (2y)^2} dxdy$$

$$= \iint_D \sqrt{1 + 4(x^2 + y^2)} dxdy.$$

Converting to polar coordinates in order to simplify this integration, we obtain

$$A = \int_0^{2\pi} \int_0^{\sqrt{2}} \sqrt{1 + 4r^2} \, r dr d\theta = 2\pi \cdot \frac{1}{8} \cdot \frac{2}{3}[(1 + 4r^2)^{3/2}]_0^{\sqrt{2}}$$

$$= \frac{13\pi}{3}.$$

In Figure 3.26, we note that the area element $d\sigma$ in the xy-plane is exactly the projection of the area element dS in the tangent plane. This indicates that

$$dS \times \cos \gamma = d\sigma, \quad \text{where } \gamma \text{ is the acute angle between the tangent plane}$$
$$\text{and the } xy\text{-plane.}$$

Note that if a smooth surface has an equation $F(x, y, z) = 0$, then $\nabla F = \langle F_x, F_y, F_z \rangle$ is a normal vector of its tangent plane. Then

$$\text{the unit vector } \frac{\nabla F}{|\nabla F|} = \left\langle \frac{F_x}{|\nabla F|}, \frac{F_y}{|\nabla F|}, \frac{F_z}{|\nabla F|} \right\rangle = \langle \cos \alpha, \cos \beta, \cos \gamma \rangle.$$

So, $dS = \frac{1}{|\cos\gamma|}d\sigma = \frac{|\nabla F|}{|F_z|}d\sigma = \frac{|\nabla F|}{|F_z|}dxdy$. Therefore, we conclude that

$$\text{surface area } S = \iint dS = \iint_{D_{xy}} \frac{|\nabla F|}{|F_z|}dxdy. \tag{3.10}$$

If the surface is given by $z = f(x,y)$, then $F(x,y,z) = f(x,y) - z = 0$, $F_z = -1$, and $\nabla F = \langle z_x, z_y, -1\rangle$. So, $|\nabla F| = \sqrt{1 + z_x^2 + z_y^2}$ and

$$\text{surface area } S = \iint_{D_{xy}} \frac{\sqrt{1 + z_x^2 + z_y^2}}{|-1|}dxdy = \iint_{D_{xy}} \sqrt{1 + z_x^2 + z_y^2}dxdy, \tag{3.11}$$

which is the same formula as the one we derived before. However, equation (3.10) does give us options to project the surface area element dS on the tangent plane onto the other two coordinate planes. Therefore, we have

$$\text{surface area } S = \iint_{D_{xy}} \frac{|\nabla F|}{|F_z|}dxdy = \iint_{D_{xy}} \sqrt{1 + z_x^2 + z_y^2}dxdy \tag{3.12}$$

$$= \iint_{D_{yz}} \frac{|\nabla F|}{|F_x|}dzdy = \iint_{D_{yz}} \sqrt{1 + x_y^2 + x_z^2}dzdy \tag{3.13}$$

$$= \iint_{D_{xz}} \frac{|\nabla F|}{|F_y|}dxdz = \iint_{D_{xz}} \sqrt{1 + y_x^2 + y_z^2}dxdz, \tag{3.14}$$

where D_{xy}, D_{yz}, and D_{xz} are projection regions of the surface onto the xy-, yz-, and xz-planes, respectively.

Example 3.7.5. Find the surface area of the part of the surface $y = x^{\frac{3}{2}}$ for $0 \le x \le \frac{4}{3}$ and $0 \le z \le 4$.

Solution. We project the surface onto the xz-plane. Since

$$F(x,y,z) = y - x^{\frac{3}{2}} = 0, \quad \text{we have } \nabla F = \left\langle -\frac{3}{2}x^{\frac{1}{2}}, 1, 0\right\rangle.$$

The surface area is given by

$$\text{surface area } S = \iint_{D_{xz}} \frac{|\nabla F|}{|F_y|}dxdz = \iint_{D_{xz}} \sqrt{1 + \left(\frac{-3}{2}x^{\frac{1}{2}}\right)^2}dxdz$$

$$= \int_0^4 dz \int_0^{\frac{4}{3}} \sqrt{1 + \frac{9x}{4}}dx = \frac{224}{27}.$$

Example 3.7.6. Find the surface area of the part of the sphere $x^2 + y^2 + z^2 = R^2$ for $z \geq h$, where $0 < h < R$.

Solution. Since the surface has equation $x^2 + y^2 + z^2 = R^2$, we let $F(x, y, z) = x^2 + y^2 + z^2 - R^2 = 0$. Then,

$$\nabla F = \langle 2x, 2y, 2z \rangle \quad \text{and} \quad F_z = 2z.$$

Thus, the surface area is

$$S = \iint_{D_{xy}} \frac{|\nabla F|}{|F_z|} dxdy = \iint_{D_{xy}} \frac{\sqrt{(4x^2 + 4y^2 + 4z^2)}}{2z} dxdy$$

$$= \iint_{D_{xy}} \frac{R}{z} dxdy = \iint_{D_{xy}} \frac{R}{\sqrt{R^2 - x^2 - y^2}} dxdy$$

$$= \int_0^{2\pi} d\theta \int_0^{\sqrt{R^2-h^2}} \frac{R}{\sqrt{R^2 - r^2}} rdr$$

$$= 2\pi R [-(R^2 - r^2)^{\frac{1}{2}}] \Big|_0^{\sqrt{R^2-h^2}}$$

$$= 2\pi R(R - h).$$

Note. The surface area is linear in h!

3.7.2 Center of mass, moment of inertia

Center of mass

In physics, the center of mass of an object is the balance point of the object or the position towards which gravity attracts. For a rod with density function $\rho(x)$, the center of mass $\bar{x}$ can be found by using the idea of moment and is given by

$$\bar{x} = \frac{\int_a^b x\rho(x)dx}{\int_a^b \rho(x)dx}. \tag{3.15}$$

For a two-dimensional lamina with density function $\rho(x, y)$, the center of mass $(\bar{x}, \bar{y})$ is given by

$$\bar{x} = \frac{\iint_D x\rho(x,y)d\sigma}{\iint_D \rho(x,y)d\sigma} \quad \text{and} \quad \bar{y} = \frac{\iint_D y\rho(x,y)d\sigma}{\iint_D \rho(x,y)d\sigma}. \tag{3.16}$$

For a three-dimensional solid with density function $\rho(x, y, z)$, the center of mass $(\bar{x}, \bar{y}, \bar{z})$ is given by

$$\bar{x} = \frac{\iiint_\Omega x\rho(x,y,z)dV}{\iiint_\Omega \rho(x,y,z)dV}, \quad \bar{y} = \frac{\iiint_\Omega y\rho(x,y,z)dV}{\iiint_\Omega \rho(x,y,z)dV}, \quad \text{and} \quad \bar{z} = \frac{\iiint_\Omega z\rho(x,y,z)dV}{\iiint_\Omega \rho(x,y,z)dV}. \tag{3.17}$$

If an object is uniform (has constant density at each point), then its center of mass is at the **centroid**, the geometric center of the figure.

Moment of inertia
In physics, we know that the moment of inertia of a rigid body m about an axis l is

$$I = \sum_i^n \Delta m_i r_i^2, \tag{3.18}$$

where Δm_i is a mass element, while r_i is the perpendicular distance from this mass element to the axis of rotation.

By taking limits, this can be represented as an integral. If ρ stands for the density function, the moments of inertia of a region D in the xy-plane being rotated about the x-axis and y-axis are, therefore,

$$I_x = \iint_D \rho y^2 dA \quad \text{and} \quad I_y = \iint_D \rho x^2 dA, \tag{3.19}$$

respectively. Thus, by the perpendicular axis theorem for the moment of inertia of a rigid body in the plane, the moment of inertia of the body about the z-axis is

$$I_z = I_x + I_y = \iint_D \rho(x^2 + y^2) dA. \tag{3.20}$$

Similarly, for a solid R in space, the moment of inertia of the solid about an axis of rotation is

$$I = \iiint_R \rho |\mathbf{r}|^2 dV, \tag{3.21}$$

where $|\mathbf{r}|$ is the distance from the volume element dV to the axis of rotation.

Therefore, finding the coordinates for a center of mass or moment of inertia of an object about an axis of rotation becomes a job of evaluating some multiple integrals.

3.8 Review

Main concepts discussed in this chapter are listed below.
1. Definition and properties of double integrals:

$$\iint_D f(x,y) d\sigma = \lim_{|\Delta \sigma_i| \to 0} \sum f(x_i, y_i) \Delta \sigma_i.$$

2. Double integral in rectangular coordinates:

$$\iint_D f(x,y)d\sigma = \int_a^b dx \int_{\phi_1(x)}^{\phi_2(x)} f(x,y)dy \quad \text{type I region,}$$

$$\iint_D f(x,y)d\sigma = \int_c^d dy \int_{\psi_1(y)}^{\psi_2(y)} f(x,y)dx \quad \text{type II region.}$$

3. Double integral in polar coordinates:

$$\iint_D f(x,y)d\sigma = \int_\alpha^\beta d\theta \int_{r_1(\theta)}^{r_2(\theta)} f(r\cos\theta, r\sin\theta)rdr.$$

4. Change of variables in a double integral:

$$\iint_D f(x,y)d\sigma = \iint_{D_{uv}} f(x(u,v), y(u,v)) \left|\frac{\partial(x,y)}{\partial(u,v)}\right| dudv.$$

5. Triple integral in rectangular coordinates:

$$\iiint_\Omega f(x,y,z)dV = \iint_{D_{xy}} \left(\int_{z_1(x,y)}^{z_2(x,y)} f(x,y,z)dz \right) d\sigma, \quad \text{a definite integral first,}$$

$$\iiint_\Omega f(x,y,z)dV = \int_a^b \left(\iint_{D_z} f(x,y,z)d\sigma \right) dz, \quad \text{a double integral first.}$$

6. Cylindrical coordinates:

$$x = r\cos\theta, \quad y = r\sin\theta, \quad \text{and} \quad z = z.$$

7. Spherical coordinates:

$$x = \rho\sin\phi\cos\theta, \quad y = \rho\sin\phi\sin\theta, \quad \text{and} \quad z = \rho\cos\phi.$$

8. Triple integral in cylindrical coordinates:

$$\iiint_\Omega f(x,y,z)dV = \iiint_{\Omega'_{r\theta z}} f(r\cos\theta, r\sin\theta, z)rdzdrd\theta.$$

9. Triple integral in spherical coordinates:

$$\iiint_\Omega f(x,y,z)dV = \iiint_{\Omega'_{\rho\phi\theta}} f(\rho\sin\phi\cos\theta, \rho\sin\phi\sin\theta, \rho\cos\phi)\rho^2\sin\phi \, d\rho d\phi d\theta.$$

10. Change of variables in a triple integral:

$$\iiint_\Omega f(x,y,z)\,dV = \iiint_{\Omega'_{uvw}} f(x(u,v,w), y(u,v,w), z(u,v,w)) \left|\frac{\partial(x,y,z)}{\partial(u,v,w)}\right| dV'.$$

11. Surface area for

$$\mathbf{r} = \mathbf{r}(u,v) \qquad S = \iint_{D_{uv}} |\mathbf{r}_u \times \mathbf{r}_v|\,du\,dv,$$

$$z = f(x,y) \qquad S = \iint_{D_{xy}} \sqrt{1 + z_x^2 + z_y^2}\,dx\,dy,$$

$$F(x,y,z) = 0 \qquad S = \iint_{D_{xy}} \frac{|\nabla F|}{|F_z|}\,dx\,dy.$$

Similar results hold if the surface is projected onto a coordinate plane other than the xy-plane.

12. Center of mass: if $f(x)$, $f(x,y)$, and $f(x,y,z)$ are density functions, then

$$\bar{x} = \frac{\int_a^b x f(x)\,dx}{\int_a^b f(x)\,dx} \qquad \text{one dimension, thin rod,}$$

$$\bar{x} = \frac{\iint_D x f(x,y)\,d\sigma}{\iint_D f(x,y)\,d\sigma} \quad \bar{y} = \frac{\iint_D y f(x,y)\,d\sigma}{\iint_D f(x,y)\,d\sigma} \qquad \text{two dimensions, thin lamina,}$$

$$\bar{x} = \frac{\iiint_\Omega x f(x,y,z)\,dV}{\iiint_\Omega f(x,y,z)\,dV} \quad \bar{y} = \frac{\iiint_\Omega y f(x,y,z)\,dV}{\iiint_\Omega f(x,y,z)\,dV} \quad \bar{z} = \frac{\iiint_\Omega z f(x,y,z)\,dV}{\iiint_\Omega f(x,y,z)\,dV}$$

three dimensions, solid.

13. Moment of inertia: if ρ is the density function, then

$$I_x = \iint_D \rho x^2\,d\sigma, \quad I_y = \iint_D \rho y^2\,d\sigma, \quad \text{and} \quad I_z = \iint_D \rho(x^2 + y^2)\,d\sigma.$$

For a solid Ω with density function ρ,

$$I = \iiint_\Omega \rho |\mathbf{r}|^2\,dV,$$

where $|\mathbf{r}|$ is the distance from the element dV to the axis of rotation.

3.9 Exercises

3.9.1 Double integrals

1. Evaluate each of the following iterated integrals:
   (1) $\int_1^2 \int_0^3 yx^2 e^{y} dxdy$, (2) $\int_{-1}^2 \int_x^{2-x} dydx$, (3) $\int_0^\pi \int_0^{\sin y} e^{\cos y} dxdy$.
2. Evaluate each of the following double integrals:
   (1) $\iint_D (x^2 + y^2) dxdy$, where $D = \{(x,y) | |x| \leq 1, |y| \leq 1\}$,
   (2) $\iint_D xy^2 dxdy$, where D is the region bounded by the parabola $y^2 = 2x$ and the line $x = \frac{1}{2}$,
   (3) $\iint_D x\sqrt{y} dxdy$, where D is bounded by the two parabolas $y = \sqrt{x}$ and $y = x^2$.
3. If $f(x,y)$ is a continuous function, find the limit $I = \lim_{\rho \to 0} \frac{1}{\pi \rho^2} \iint_{x^2+y^2 \leq \rho^2} f(x,y) d\sigma$.
4. Use double integrals to find the area for each of the following plane regions:
   (1) region D bounded by curves $y = x^2$ and $y = 2$,
   (2) region R below the curve $y = 4 - x^2$ and the line $y = 6 - 3x$ and above the line $y = 2 - x$.
5. Find the volume for each of the following solids:
   (1) solid Ω beneath the paraboloid $z = 12 - x^2 - 2y^2$ and above the square region $D = \{(x,y,0) | 0 \leq x \leq 1, 0 \leq y \leq 1\}$,
   (2) solid R bounded by $z = 1 + x + y$, $z = 0$, $x + y = 1$, $x = 0$, and $y = 0$,
   (3) solid R below the surface of $f(x,y) = 2 + \frac{1}{y}$ and above the rectangular region in the xy-plane with vertices $(1,1)$, $(1,3)$, $(2,1)$, and $(2,3)$,
   (4) solid R bounded by $z = 0$, $z = xy$, and $x + y + z = 1$.
6. Sketch the region of each of the following integrations and change the order of each integration:
   (1) $\int_1^2 dx \int_x^{2x} f(x,y) dy$, (2) $\int_0^1 dx \int_0^x f(x,y) dy + \int_1^2 dx \int_0^{2-x} f(x,y) dy$,
   (3) $\int_0^3 \int_{-\sqrt{9-y^2}}^{\sqrt{9-y^2}} f(x,y) dxdy$.
7. Evaluate each of the following integrals by reversing the order of integration:
   (1) $\int_0^3 dy \int_{y^2}^9 y\cos(x^2) dx$, (2) $\int_0^{\sqrt{\pi}} \int_y^{\sqrt{\pi}} e^{x^2} dxdy$.
8. Evaluate the integral $\int_0^1 \frac{x^b - x^a}{\ln x} dx$, where $0 < a < b$.
9. The **average value** of a function $f(x,y)$ over a plane region D is defined as
$$f_{ave} = \frac{1}{A(D)} \iint_D f(x,y) d\sigma, \quad \text{where } A(D) \text{ is the area of the region } D.$$
   Find the average of the function $f(x,y) = \sin(2x - 3y)$ over the region D bounded by $0 \leq x \leq \frac{\pi}{2}$ and $x \leq y \leq \frac{\pi}{2}$.
10. Use polar coordinates to combine the sum
$$\int_{1/\sqrt{2}}^1 dx \int_{\sqrt{1-x^2}}^x xy\,dy + \int_1^{\sqrt{2}} dx \int_0^x xy\,dy + \int_{\sqrt{2}}^2 dx \int_0^{\sqrt{4-x^2}} xy\,dy.$$

11. Find the volume for each of the following solids:
    (1) solid R enclosed by two cylinders $x^2 + y^2 = 4$ and $x^2 + z^2 = 4$,
    (2) solid R inside both the sphere $x^2 + y^2 + z^2 = 16$ and the cylinder $x^2 + y^2 = 4$,
    (3) solid R that is below the surface $z = \frac{18}{2+x^2+y^2} - 3$ and above the plane $z = 0$,
    (4) solid R between two surfaces $z = x^2 + y^2$ and $z = 2 - x^2 - y^2$.
12. Find the volume of the solid bounded by $z = 25 - (x-1)^2 - (y+2)^2$ and $z \geq 0$.
13. Use change of variables in a double integral to
    (1) find the area of the region bounded by the lines $x + y = c$, $x + y = d$, $y = ax$, and $y = bx$ ($0 < c < d$, $0 < a < b$).
    (2) evaluate the double integral $\iint_D x \sin(y - 2x) d\sigma$, where D is region bounded by the parallelogram with vertices $(1, 0)$, $(1, 3)$, $(3, 7)$, $(3, 4)$.
    (3) find $\iint_D \sqrt{1 - \frac{x^2}{a^2} - \frac{y^2}{b^2}} dxdy$, where D is the region enclosed by the ellipse $\frac{x^2}{a^2} + \frac{y^2}{b^2} = 1$.
14. We can define the improper integral

$$I = \int_{-\infty}^{\infty} \int_{-\infty}^{\infty} e^{-(x^2+y^2)} dxdy = \lim_{r \to +\infty} \iint_{D_r} e^{-(x^2+y^2)} dxdy,$$

where D_r is the disk with radius r and center the origin. Show that

$$\int_{-\infty}^{\infty} \int_{-\infty}^{\infty} e^{-(x^2+y^2)} dxdy = \pi$$

and deduce

$$\int_{-\infty}^{\infty} e^{-x^2} dx = \sqrt{\pi} \quad \text{and} \quad \int_{-\infty}^{\infty} e^{-x^2/2} dx = \sqrt{2\pi}.$$

3.9.2 Triple integrals

1. Find the region Ω for which the triple integral

$$\iiint_\Omega 1 - x^2 - 3y^2 - 2z^2 dV$$

is a maximum.
2. Evaluate the iterated integral $\int_0^1 \int_0^z \int_0^{x-z} 6xz \, dy \, dx \, dz$.
3. If $\Omega_2 = \{(x, y, z) | x^2 + y^2 + z^2 \leq R^2\}$ and

$$\Omega_1 = \{(x, y, z) | x^2 + y^2 + z^2 \leq R^2, x \geq 0, y \geq 0, z \geq 0\},$$

find (1) $\iiint_{\Omega_1} 3 dV$ and (2) $\iiint_{\Omega_2} xyz \, dV$.

4. Evaluate the triple integral $\iiint_\Omega 2x\,dV$, where

$$\Omega = \{(x,y,z) \mid 0 \le y \le 2, 0 \le x \le \sqrt{4-y^2}, 0 \le z \le y\}.$$

5. The average value of a function $f(x,y,z)$ over a solid region Ω is defined to be

$$f_{ave} = \frac{1}{V(\Omega)} \iiint_\Omega f(x,y,z)\,dV, \quad \text{where } V(\Omega) \text{ is the volume of } \Omega.$$

Find the average value of the function $f(x,y,z) = x^2z + y^2z$ over the region enclosed by the paraboloid $z = 1 - x^2 - y^2$ and the plane $z = \frac{3}{4}$.

6. Evaluate the following triple integrals by converting to cylindrical or spherical coordinates:
   (1) $\iiint_\Omega (x^2 + y^2)\,dxdydz$, where Ω is the region enclosed by the paraboloid $x^2 + y^2 = 2z$ and the plane $z = 2$,
   (2) $\int_0^3 \int_0^{\sqrt{9-x^2}} \int_0^{\sqrt{x^2+y^2}} \sqrt{x^2+y^2}\,dzdydx$,
   (3) $\iiint_\Omega \sqrt{x^2+y^2+z^2}\,dxdydz$, where Ω is the ball $x^2+y^2+z^2 \le z$,
   (4) $\int_0^1 dx \int_0^{\sqrt{1-x^2}} dy \int_{\sqrt{x^2+y^2}}^{\sqrt{2-x^2-y^2}} z^2\,dz$.

7. Use triple integrals to find the volume of the solid
   (1) bounded by planes $x = 0$, $y = 0$, $z = 0$, $x = 4$, and $y = 4$ and the paraboloid $z = x^2 + y^2 + 1$.
   (2) enclosed by the cone $z = \sqrt{x^2+y^2}$ and the paraboloid $az = x^2 + y^2$, where $a > 0$.
   (3) enclosed by the spherical surfaces $\rho = 4\sin\phi$.

8. Use the change of variables in a triple integral to evaluate $\iiint_\Omega z\,dV$, where Ω is bounded by the planes $y = x$, $y = x+2$, $z = x$, $z = x+2$, $z = 0$, and $z = 6$.

9. Let $f(x) > 0$ be a continuous function, $\Omega = \{(x,y,z) \mid x^2+y^2+z^2 \le t^2\}$, and $D = \{(x,y) \mid x^2+y^2 \le t^2\}$. Prove that the function

$$F(t) = \frac{\iiint_\Omega f(x^2+y^2+z^2)\,dV}{\iint_D f(x^2+y^2)\,d\sigma}$$

is increasing for $t > 0$.

3.9.3 Other applications of multiple integrals

1. Find the area of each of the following surfaces:
   (1) the part of the hyperbolic paraboloid $z = y^2 - x^2$ that lies between the cylinders $x^2 + y^2 = 1$ and $x^2 + y^2 = 4$,
   (2) the part of the surface $z = xy$ that lies within the cylinder $x^2 + y^2 = 1$,

(3) the part of the cylinder $x^2 + y^2 = x$ that lies within the sphere $x^2 + y^2 + z^2 = 1$,
(4) the part of the sphere $x^2 + y^2 + z^2 = a^2$ that lies within the cylinder $x^2 + y^2 = ax$ and above the xy-plane,
(5) the part of the cylinder $x^2 + y^2 = 16$ that is between the planes $z = 0$ and $z = 16 - 2x$,
(6) the part of the trough $z = x^2$ for $-3 \le x \le 3$ and $1 \le y \le 4$.

2. A lamina is represented by the part of the disk $x^2 + y^2 \le 1$ in the first and second quadrants. Find the mass of the lamina if the density at any point is proportional to its distance from the x-axis with constant of proportionality equal to K.

3. A solid has the shape of a half-cylinder bounded by $-3 \le x \le 3$, $0 \le y \le \sqrt{9-x^2}$, and $0 \le z \le 2$. Each point (x,y,z) in the solid has density given by the function $f(x,y,z) = \frac{1}{1+x^2+y^2}$. Find the total mass of this solid.

4. Find the coordinates of the centroid of the constant-density cone D bounded by $z = 4 - \sqrt{x^2 + y^2}$ and $z = 0$.

5. A **torus** is a surface obtained by rotating a closed plane curve (for example a circle) about an axis, where the axis usually does not intersect the curve. Considered the torus obtained by rotating the circle $(x-R)^2 + z^2 = r^2$ ($r < R$) in the xz-plane about the z-axis.
   (a) Parameterize the torus.
   (b) Find its volume.
   (c) Find the centroid of the half-torus ($z \ge 0$).
   (d) Find the surface area and the moment of inertia of the torus.

6. Find the moment of inertia for:
   (a) a uniform semidisk about its straight edge (the diameter),
   (b) a uniform semidisk about the axis that is the perpendicular bisector of its straight edge (in the same plane),
   (c) a ball with radius R center at the origin, and density function $\sqrt{x^2+y^2+z^2}$ about the z-axis.

4 Line and surface integrals

4.1 Line integral with respect to arc length

The definite integral

$$\int_a^b f(t)\,dt \quad \text{or equivalently} \quad \int_a^b f(x)\,dx \tag{4.1}$$

of a one-variable function f defined on an interval $[a, b] \subset \mathbb{R}$ is used to model many physical phenomena. For example, if $f(t)$ is the speed of an object at time t, then the first integral gives the total distance traveled by the object between times $t = a$ and $t = b$. If $f(x)$ is the density of a wire at distance x along the wire, then the second integral of (4.1) is the mass of the wire. If $f(x) \geq 0$ on $a \leq x \leq b$, then the second form of (4.1) is the area of the region under the graph of f and above the x-axis, between $x = a$ and $x = b$.

It is useful to extend the integral to functions defined on domains other than an interval $[a, b]$ in $\mathbb{R}$. For example, how can we find the mass of a wire represented by a curve $C \subset \mathbb{R}^2$, with $f(x, y)$ the density function of the wire at a point (x, y) on C? As shown in Figure 4.1, what is the area of a "curtain" that is part of a cylinder with base C, a curve in the xy-plane, and at each point (x, y) the height is given by $f(x, y)$? We can use the element method like those we used for other integrals. We can split the curve into many small pieces, and for each piece, say, Δs_i, we arbitrarily choose a sample point, say, (x_i^*, y_i^*), so that $f(x_i^*, y_i^*)\Delta s_i$ approximates the mass element (or area element). Then we can form a Riemann sum and take the limit of this Riemann sum as Δs_i tends to 0. This leads to the ideas of a line integral. We now give a formal definition of a line integral with respect to arc length.

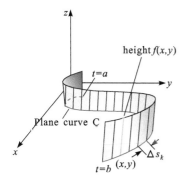

Figure 4.1: Area of a curtain.

4.1.1 Definition and properties

Definition 4.1.1 (Line integral with respect to arc length). Let C be a piecewise smooth curve in the plane $\mathbb{R}^2$, connecting two fixed points A and B. Let s be the distance along C measured from $A = S_0$. Subdivide C between A and B (Figure 4.2(a)) by points $A = S_0, S_1, S_2, \ldots, S_n = B$ and let s_i for each $i = 1, 2, \ldots, n$ be the distance along the curve from S_0 to S_i. Arbitrarily choose $(x_i^*, y_i^*) \in \mathbb{R}^2$ on C between S_{i-1} and S_i for $i = 1, 2, \ldots, n$. Let f be a real-valued function defined on C. If the limit

$$\lim_{\max |\Delta s_i| \to 0, \, n \to \infty} \sum_{i=1}^{n} f(x_i^*, y_i^*) \Delta s_i$$

exists for all possible subdivisions and choices of the (x_i^*, y_i^*), then we define this to be the line integral of f on the curve C with respect to arc length, and we write

$$\lim_{\max |\Delta s_i| \to 0, \, n \to \infty} \sum_{i=1}^{n} f(x_i^*, y_i^*) \Delta s_i = \int_C f(x, y) \, ds.$$

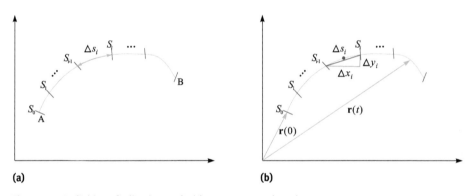

Figure 4.2: Definition of a line integral with respect to arc length.

Note.
1. It can be proved that this line integral exists when f is continuous, or *piecewise continuous*, provided C is finite and *piecewise smooth*. A curve $C: \mathbf{r}(t)$, $a \leq t \leq b$, is piecewise smooth if there is a subdivision $a = t_0 < t_1 < t_2 < \cdots < t_n = b$ such that $\mathbf{r}'(t)$ is continuous on each subinterval $t_{i-1} < t < t_i$.
2. Sometimes we also use the notation $\int_L f(x, y) \, ds$ for a line integral.

Properties of the line integral

The definition can be used to prove, without difficulty, the following properties for line integrals of the function $f(x, y)$ and $g(x, y)$ in $\mathbb{R}^2$, provided all line integrals involved exist:

(1) if $f(x,y) = 1$ for all $(x,y) \in C$ in $\mathbb{R}^2$, then $\int_C 1 ds$ is the length of the curve C,
(2) $\int_C kf(x,y) + mg(x,y) ds = k \int_C f(x,y) ds + m \int_C g(x,y) ds$ if $k, m \in \mathbb{R}$ (linearity),
(3) $\int_{C_1+C_2} f(x,y) ds = \int_{C_1} f(x,y) ds + \int_{C_2} f(x,y) ds$, where $C_1 + C_2$ means the curve formed by combining curves C_1 and C_2 into one curve,
(4) if $f(x,y) \le g(x,y)$, then $\int_C f(x,y) ds \le \int_C g(x,y) ds$,
(5) if $f(x,y)$ is continuous, then there is a point (η, ξ) such that

$$\int_C f(x,y) ds = f(\eta, \xi) \times \text{length of } C.$$

Example 4.1.1. Evaluate the line integral $\int_C (2x \ln(1+y^2) + \pi) ds$, where C is the half-circle $y = \sqrt{4-x^2}$.

Solution. By the linearity property, we have

$$\int_C (2x \ln(1+y^2) + \pi) ds = 2 \int_C x \ln(1+y^2) ds + \pi \int_C 1 ds.$$

By symmetry, as we saw in the discussion of multiple integrals, $\int_C x \ln(1+y^2) ds$ is equal to 0 since the curve is symmetrical about the y-axis, and $x \ln(1+y^2)$ is odd with respect to x. Since $\int_C 1 ds$ gives the length of C, we obtain

$$\int_C (2x \ln(1+y^2) + \pi) ds = 2 \int_C x \ln(1+y^2) ds + \pi \int_C 1 ds$$

$$= 0 + \pi \times \pi \times 2 = 2\pi^2.$$

4.1.2 Evaluating a line integral, $\int_C f(x,y) ds$, in $\mathbb{R}^2$

Suppose that a curve C is given by a vector-valued function (in a vector parametric form)

$$\mathbf{r}(t) = \langle x(t), y(t) \rangle, \quad \text{for } a \le t \le b,$$

where the functions $x(t)$ and $y(t)$ both have continuous first-order derivatives, and where $\mathbf{r}(a)$ and $\mathbf{r}(b)$ correspond to the points A and B, respectively. We approximate $\Delta s_i \approx \sqrt{\Delta x_i^2 + \Delta y_i^2}$ (see Figure 4.1.2(b)). Let $t = t_i^*$ be the value of t that corresponds to the point (x_i^*, y_i^*), and let $t = t_i$ be the value of t that corresponds to the point S_i on C, for $i = 1, 2, \ldots, n$. Writing $\Delta t_i = t_i - t_{i-1}$ and noting that $\max |\Delta t_i| \to 0 \iff \max |\Delta s_i| \to 0$, it follows that

$$\int_C f(x,y) ds = \lim_{\max |\Delta s_i| \to 0 \, n \to \infty} \sum_{i=1}^n f(x_i^*, y_i^*) \Delta s_i$$

$$= \lim_{\max |\Delta s_i| \to 0\ n \to \infty} \sum_{i=1}^{n} f(x(t_i^*), y(t_i^*)) \sqrt{\Delta x_i^2 + \Delta y_i^2}$$

$$= \lim_{\max |\Delta t_i| \to 0\ n \to \infty} \sum_{i=1}^{n} f(x(t_i^*), y(t_i^*)) \sqrt{\frac{\Delta x_i^2}{\Delta t_i^2} + \frac{\Delta y_i^2}{\Delta t_i^2}} \Delta t_i.$$

The last Riemann sum is an ordinary one-variable Riemann sum. Hence, if the limit exists, we can express the line integral as an ordinary one-variable definite integral, i.e.,

$$\int_C f(x,y)ds = \int_a^b f(x(t), y(t)) \sqrt{\left(\frac{dx}{dt}\right)^2 + \left(\frac{dy}{dt}\right)^2}\, dt. \tag{4.2}$$

We can also write this in a vector form,

$$\int_C f(x,y)ds = \int_a^b f(\mathbf{r}(t))|\mathbf{r}'(t)|dt. \tag{4.3}$$

Note that we can interpret ds, the arc length element, as $|\mathbf{r}'(t)|dt$. This is consistent with $\frac{ds}{dt} = |\mathbf{r}'(t)|$.

Example 4.1.2. Compute the circumference of the circle $x^2 + y^2 = R^2$ using a line integral.

Solution. The circumference is $\int_C 1 ds$, where C is the circle. Choose the standard parameterization,

$$x = R\cos t, \quad y = R\sin t, \quad 0 \le t \le 2\pi.$$

Evaluating the integral using equation (4.2), we have

$$\int_C 1ds = \int_0^{2\pi} \sqrt{\left(\frac{dx}{dt}\right)^2 + \left(\frac{dy}{dt}\right)^2}\, dt = \int_0^{2\pi} \sqrt{(-R\sin t)^2 + (R\cos t)^2}\, dt$$

$$= \int_0^{2\pi} R\sqrt{\sin^2 t + \cos^2 t}\, dt = 2\pi R.$$

Example 4.1.3. Find the total mass of the wire given by the curve C: $y = x^2$, $-2 \le x \le 2$ with density function $f(x,y) = x + \sqrt{y}$.

Solution. We choose a parameterization $\mathbf{r}(t) = \langle t, t^2 \rangle$, and then $|\mathbf{r}'(t)| = \sqrt{1^2 + (2t)^2}$. So we have

$$\text{total mass } m = \int_C f(x,y)ds = \int_C x + \sqrt{y}\, ds$$

$$= \int_{-2}^{2}(t+\sqrt{t^2})|\mathbf{r}'(t)|dt = \int_{-2}^{2}(t+\sqrt{t^2})\sqrt{1^2+(2t)^2}dt$$

$$= \int_{-2}^{2} t\sqrt{1^2+(2t)^2}dt + \int_{-2}^{2} \sqrt{t^2}\sqrt{1^2+(2t)^2}dt$$

$$= 0 + 2\int_{0}^{2} t\sqrt{1+4t^2}dt$$

$$= \frac{1}{4}\frac{2}{3}(1+4t^2)^{\frac{3}{2}}\Big|_{0}^{2} = \frac{17}{6}\sqrt{17} - \frac{1}{6}.$$

4.1.3 Line integrals $\int_C f(x,y,z)ds$ in $\mathbb{R}^3$

The process for expressing a line integral in $\mathbb{R}^3$ as a definite integral is very similar to the one used in $\mathbb{R}^2$, and is, therefore, not repeated here. When the curve C is, in a vector parametric form, $\mathbf{r}(t) = \langle x(t), y(t), z(t)\rangle$, for $a \le t \le b$, then the result is

$$\int_C f(x,y,z)ds = \int_a^b f(x(t),y(t),z(t))\sqrt{\left(\frac{dx}{dt}\right)^2 + \left(\frac{dy}{dt}\right)^2 + \left(\frac{dz}{dt}\right)^2}dt \quad (4.4)$$

or $\int_C f(x,y,z)ds = \int_a^b f(\mathbf{r}(t))|\mathbf{r}'(t)|dt.$

Note. A curve can have many different parametric forms, but the line integral will always have the same value because the parametric form of the line integral will always be the same as the original definition of the line integral in equation (4.2).

Example 4.1.4. Find the mass of the wire represented by the helical curve $C: \vec{r}(t) = \langle 2\cos t, 2\sin t, t\rangle$, $\pi \le t \le 2\pi$, when the density at a point (x,y,z) on the wire is given by $\delta(x,y,z) = x^2 + y^2 + z^2$.

Solution. The mass is given by the following line integral:

$$\int_C \delta(x,y,z)ds = \int_\pi^{2\pi}(x^2(t)+y^2(t)+z^2(t))\sqrt{\left(\frac{dx}{dt}\right)^2+\left(\frac{dy}{dt}\right)^2+\left(\frac{dz}{dt}\right)^2}dt$$

$$= \int_\pi^{2\pi}(4\cos^2 t + 4\sin^2 t + t^2)\sqrt{4\sin^2 t + 4\cos^2 t + 1}\,dt$$

$$= \sqrt{5}\int_\pi^{2\pi}(4+t^2)dt = \sqrt{5}\left(4\pi + \frac{7}{3}\pi^3\right).$$

Example 4.1.5. Evaluate $\int_C f ds$ when $f(x,y,z) = x + yz$ and $C = C_1 + C_2$, where C_1 is the line segment from $(0,0,2)$ to $(2,0,2)$ and C_2 is the line segment from $(2,0,2)$ to $(1,1,1)$.

Solution. The parametric form of the line through $(0,0,2)$ and $(2,0,2)$ is

$$\mathbf{r}(t) = \langle 0,0,2\rangle + t(\langle 2,0,2\rangle - \langle 0,0,2\rangle)$$
$$= (1-t)\langle 0,0,2\rangle + t\langle 2,0,2\rangle.$$

Since $\mathbf{r}(0) = \langle 0,0,2\rangle$ and $\mathbf{r}(1) = \langle 2,0,2\rangle$, the line segment C_1 has parameterization

$$C_1 : \mathbf{r}_1(t) = (1-t)\langle 0,0,2\rangle + t\langle 2,0,2\rangle = \langle 2t, 0, 2\rangle, \; 0 \le t \le 1.$$

Similarly, line segment C_2 has parameterization

$$C_2 : \mathbf{r}_2(t) = (1-t)\langle 2,0,2\rangle + t\langle 1,1,1\rangle = \langle 2-t, t, 2-t\rangle, \; 0 \le t \le 1.$$

Using properties of line integrals, we have

$$\int_C f ds = \int_{C_1} f ds + \int_{C_2} f ds = \int_0^1 f(\mathbf{r}_1(t))|\mathbf{r}_1'(t)|dt + \int_0^1 f(\mathbf{r}_2(t))|\mathbf{r}_2'(t)|dt$$

$$= \int_0^1 (2t)\sqrt{2^2}dt + \int_0^1 (2-t+t(2-t))\sqrt{(-1)^2+1+(-1)^2}dt$$

$$= [2t^2]_0^1 + \sqrt{3}\left[2t + \frac{t^2}{2} - \frac{t^3}{3}\right]_0^1 = 2 + \sqrt{3}\frac{13}{6}.$$

Figure 4.3 shows the graph of the curve C.

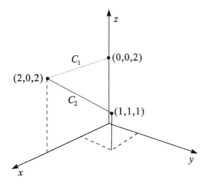

Figure 4.3: Line integral with respect to arc length, Example 4.1.5.

4.2 Line integral of a vector field

4.2.1 Vector fields

Many physical phenomena can be modeled by associating a vector with each point in space. Examples include electric fields, magnetic fields, gravitational fields, and the velocity field for a fluid. A function that associates a vector with each point is called a *vector field*.

More precisely, it is defined as follows.

Definition 4.2.1. A function **F** with domain set $D \subset \mathbb{R}^n$ and with range a set of vectors in $\mathbb{R}^n$ is called a vector field.

For example, the functions **F**, **G**, and **H** defined as follows are vector fields:

$$\mathbf{F}(x,y) = x^2 y\, \vec{i} + 2xy e^x\, \vec{j},$$
$$\mathbf{G}(x,y) = \langle -x, -y, -z \rangle,$$
$$\mathbf{H}(\rho, \theta, \phi) = \langle \rho \cos\theta \sin\phi, \rho \sin\theta \sin\phi, \cos\phi \rangle.$$

The gradient of a function f is also a vector field, i.e.,

$$\operatorname{grad} f(x,y) = \nabla f = \frac{\partial f}{\partial x}\mathbf{i} + \frac{\partial f}{\partial y}\mathbf{j} = \left\langle \frac{\partial f}{\partial x}, \frac{\partial f}{\partial y} \right\rangle,$$

$$\operatorname{grad} f(x,y,z) = \nabla f = \frac{\partial f}{\partial x}\mathbf{i} + \frac{\partial f}{\partial y}\mathbf{j} + \frac{\partial f}{\partial z}\mathbf{k}.$$

We can graph a vector field by drawing arrows at some selected points in $\mathbb{R}^2$ or $\mathbb{R}^3$. Some examples are shown in Figure 4.4 and Figure 4.5.

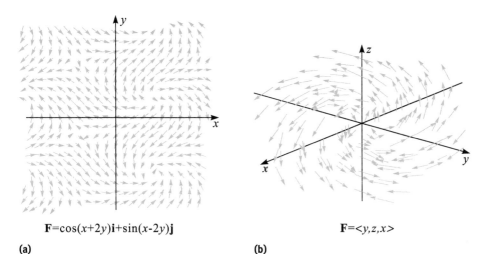

(a) F=cos(x+2y)i+sin(x−2y)j

(b) F=⟨y,z,x⟩

Figure 4.4: Some vector fields.

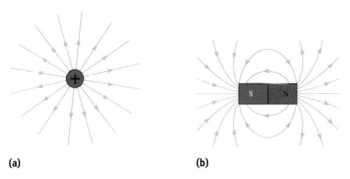

(a) (b)

Figure 4.5: Electronic and magnetic fields.

Example 4.2.1. The **gravitational field** of the Earth (of mass M_e) acting on an object of mass m is a vector field. Find the formula for this force field.

Solution. The gravitational force F is given by the inverse square law formula

$$F = G\frac{M_e m}{r^2},$$

where G is the gravitational constant, r is the distance between the object and the center of the Earth, and m and M_e are the mass of the object and the Earth, respectively. In order to find the gravitational field, let $\vec{r}$ be the vector from the center of the Earth to the object. The force on the object acts along the same line, but in the opposite direction to $\vec{r}$, and so the field becomes

$$\mathbf{F} = -G\frac{M_e m \vec{r}}{r^3}.$$

If we define a Euclidean coordinate system with origin at the center of the Earth, then this becomes

$$\mathbf{F} = -G\frac{M_e m}{r^2}\frac{\vec{r}}{|\vec{r}|} = -G\frac{M_e m(x\vec{i} + y\vec{j} + z\vec{k})}{(x^2 + y^2 + z^2)^{\frac{3}{2}}},$$

where (x,y,z) is the location of the object.

4.2.2 The line Integral of a vector field along a curve C

Suppose an object moves from point A to point B along a smooth curve C in $\mathbb{R}^2$ in a force field $\mathbf{F} = \langle f, g \rangle$, where $f = f(x,y)$ and $g = g(x,y)$ are two differentiable functions of two variables. How much work does the force field do? Note that the force acting on the object varies from point to point, maybe both in direction and in magnitude.

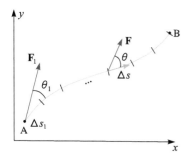

Figure 4.6: Work integral.

To find the total work, we again use the element method to break down the problem. Thinking about a very small piece of the curve, say, Δs, the work element ΔW can be obtained by using $|\mathbf{F}|\cos\theta \cdot \Delta s$, as shown in Figure 4.6. Suppose $\mathbf{T}$ is the unit tangent vector of the curve. Then

$$\Delta W \approx |\mathbf{F}|\cos\theta \cdot \Delta s = |\mathbf{F}|\cos\theta|\mathbf{T}| \cdot \Delta s = (\mathbf{F} \cdot \mathbf{T})\Delta s.$$

Adding them up and taking a limit, we will have a line integral similar to the one in the previous section. We now give a formal definition of a line integral of a vector field along a curve C in $\mathbb{R}^2$.

Definition 4.2.2 (Line integral of a vector field). Let $\mathbf{F} = \langle f(x,y), g(x,y) \rangle$ be a vector field in $\mathbb{R}^2$, and let $C: \mathbf{r}(t)$, $a \leq t \leq b$, be a curve in the xy-plane. If the limit

$$\lim_{\max|\Delta s_i| \to 0} \sum_{i=1}^{n} \langle f(x_i^*, y_i^*), g(x_i^*, y_i^*) \rangle \cdot \frac{\frac{\Delta \mathbf{r}_i}{\Delta t_i}}{\left|\frac{\Delta \mathbf{r}_i}{\Delta t_i}\right|} \Delta s_i$$

exists and is the same for any possible choice of sample points (x_i^*, y_i^*) between S_{i-1} and S_i on C and for any subdivision $S_0, S_1, \ldots, S_n$ of the curve C, we define this limit to be the line integral of the vector $\mathbf{F}$ along the curve C from point $A = \mathbf{r}(a)$ to point $B = \mathbf{r}(b)$, and we write

$$\int_C (\mathbf{F} \cdot \mathbf{T}) ds = \lim_{\max|\Delta s_i| \to 0} \sum_{i=1}^{n} \langle f(x_i^*, y_i^*), g(x_i^*, y_i^*) \rangle \cdot \frac{\frac{\Delta \mathbf{r}_i}{\Delta t_i}}{\left|\frac{\Delta \mathbf{r}_i}{\Delta t_i}\right|} \Delta s_i.$$

Note.
1. If both components f and g of $\mathbf{F}$ are continuous or piecewise continuous and the curve C is also piecewise smooth, then the above line integral of the vector field $\mathbf{F}$ along C always exists.
2. Unlike the line integral with respect to arc length, the orientation of the curve C does make a difference in the line integral of a vector field. We usually define the positive orientation of a parameterized curve C to be the direction that is consistent with the increasing value of its parameter. If we use $-C$ to denote the negative

direction of C, then we define

$$\int_C (\mathbf{F} \cdot \mathbf{T}) ds = -\int_{-C} (\mathbf{F} \cdot \mathbf{T}) ds.$$

3. If the curve is closed, that is, $\mathbf{r}(a) = \mathbf{r}(b)$, then we also adopt the notation $\oint_C (\mathbf{F} \cdot \mathbf{T}) ds$.

Notations for line integrals of a vector field

The notation $\int_C (\mathbf{F} \cdot \mathbf{T}) ds$ for the line integral of the vector field $\mathbf{F} = \langle f, g \rangle$ along the curve $C : \mathbf{r}(t) = \langle x(t), y(t) \rangle$, $a \le t \le b$, is easy to understand, although it can be mathematically hard to manipulate. Fortunately, we have some other equivalent notations. Since

$$\int_C (\mathbf{F} \cdot \mathbf{T}) ds = \int_C \langle f, g \rangle \cdot \frac{\mathbf{r}'(t)}{|\mathbf{r}'(t)|} ds$$

$$= \int_C \langle f, g \rangle \cdot \frac{\mathbf{r}'(t)}{|\mathbf{r}'(t)|} |\mathbf{r}'(t)| dt = \int_C \langle f, g \rangle \cdot \mathbf{r}'(t) dt$$

$$= \int_C \langle f, g \rangle \cdot \left\langle \frac{dx}{dt}, \frac{dy}{dt} \right\rangle dt$$

$$= \int_C f dx + g dy,$$

we have

$$\int_C (\mathbf{F} \cdot \mathbf{T}) ds = \int_C f dx + g dy. \tag{4.5}$$

If we denote $d\mathbf{r} = \langle dx, dy \rangle$, then we also have

$$\int_C (\mathbf{F} \cdot \mathbf{T}) ds = \int_C \mathbf{F} \cdot d\mathbf{r} = \int_C \mathbf{F} \cdot \mathbf{r}'(t) dt. \tag{4.6}$$

Note. We can write $\int_C \mathbf{F} \cdot d\mathbf{r} = \int_C (\mathbf{F} \cdot \mathbf{T}) ds$, where $\mathbf{T}$ is a unit tangent to C, and the integrals with respect to arc length s are unchanged when the orientation of C is reversed. However, the orientation property still holds true because the unit vector $\mathbf{T}$ is replaced by its negative, $-\mathbf{T}$, when C is replaced by $-C$.

Example 4.2.2. Evaluate $\int_C x dy - y dx$, where C is the arc of the ellipse $\frac{x^2}{a^2} + \frac{y^2}{b^2} = 1$ from $(a, 0)$ to $(0, b)$.

Solution. A vector parametric form of the curve is

$$\mathbf{r}(t) = \langle a\cos t, b\sin t\rangle, \quad \text{for } 0 \le t \le \frac{\pi}{2},$$

and $\mathbf{r}'(t) = \langle -a\sin t, b\cos t\rangle$. The vector field is $\mathbf{F} = \langle -y, x\rangle$, so

$$\int_C x\,dy - y\,dx = \int_C \mathbf{F}\cdot\mathbf{r}'(t)\,dt = \int_C \langle -y, x\rangle \cdot \langle -a\sin t, b\cos t\rangle\,dt$$

$$= \int_0^{\frac{\pi}{2}} \langle -b\sin t, a\cos t\rangle \cdot \langle -a\sin t, b\cos t\rangle\,dt$$

$$= \int_0^{\frac{\pi}{2}} (-b\sin t)(-a\sin t) + (a\cos t \times b\cos t)\,dt$$

$$= \int_0^{\frac{\pi}{2}} (ab\cos^2 t + ab\sin^2 t)\,dt = \frac{\pi}{2}ab.$$

Example 4.2.3. Find the line integral $\int_C x^2\,dx - xy\,dy$, where C consists of the line segment C_1 from the point $(1, 0)$ to the point $(0, 0)$ followed by the vertical line segment C_2 from the point $(0, 0)$ to $(0, 1)$.

Solution. The path and the vector field are shown in Figure 4.7(a). Along C_1, we have $y = 0$ and $dy = 0$, so

$$\int_{C_1} x^2\,dx - xy\,dy = \int_1^0 x^2\,dx = -\frac{1}{3}.$$

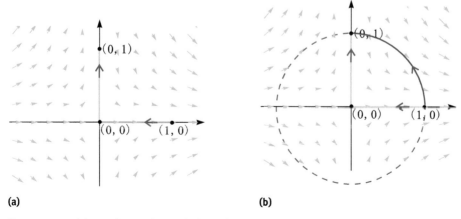

(a) (b)

Figure 4.7: Work integral examples, path dependence.

Along C_2, $x = 0$, so $dx = 0$ and

$$\int_{C_2} x^2 dx - xy dy = 0.$$

Altogether, we have

$$\int_C x^2 dx - xy dy = \int_{C_1} x^2 dx - xy dy + \int_{C_2} x^2 dx - xy dy = -\frac{1}{3}.$$

Example 4.2.4. Find the work done by the force field $\vec{F}(x,y) = x^2 \vec{i} - xy \vec{j}$ acting on a particle moving along the quarter-circle $\vec{r}(t) = \cos t \, \vec{i} + \sin t \, \vec{j}$ for $0 \le t \le \frac{\pi}{2}$.

Solution. The path and the vector field are shown in Figure 4.7(b). Since $x = \cos t$ and $y = \sin t$, we have

$$\vec{F}(x,y) = \langle x^2, -xy \rangle = \langle \cos^2 t, -\cos t \sin t \rangle$$

and $\frac{d\vec{r}(t)}{dt} = \langle -\sin t, \cos t \rangle$. Therefore, the work done is

$$\int_C \mathbf{F} \cdot d\mathbf{r} = \int_0^{\pi/2} \langle \cos^2 t, -\cos t \sin t \rangle \cdot \langle -\sin t, \cos t \rangle dt$$

$$= \int_0^{\pi/2} (-2\cos^2 t \sin t) dt = \left[2\frac{\cos^3 t}{3} \right]_0^{\pi/2} = -\frac{2}{3}.$$

Note. The two previous examples both compute the work done by the same force field, $x^2 \mathbf{i} - xy \mathbf{j}$, between the same two points, but reach different answers. This shows that the work done by the same vector field may be different when calculated along different routes. This is not true of so-called conservative force fields, such as gravity, where the work done is the same, no matter what path is taken, so long as the starting point and ending point are fixed. The next example with a conservative force field illustrates this point.

Example 4.2.5. Evaluate the line integral $\int_C xy dx + \frac{x^2}{2} dy$ between $O(0,0)$ and $B(1,1)$ along the following curves:
(1) the line segment from $O(0,0)$ to $A(1,0)$, and then to $B(1,1)$.
(2) the line segment from $O(0,0)$ to $B(1,1)$.
(3) an arc of the parabola $x = y^2$.

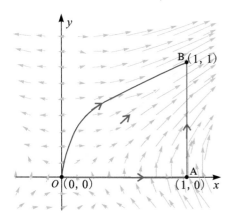

Figure 4.8: Work integral examples, path independence.

Solution. The vector field and the paths are shown in Figure 4.8.
(1) Along the line segment from $O(0, 0)$ to $A(1, 0)$, $y = 0$, and then

$$\int_C xy\,dx + \frac{x^2}{2}dy = 0.$$

Along the line segment from $A(1, 0)$ to $B(1, 1)$, $x = 1$ and $dx = 0$. Thus

$$\int_C xy\,dx + \frac{x^2}{2}dy = \int_0^1 \frac{1^2}{2}dy = \frac{1}{2}.$$

All together,

$$\int_C xy\,dx + \frac{x^2}{2}dy = \int_{OA} xy\,dx + \frac{x^2}{2}dy + \int_{AB} xy\,dx + \frac{x^2}{2}dy = 0 + \frac{1}{2} = \frac{1}{2}.$$

(2) Along the line segment from $O(0, 0)$ to $B(1, 1)$, a parametric form is $x = x, y = x$ for $0 \le x \le 1$, so

$$\int_C xy\,dx + \frac{x^2}{2}dy = \int_0^1 x^2 dx + \frac{x^2}{2}dx = \int_0^1 \frac{3x^2}{2}dx = \frac{1}{2}.$$

(3) Along the parabola $x = y^2$, a parametric form is $x = y^2, y = y$ for $0 \le y \le 1$, so

$$\int_C xy\,dx + \frac{x^2}{2}dy = \int_0^1 y^3 d(y^2) + \frac{y^4}{2}dy = \int_0^1 \frac{5y^4}{2}dy = \frac{1}{2}.$$

Note. In this example, you can see that the line integral of the vector field $\langle xy, \frac{x^2}{2}\rangle$ from $(0, 0)$ to $(1, 1)$ is the same along three different curves. In fact it would be the same along any curve joining the two points. This is an example of a conservative vector field. We will discuss this kind of field in more details in the coming section.

Line integral of a vector field in $\mathbb{R}^3$

The work integral $\int_C (\mathbf{F} \cdot \mathbf{T})ds = \int_C f dx + g dy$ can be generalized to a three-dimensional vector field $\mathbf{F} = \langle f, g, h \rangle$, where $f = f(x, y, z)$, $g = g(x, y, z)$, and $h = h(x, y, z)$ are three piecewise continuous functions, and C is a piecewise smooth curve in space. In this case, we have

$$\int_C (\mathbf{F} \cdot \mathbf{T})ds = \int_C \mathbf{F} \cdot d\mathbf{r} = \int_C f dx + g dy + h dz.$$

Example 4.2.6. Evaluate $\int_C (y - x)dx + xdy + (x + z)dz$, where C consists of the line segment C_1 from $(2, 0, 0)$ to $(3, 4, 5)$ followed by the vertical line segment C_2 from $(3, 4, 5)$ to $(-1, 4, 1)$.

Solution. The line segments C_1 and C_2 can be written parametrically for $0 \leq t \leq 1$ as

$$C_1 : \mathbf{r}_1(t) = (1 - t)\langle 2, 0, 0 \rangle + t\langle 3, 4, 5 \rangle = \langle 2 + t, \ 4t, \ 5t \rangle,$$
$$C_2 : \mathbf{r}_2(t) = (1 - t)\langle 3, 4, 5 \rangle + t\langle -1, 4, 1 \rangle = \langle 3 - 4t, \ 4, \ 5 - 4t \rangle.$$

Thus,

$$\int_{C_1+C_2} (y - x)dx + xdy + (x + z)dz = \int_0^1 (4t - (2 + t))dt + (2 + t)d(4t) + (2 + t + 5t)d(5t)$$

$$+ \int_0^1 (4 - (3 - 4t))d(3 - 4t) + (3 - 4t)d(4)$$

$$+ (3 - 4t + 5 - 4t)d(5 - 4t)$$

$$= \int_0^1 (-20 + 47t)dt = -20t + \frac{47t^2}{2}\Big|_0^1 = \frac{7}{2}.$$

4.3 The fundamental theorem of line integrals

As you may have already noted in Example 4.2.5, the line integral of a vector field from point A to point B may not depend on the path that connects A and B. The property is called path independence, and its formal definition is given below. A conservative field has this property. For example, we know that the gravitational force field is conservative and path independent.

Definition 4.3.1 (Path independence). Let $\mathbf{F} = \langle f, g \rangle$ be a vector field defined on a region D in $\mathbb{R}^2$. Let A and B be any two points in D, and C be any path with endpoints A and B. If the value of the line integral $\int_C f dx + g dy$ is independent of the path that connects A and B, then we say the line integral $\int_C f dx + g dy$ is path-independent and that the vector field $\mathbf{F}$ is path-independent in D.

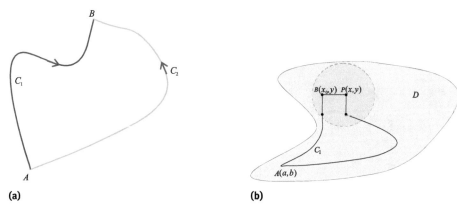

Figure 4.9: Path independence.

From the definition of path independence, we know that if $\mathbf{F} = \langle f, g \rangle$ is path-independent in D, then for any given two points A and B, where C_1 is a path (curve) in D from A to B and C_2 is any other path (curve) in D from A to B (see Figure 4.9(a)), we must have

$$\int_{C_1} \mathbf{F} \cdot d\mathbf{r} = \int_{C_2} \mathbf{F} \cdot d\mathbf{r}.$$

This means

$$0 = \int_{C_1} \mathbf{F} \cdot d\mathbf{r} - \int_{C_2} \mathbf{F} \cdot d\mathbf{r} = \int_{C_1} \mathbf{F} \cdot d\mathbf{r} + \int_{-C_2} \mathbf{F} \cdot d\mathbf{r} = \oint_{C_1 + (-C_2)} \mathbf{F} \cdot d\mathbf{r}.$$

Note that $C_1 + (-C_2)$ is a closed curve in D. (We say a parameterized curve $\mathbf{r}(t)$, $a \leq t \leq b$, is **closed** if $\mathbf{r}(a) = \mathbf{r}(b)$).

On the other hand, if $\mathbf{F}$ is defined in D and

$$\oint_C \mathbf{F} \cdot d\mathbf{r} = 0$$

for any closed curve C in D, then we claim that the vector field $\mathbf{F}$ must be path-independent in D. This is because for any two points A and B in D, and any two different paths C_1 and C_2 from A to B, we have

$$0 = \oint_C \mathbf{F} \cdot d\mathbf{r} = \int_{C_1} \mathbf{F} \cdot d\mathbf{r} + \int_{-C_2} \mathbf{F} \cdot d\mathbf{r}$$

$$= \int_{C_1} \mathbf{F} \cdot d\mathbf{r} - \int_{C_2} \mathbf{F} \cdot d\mathbf{r}.$$

This means

$$\int_{C_1} \mathbf{F} \cdot d\mathbf{r} = \int_{C_2} \mathbf{F} \cdot d\mathbf{r}.$$

We summarize the above arguments in the following theorem.

Theorem 4.3.1. *$\mathbf{F} = \langle f, g \rangle$ is path-independent in D, if and only if for any closed curve C in D, we must have*

$$\oint_C \mathbf{F} \cdot d\mathbf{r} = \oint_C f dx + g dy = 0.$$

But how do we know that a line integral is path-independent? Recall the fundamental theorem of calculus

$$\int_a^b f'(x) dx = f(b) - f(a).$$

The integration of the rate of change of a function is equal to the net change of the function over the interval $[a, b]$. For functions of two or more variables, we have seen that the gradient of a function plays much the same role as the derivative of a function does for functions of one variable. We consider the integral $\int_C \nabla \varphi \cdot d\mathbf{r}$. This means the vector field $\mathbf{F}$ is the gradient of some scalar function φ. This indeed works, and we now state the fundamental theorem of line integrals.

Theorem 4.3.2. *Let $\varphi(x, y)$ be a differentiable function and $\mathbf{F} = \nabla \varphi$. Suppose that C is any curve that has a parameterization $\mathbf{r}(t) = \langle x(t), y(t) \rangle$, $a \leq t \leq b$, where $\mathbf{r}(a)$ and $\mathbf{r}(b)$ correspond to the points A and B, respectively. Then $\mathbf{F}$ is path-independent and*

$$\int_C \mathbf{F} \cdot d\mathbf{r} = \int_C \nabla \varphi \cdot d\mathbf{r} = \varphi(B) - \varphi(A).$$

Proof. Since $\nabla \varphi = \langle \frac{\partial \varphi}{\partial x}, \frac{\partial \varphi}{\partial y} \rangle$ and $\frac{d\mathbf{r}}{dt} = \langle \frac{dx}{dt}, \frac{dy}{dt} \rangle$,

$$\int_C \nabla \varphi \cdot d\mathbf{r} = \int_C \nabla \varphi \cdot \frac{d\mathbf{r}}{dt} dt = \int_C \langle \frac{\partial \varphi}{\partial x}, \frac{\partial \varphi}{\partial y} \rangle \cdot \langle \frac{dx}{dt}, \frac{dy}{dt} \rangle dt$$

$$= \int_a^b \left(\frac{\partial \varphi}{\partial x} \frac{dx}{dt} + \frac{\partial \varphi}{\partial y} \frac{dy}{dt} \right) dt$$

$$= \int_a^b \left(\frac{d\varphi}{dt} \right) dt = \varphi(x(b), y(b)) - \varphi(x(a), y(a))$$

$$= \varphi(B) - \varphi(A).$$

This completes the proof. □

Due to the fundamental theorem of line integrals, we now give the definition of a conservative vector field and a *potential function* of the vector field **F**.

Definition 4.3.2 (Conservative field and potential function). A vector field **F** is called conservative if it is the gradient of a scalar function φ, that is, $\mathbf{F} = \nabla \varphi$. Such a φ is called a potential function of the vector field **F**.

Note that if $\mathbf{F} = \langle f, g \rangle = \nabla \varphi$, then

$$f dx + g dy = \frac{\partial \varphi}{\partial x} dx + \frac{\partial \varphi}{\partial y} dy = d\varphi.$$

This means

$$\mathbf{F} = \langle f, g \rangle \text{ is conservative} \iff \text{there is a function } \varphi(x,y) \text{ such that } \mathbf{F} = \nabla \varphi$$
$$\iff \text{there is a function } \varphi(x,y) \text{ such that } f dx + g dy = d\varphi.$$

Also in this case, we can write

$$\int_C f dx + g dy = \int_C \frac{\partial \varphi}{\partial x} dx + \frac{\partial \varphi}{\partial y} dy = \int_C d\varphi(x,y) = \varphi(B) - \varphi(A).$$

Therefore, if we can find a potential function for the vector field **F**, then we can evaluate a work integral along a curve C by directly finding the difference of the potential function at the initial and terminal points of the curve C. In this case we also write

$$\int_C f dx + g dy = \int_A^B f dx + g dy$$

to indicate that the line integral is path-independent.

Example 4.3.1. Note from Example 4.2.5 that

$$xy dx + \frac{x^2}{2} dy = d\left(\frac{x^2 y}{2}\right),$$

so $\frac{x^2 y}{2}$ is a potential function of $\langle xy, \frac{x^2}{2} \rangle$. Therefore, the line integral $\int_C xy dx + \frac{x^2}{2} dy$ from $A(0,0)$ to $B(1,1)$ along any curve C is obtained by

$$\int_C xy dx + \frac{x^2}{2} dy = \left.\frac{x^2 y}{2}\right|_{(0,0)}^{(1,1)} = \frac{1^2 \cdot 1}{2} - \frac{0^2 \cdot 0}{2} = \frac{1}{2}.$$

We now establish the famous principle of *conservation of energy*. Let us assume that a continuous force field **F** acts on an object moving along a path C. The path C is given parametrically by $\mathbf{r}(t)$, $a \leq t \leq b$, $\mathbf{r}(a) = A$ is the initial point, and $\mathbf{r}(b) = B$ is the terminal point of C. By Newton's second law of motion, the relation between the force $\mathbf{F}$ at a point on C and the acceleration, $\mathbf{a}(t)$ or $\mathbf{r}''(t)$, is given by

$$\mathbf{F}(\mathbf{r}(t)) = m\mathbf{r}''(t).$$

The line integral giving the work done by the force acting on the object while it is moving along C from A to B is

$$W = \int_C \mathbf{F} \cdot d\mathbf{r} = \int_a^b \mathbf{F}(\mathbf{r}(t)) \cdot \mathbf{r}'(t) dt$$

$$= \int_a^b m\mathbf{r}''(t) \cdot \mathbf{r}'(t) dt$$

$$= \frac{m}{2} \int_a^b \frac{d}{dt}[\mathbf{r}'(t) \cdot \mathbf{r}'(t)] dt$$

$$= \frac{m}{2} \int_a^b \frac{d}{dt}|\mathbf{r}'(t)|^2 dt$$

$$= \frac{m}{2}[|\mathbf{r}'(t)|^2]_a^b$$

$$= \frac{m}{2}(|\mathbf{r}'(b)|^2 - |\mathbf{r}'(a)|^2).$$

Therefore, if we write $\mathbf{v} = \mathbf{r}'(t)$ (the velocity vector), then

$$W = \frac{1}{2}m|\mathbf{v}(b)|^2 - \frac{1}{2}m|\mathbf{v}(a)|^2.$$

The quantity $\frac{1}{2}m|\mathbf{v}(t)|^2$ is called the *kinetic energy* of the object and is denoted by $K(t)$. Thus, we can rewrite the above equation as

$$W = K(B) - K(A).$$

This means that the work done by the force acting on the object as it moves along C is equal to the change in kinetic energy between $t = a$ and $t = b$, regardless of the path that C takes between A and B.

Now, if $\mathbf{F}$ is also a *conservative force field*, then we have $\mathbf{F} = \nabla \varphi$ for some function φ. Note that in physics, the *potential energy* P of an object at the point (x, y, z) is defined as the negative of a potential function, i. e., $P(x, y, z) = -\varphi(x, y, z)$. Applying this here, we have $\mathbf{F} = -\nabla P$, and then the work done in moving along C from A to B can be written

$$W = \int_C \mathbf{F} \cdot d\mathbf{r} = -\int_C \nabla P \cdot d\mathbf{r}$$

$$= -\int_a^b \left(\frac{\partial P}{\partial x}\frac{dx}{dt} + \frac{\partial P}{\partial y}\frac{dy}{dt} + \frac{\partial P}{\partial z}\frac{dz}{dt} \right) dt$$

$$= -\int_a^b \frac{d}{dt}(P(x(t), y(t), z(t)))dt$$
$$= -[P(\mathbf{r}(b)) - P(\mathbf{r}(a))]$$
$$= P(A) - P(B).$$

Comparing this equation with the previous expression in terms of kinetic energy we see that

$$P(A) + K(A) = P(B) + K(B).$$

This means that if an object moves from point A to point B under the influence of a conservative force field, then the sum of its potential energy and its kinetic energy remains constant. That is called the *law of conservation of (mechanical) energy*, and it is why such a vector field is called *conservative*.

By the fundamental theorem of line integrals, we know that a conservative field in a region D must also be path-independent in D. Conversely, if a vector field is path-independent in D, do we know whether it is conservative? If the region D is open and connected, the answer is yes. We say a region D is connected if for any two points in D there exists a line/curve that lies entirely in D and connects the two points. We have the following theorem.

Theorem 4.3.3. *If D is an open and connected region, and a continuous vector field $\mathbf{F} = \langle f, g \rangle$ is path-independent in D, then $\mathbf{F}$ is also conservative in D. That is, there exists a function $\varphi(x, y)$ such that $\mathbf{F} = \nabla \varphi$.*

Proof. We construct such a φ in the following way.

Suppose $A(a, b)$ is a fixed point in D and $P(x, y)$ is any point in D. Since $\mathbf{F} = \langle f, g \rangle$ is path-independent in D, the function

$$\varphi(x, y) = \int_{(a,b)}^{(x,y)} \mathbf{F} \cdot d\mathbf{r}$$

is independent of the path that connects $A(a, b)$ and the point $P(x, y)$. We are going to show that

$$\nabla \varphi(x, y) = \mathbf{F}, \quad \text{or equivalently,} \quad \frac{\partial \varphi}{\partial x} = f \quad \text{and} \quad \frac{\partial \varphi}{\partial y} = g.$$

Since D is open, there exists a disk centered at (x, y) that lies entirely in D. We choose a point $B(x_0, y)$ that is also in this disk, and then $B(x_0, y)$ and $P(x, y)$ are the two endpoints of the horizontal line segment $\overline{BP}$. As shown in Figure 4.9(b), since D is connected, there is a path C_1 in D that connects A and B, and

$$\varphi(x, y) = \int_{(a,b)}^{(x,y)} \mathbf{F} \cdot d\mathbf{r} = \int_{A(a,b)}^{B(x_0,y)} \mathbf{F} \cdot d\mathbf{r} + \int_{B(x_0,y)}^{P(x,y)} \mathbf{F} \cdot d\mathbf{r}.$$

Note that $\int_{A(a,b)}^{B(x_0,y)} \mathbf{F}\cdot d\mathbf{r}$ is independent of x and along the line segment $\overline{BP}$, $dy = 0$. Therefore,

$$\varphi(x,y) = \int_{A(a,b)}^{B(x_0,y)} \mathbf{F}\cdot d\mathbf{r} + \int_{x_0}^{x} f(t,y)dt, \quad \text{where we changed the dummy variable to } t,$$

and

$$\frac{\partial \varphi}{\partial x} = 0 + f(x,y) = f(x,y).$$

In a very similar way, with the aid of a vertical line segment, we could prove

$$\frac{\partial \varphi}{\partial y} = g(x,y).$$

Therefore, $\varphi(x,y)$ is a function such that $\mathbf{F} = \nabla \varphi$, so $\mathbf{F}$ is conservative. □

Note. If we choose the point $A(a,b)$ such that $\overline{AP}$ lies in a disk centered at $P(x,y)$ and lies in D, then by integration along a horizontal line segment followed by a vertical line segment, we would have a nice formula for finding a potential function,

$$\varphi(x,y) = \int_{a}^{x} f(t,b)dt + \int_{b}^{y} g(x,t)dt, \tag{4.7}$$

or by integration along a vertical line segment first followed by a horizontal one, we have

$$\varphi(x,y) = \int_{b}^{y} g(a,t)dt + \int_{a}^{x} f(t,y)dt. \tag{4.8}$$

Now, there are still some questions. For example, how do we know whether a potential function exists? Observing again, if

$$\mathbf{F} = \langle f, g \rangle = \nabla \varphi = \langle \varphi_x, \varphi_y \rangle,$$

then $\varphi_x = f$ and $\varphi_y = g$. If f and g both have continuous partial derivatives, then we would have

$$f_y = \varphi_{xy} = \varphi_{yx} = g_x.$$

This means that if f and g both have continuous partial derivatives, then a necessary condition for it to be a conservative field is

$$\frac{\partial g}{\partial x} = \frac{\partial f}{\partial y}.$$

4.3 The fundamental theorem of line integrals

Remarkably, it turns out this is also a sufficiently condition if the region D is simply connected, and the curve C is simple and closed. This is proved in the next section by using Green's theorem. We now adopt this fact and demonstrate how to find a potential function in two different ways.

Example 4.3.2. Given a vector field $\mathbf{F} = \langle xy^2, x^2y + y \rangle$, find a function $\varphi(x,y)$ such that $\mathbf{F} - \nabla\varphi = \langle xy^2, x^2y + y \rangle$.

Solution. We first note that

$$\frac{\partial(x^2y+y)}{\partial x} = 2xy = \frac{\partial(xy^2)}{\partial y}.$$

So there may exist a potential function φ.

Method 1: If $\mathbf{F} = \nabla\varphi = \langle xy^2, x^2y + y \rangle$, then we must have

$$\frac{\partial\varphi}{\partial x} = xy^2 \quad \text{and} \quad \frac{\partial\varphi}{\partial y} = x^2y + y.$$

From $\frac{\partial\varphi}{\partial x} = xy^2$, we integrate with respect to x to obtain

$$\varphi = \int xy^2 \, dx = \frac{x^2y^2}{2} + C(y).$$

We need to remember that the integration is with respect to x, so the arbitrary constant may be a function of y. To determine $C(y)$, we take the partial derivative with respect to y to obtain

$$\frac{\partial\varphi}{\partial y} = \left(\frac{x^2y^2}{2} + C(y)\right)'_y = x^2y + C'(y); \quad \text{but we also have} \quad \frac{\partial\varphi}{\partial y} = x^2y + y.$$

Thus,

$$x^2y + C'(y) = x^2y + y.$$

So, $C'(y) = y$. Then, $C'(y) = \frac{y^2}{2} + C$, where C is an arbitrary constant. Now we can conclude that

$$\varphi(x,y) = \frac{x^2y^2}{2} + \frac{y^2}{2} + C.$$

Method 2: We can also try either of equation (4.7) or equation (4.8). Then we have

$$\varphi(x,y) = \int_a^x f(t,b)\,dt + \int_b^y g(x,t)\,dt$$

$$= \int_a^x (tb^2)dt + \int_b^y (x^2 t + t)dt = \left(\frac{t^2 b^2}{2}\right)_a^x + \left(\frac{x^2 t^2}{2} + \frac{t^2}{2}\right)_b^y$$

$$= \frac{x^2 b^2}{2} - \frac{a^2 b^2}{2} + \frac{x^2 y^2}{2} + \frac{y^2}{2} - \frac{x^2 b^2}{2} - \frac{b^2}{2}$$

$$= \frac{x^2 y^2}{2} + \frac{y^2}{2} + C.$$

4.4 Green's theorem: circulation-curl form

4.4.1 Positive oriented simple curve and simply connected region

A *simple curve* is a curve which does not intersect itself except possibly at its endpoints. The set $D \subset \mathbb{R}^2$ is *connected* if for any two points in D there exists a line (curve) that lies entirely in D and connects the two points; D is called *simply connected* if for every *simple* (i.e., nonself-intersecting) closed curve C composed of points of D, the region inside of C is also part of D, that is, D has no holes and does not consist of separate parts. In other words, one can continuously shrink any simple closed curve to a point while remaining in the domain. Figure 4.10 shows some such curves and regions. The boundary curve of a region D has a *positive orientation* if, as you walk along the boundary, the region D is on your left-hand side. Figure 4.11 shows positive orientation for two connected regions.

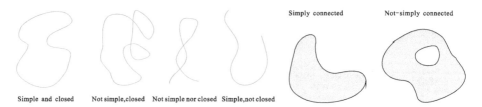

Figure 4.10: Simple curve and simply connected region.

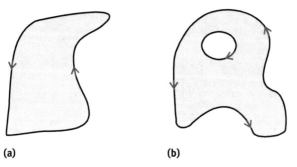

Figure 4.11: Positive orientation.

4.4.2 Circulation around a closed curve

The line integral $\int_C \mathbf{F} \cdot d\mathbf{r}$ along an oriented curve C "adds up" the component of the vector field that is tangent to the curve C. In this sense, the line integral measures how much the vector field is aligned with the curve. If the curve C is a closed curve, then the line integral indicates how much the vector field tends to circulate around the curve C. We call the line integral the "circulation" of $\mathbf{F}$ around C, i. e.,

$$\text{circulation integral} = \oint_C (\mathbf{F} \cdot \mathbf{T}) ds = \oint_C f dx + g dy. \tag{4.9}$$

Note. The symbol "$\oint_C$" is used to indicate a line integral along a closed oriented curve C.

Example 4.4.1. Investigate the circulation integral

$$\oint_C \frac{xdy - ydx}{x^2 + y^2}$$

for any circle C with center at the origin and radius $R > 0$ with counterclockwise orientation.

Solution. Figure 4.12 shows the vector field and the circle. Note that $\mathbf{F} = \langle f, g \rangle$ with $f = -\frac{y}{x^2+y^2}$, $g = \frac{x}{x^2+y^2}$, and C has a parametric form $\mathbf{r}(t) = \langle R\cos t, R\sin t \rangle$. Then

$$\oint_C -\frac{y}{x^2+y^2} dx + \frac{x}{x^2+y^2} dy = \oint_C \left\langle -\frac{y}{x^2+y^2}, \frac{x}{x^2+y^2} \right\rangle \cdot \mathbf{r}'(t) dt$$

$$= \int_0^{2\pi} \left\langle -\frac{R\sin t}{R^2}, \frac{R\cos t}{R^2} \right\rangle \cdot \langle -R\sin t, R\cos t \rangle dt$$

$$= \int_0^{2\pi} (\sin^2 t + \cos^2 t) dt$$

$$= \int_0^{2\pi} 1 dt = 2\pi.$$

4.4.3 Circulation density

We first introduce the concept of *circulation density*. As shown in Figure 4.13, we consider the counterclockwise circulation along the rectangle. We start with writing the following:

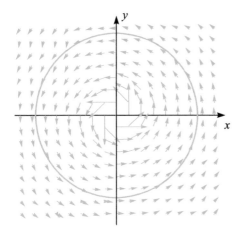

Figure 4.12: Circulation integral, Example 4.4.1.

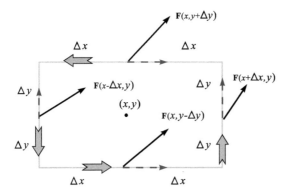

Figure 4.13: Circulation density, the curl.

the bottom circulation is $\mathbf{F}(x, y - \Delta y) \cdot \mathbf{i} \times 2\Delta x = f(x, y - \Delta y) \times 2\Delta x$,
the top circulation is $\mathbf{F}(x, y + \Delta y) \cdot -\mathbf{i} \times 2\Delta x = -f(x, y + \Delta y) \times 2\Delta x$,
the right circulation is $\mathbf{F}(x + \Delta x, y) \cdot \mathbf{j} \times 2\Delta y = g(x + \Delta x, y) \times 2\Delta y$,
the left circulation is $\mathbf{F}(x - \Delta x, y) \cdot -\mathbf{j} \times 2\Delta x = -g(x - \Delta x, y) \times 2\Delta y$.
So the total counterclockwise circulation along the rectangle is

$$f(x, y - \Delta y) \times 2\Delta x - f(x, y + \Delta y) \times 2\Delta x + g(x + \Delta x, y) \times 2\Delta y - g(x - \Delta x, y) \times 2\Delta y,$$

which we can rearrange to obtain

$$(g(x + \Delta x, y) - g(x - \Delta x, y))2\Delta y - (f(x, y + \Delta y) - f(x, y - \Delta y))2\Delta x.$$

The density over the rectangle is, therefore,

$$\frac{(g(x + \Delta x, y) - g(x - \Delta x, y))2\Delta y - (f(x, y + \Delta y) - f(x, y - \Delta y))2\Delta x}{2\Delta x \times 2\Delta y}.$$

Simplifying this, we have
$$\frac{g(x+\Delta x, y) - g(x-\Delta x, y)}{2\Delta x} - \frac{f(x, y+\Delta y) - f(x, y-\Delta y)}{2\Delta y}.$$

Taking the limit as $\Delta x \to 0$ and $\Delta y \to 0$, and noting that $\frac{f(x+h)-f(x-h)}{2h} \to f'(x)$ as $h \to 0$, we obtain the circulation density at (x, y)

$$\frac{\partial g}{\partial x} - \frac{\partial f}{\partial y}. \tag{4.10}$$

Then an integration of $\frac{\partial g}{\partial x} - \frac{\partial f}{\partial y}$ over the region D will give a cumulated circulation effect on the boundary the region D, i.e.,

$$\oint_C f dx + g dy = \iint_D \left(\frac{\partial g}{\partial x} - \frac{\partial f}{\partial y} \right) d\sigma.$$

This is exactly Green's theorem. This is illustrated in Figure 4.14.

The term $\frac{\partial g}{\partial x} - \frac{\partial f}{\partial y}$ is called *the **k**-component of the curl* of the vector field **F**. The curl is a vector relating to the rotational effect of **F** as shown in Figure 4.15. We will discuss the curl vector again in a later section.

4.4.4 Green's theorem: circulation-curl form

Recall the fundamental theorem of calculus

$$\int_a^b f'(x) dx = f(b) - f(a),$$

which says that the integration of the rate of change of a function is related to the value of the function at the boundary points of the integration interval. There is indeed a similar relationship between the integration of rate of change of some functions on the region that is bounded by a closed curve and a line integral along the curve. The remarkable *Green's theorem* is, therefore, sometimes called the *fundamental theorem of calculus for double integrals*, as it reveals the relationship between a double integral over the planar region D and a line integral on the boundary of D.

With the ideas of circulation integral and circulation density, we now state the great Green's theorem and give a partial proof of it.

Theorem 4.4.1 (Green's theorem: circulation or tangential form). *Let L be a piecewise smooth simple closed curve in $\mathbb{R}^2$ having **positive orientation** (the interior is on the left as you travel around L) and the region D inside of L is simply connected. Let $\vec{F} = f\vec{i} + g\vec{j}$ be a vector field for which f and g have continuous partial derivatives in a region containing C and D. Then*

$$\oint_L f(x, y) dx + g(x, y) dy = \iint_D \left(\frac{\partial g}{\partial x} - \frac{\partial f}{\partial y} \right) d\sigma. \tag{4.11}$$

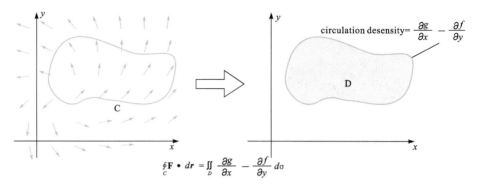

Figure 4.14: Green's theorem: the circulation-curl form.

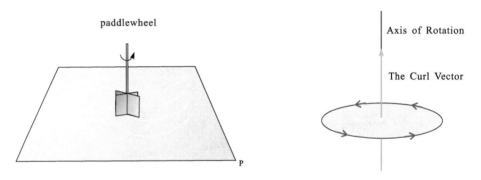

Figure 4.15: The curl vector, axis of rotation.

Verifying Green's theorem

We first give the following example to demonstrate that the two integrals in Green's theorem are equal.

Example 4.4.2. For the vector field $\mathbf{F} = \langle -y, x \rangle$ and the closed curve $x^2 + y^2 = 1$, evaluate both integrals in Green's theorem and check that they are equal.

Solution. Since $f = -y$, $g = x$, and the circle has a parametric representation $\mathbf{r} = \langle \cos t, \sin t \rangle$ with $r'(t) = \langle -\sin t, \cos t \rangle$, the line integral is

$$\oint_C f dx + g dy = \oint_C \langle -y, x \rangle \cdot \mathbf{r}'(t) dt$$

$$= \int_0^{2\pi} \langle -\sin t, \cos t \rangle \langle -\sin t, \cos t \rangle dt$$

$$= \int_0^{2\pi} (\sin^2 t + \cos^2 t) dt = 2\pi.$$

The double integral is

$$\iint_D \left(\frac{\partial g}{\partial x} - \frac{\partial f}{\partial y}\right) d\sigma = \iint_D \left(\frac{\partial x}{\partial x} - \frac{\partial (-y)}{\partial y}\right) d\sigma$$

$$= \iint_D (1 - (-1)) d\sigma = 2 \iint_D 1 d\sigma$$

$$= 2 \cdot A(D) = 2\pi.$$

Therefore, the two integrals in Green's theorem are equal in this example.

Proof of Green's theorem in a simple case

Proof. Suppose D is both a type I and a type II region, as shown in Figure 4.16, and the type I form is

$$D = \{(x, y) | \phi_1(x) \le y \le \phi_2(x), \ a \le x \le b\},$$

where $L_1 : y = \phi_1(x)$ and $L_2 : y = \phi_2(x)$, for $a \le x \le b$, are two curves comprising L (see Figure 4.16(a)). □

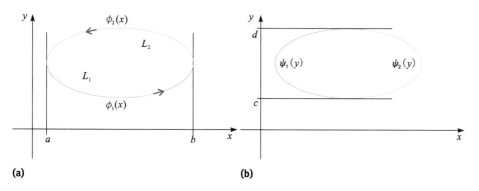

Figure 4.16: Green's theorem: a proof in a simple case.

Since $\frac{\partial f}{\partial y}$ is continuous, by applying the integration formula for double integrals, we have

$$\iint_D \frac{\partial f}{\partial y} d\sigma = \int_a^b \left\{ \int_{\phi_1(x)}^{\phi_2(x)} \frac{\partial f(x, y)}{\partial y} dy \right\} dx$$

$$= \int_a^b \{f(x, \phi_2(x)) - f(x, \phi_1(x))\} dx.$$

On the other hand, the line integral

$$\oint_C f(x,y)dx = \int_{L_1} f(x,y)dx + \int_{L_2} f(x,y)dx$$

$$= \int_a^b f(x, \phi_1(x))dx + \int_b^a f(x, \phi_2(x))dx$$

$$= -\int_a^b \{f(x, \phi_2(x)) - f(x, \phi_1(x))\} dx.$$

So

$$-\iint_D \frac{\partial f}{\partial y} d\sigma = \oint_L f dx.$$

Since D is also a type II region of form $D = \{(x,y) | \psi_1(y) \le x \le \psi_2(y), c \le y \le d\}$, a similar argument leads to a proof that

$$\iint_D \frac{\partial g}{\partial x} d\sigma = \oint_C g dy.$$

By the linearity property of integrals, we obtain the following result:

$$\oint_C f dx + g dy = \iint_D \left(\frac{\partial g}{\partial x} - \frac{\partial f}{\partial y} \right) d\sigma.$$

Note. If a simply connected region is not both type I and type II, then we can partition it into several subregions which are both type I and type II. Green's theorem can then be still proved.

4.4.5 Applications of Green's theorem in circulation-curl form

Determination of a conservative field
Recall that if a vector field $\mathbf{F} = \langle f, g \rangle$ is path-independent in D, this means that for any closed curve C in D we have

$$\oint_C f dx + g dy = 0.$$

Now, from Green's theorem in circulation-curl form, if D is simply connected, we know that $\oint_C f dx + g dy = 0$ for every simple closed curve C in D if and only if $\frac{\partial g}{\partial x} = \frac{\partial f}{\partial y}$ everywhere on D. We summarize properties about a conservative field as follows.

Theorem 4.4.2. *Let $\mathbf{F} = \langle f, g \rangle$ be a continuous vector field defined in a simply connected region $R \subset \mathbb{R}^2$ and f and g both have continuous partial derivatives. Then we have*

$\mathbf{F} = \langle f, g \rangle$ *is conservative*

$\iff$ *there is a function φ such that $\mathbf{F} = \nabla \varphi$ or $d\varphi = f dx + g dy$*

$\iff$ $\int_C f dx + g dy$ *is path-independent for every piecewise smooth C in R*

$\iff$ $\oint_C f dx + g dy = 0$ *for every simple piecewise smooth closed curve C in R*

$\iff$ $\frac{\partial f}{\partial y} = \frac{\partial g}{\partial x}$ *means $\varphi_{xy} = \varphi_{yx}$ throughout R.*

So $\frac{\partial f}{\partial y} = \frac{\partial g}{\partial x}$ is a simple criterion to determine whether or not a vector field is conservative. In this case, we can use the method introduced in Example 4.3.2 to find a potential function. We can also find a potential function for the vector field by the method introduced in the following example.

Example 4.4.3. Given a vector field $\mathbf{F} = \langle e^x + y^2, 2xy \rangle$:
(1) Determine whether $\mathbf{F}$ is conservative.
(2) Find the line integral $\int_C \mathbf{F} \cdot d\mathbf{r}$, where C is the part of the curve $y = \sin x^2$, $0 \le x \le \sqrt{\pi}$.
(3) Find a potential function $\mathbf{F}$ if it exists.

Solution. The graph of the vector field is shown in Figure 4.17(a). We now use the simple criterion derived in the proof of Green's theorem to determine whether it is conservative or not.
(1) Since $\mathbf{F}$ has continuous partial derivatives in $\mathbb{R}^2$ and

$$\frac{\partial g}{\partial x} = \frac{\partial (2xy)}{\partial x} = 2y = \frac{\partial (e^x + y^2)}{\partial y} = \frac{\partial f}{\partial y},$$

this field is conservative.
(2) Since the field is conservative, the line integral $\int_C \mathbf{F} \cdot d\mathbf{r}$ is path-independent. Therefore, we will not use the original path where it is hard to evaluate the integral, and, instead, we try the line segment from $(0, 0)$ to $(\sqrt{\pi}, 0)$, as shown in Figure 4.17(b). Note that along this new path, $y = 0$. We have

$$\int_C \mathbf{F} \cdot d\mathbf{r} = \int_C (e^x + y^2) dx + 2xy dy = \int_0^{\sqrt{\pi}} (e^x + 0^2) dx + 0$$

$$= e^x \big|_{x=0}^{x=\sqrt{\pi}} = e^{\sqrt{\pi}} - 1.$$

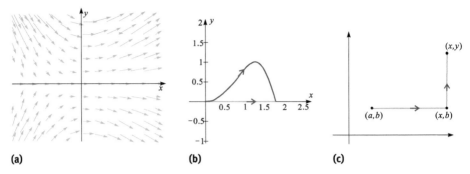

Figure 4.17: Green's theorem, Example 4.4.3.

(3) We could have used equation (4.7) or equation (4.8). To enhance our understanding, we use the idea of path independence again. Assume we are going to evaluate the line integral from $(0,0)$ to any point, say, (u,v). Since the line integral is path-independent in this field, if $\varphi(x,y)$ is a potential function of **F**, then $\int_{(0,0)}^{(u,v)} (e^x + y^2)dx + 2xy\,dy = \varphi(u,v) - \varphi(0,0)$. We now choose the line segments from $(0,0)$ to $(u,0)$ and from $(u,0)$ to (u,v). Note that on the first line segment $y = 0$ and along the second line segment $x = u$ and $dx = 0$. Then

$$\int_{(0,0)}^{(u,v)} (e^x + y^2)dx + 2xy\,dy = \int_{(0,0)}^{(u,0)} (e^x + 0^2)dx + 2x0dy + \int_{(u,0)}^{(u,v)} (e^u + y^2)dx + 2uy\,dy$$

$$= \int_0^u e^x dx + 2u \int_0^v y\,dy$$

$$= e^u - 1 + u(v^2 - 0^2)$$

$$= e^u + uv^2 - 1.$$

This means $\varphi(u,v) - \varphi(0,0) = e^u + uv^2 - 1$, so

$$\varphi(u,v) = e^u + uv^2 - 1 + \varphi(0,0).$$

Since (u,v) is any point in D, we have a potential function $\varphi(x,y) = e^x + xy^2$ (note again that any two potential functions could differ by a constant).

Note. If we switch the letters (u,v) and (x,y), then the equation

$$\varphi(x,y) = \int_0^x f(u,0)du + \int_0^y g(x,v)dv$$

gives a potential function of a conservative field $\mathbf{F} = \langle f, g \rangle$. Of course, the initial point is not necessarily $(0,0)$; it could be any qualifying point, say, (a,b), as shown in Figure 4.17(c). Then

$$\varphi(x,y) = \int_a^x f(u,b)\,du + \int_b^y g(x,v)\,dv \tag{4.12}$$

still gives a potential function of the conservative field $\mathbf{F} = \langle f, g \rangle$. These are the same as equation (4.7) and equation (4.8).

Finding the circulation integral

Green's theorem in circulation-curl form offers us an alternative way to evaluate a line integral by evaluating a double integral. This may save a lot of work, especially when $\frac{\partial g}{\partial x} - \frac{\partial f}{\partial y}$ has a simple expression and $\iint_D (\frac{\partial g}{\partial x} - \frac{\partial f}{\partial y})\,d\sigma$ is easy to compute.

Example 4.4.4. Evaluate $\int_C (x^2 - 2y)\,dx + (3xy + ye^y)\,dy$, where C is the closed triangular curve consisting of the line segments from $(0,0)$ to $(1,0)$, from $(1,0)$ to $(0,1)$, and from $(0,1)$ to $(0,0)$.

Solution. The graph of the vector field and the triangular curve are shown in Figure 4.18. The given line integral could be evaluated directly by integrating along each line segment. But instead we use Green's theorem. Note that the region D enclosed by C is simply connected, and C has positive orientation. If we let $f(x,y) = x^2 - 2y$ and $g(x,y) = 3xy + ye^y$, then we have

$$\int_C (x^2 - 2y)\,dx + (3xy + ye^y)\,dy = \iint_D \left(\frac{\partial g}{\partial x} - \frac{\partial f}{\partial y}\right) d\sigma = \iint_D (3y - (-2))\,d\sigma$$

$$= \iint_D (3y + 2)\,d\sigma = \int_0^1 dx \int_0^{1-x} (3y + 2)\,dy$$

$$= \int_0^1 \left(\frac{3}{2}x^2 - 5x + \frac{7}{2}\right) dx = \frac{3}{2}.$$

Example 4.4.5. Use Green's theorem to find $\int_C (x^2 - y)\,dx - (x + \sin^2 y)\,dy$, where C is the arc of the circle $y = \sqrt{2x - x^2}$ from the point $O(0,0)$ to $A(1,1)$.

Solution. The graph of the vector field and the curve C are shown in Figure 4.19. Let $f = x^2 - y$ and $g = -(x + \sin^2 y)$. Note that $\frac{\partial f}{\partial y} = \frac{\partial g}{\partial x}$ for all x and y. The field is path-independent. We choose another path, O to B, and then B to A, where B is the point $(1,0)$.

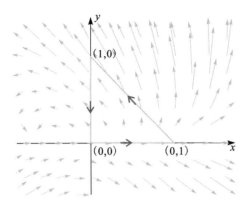

Figure 4.18: Green's theorem, Example 4.4.4.

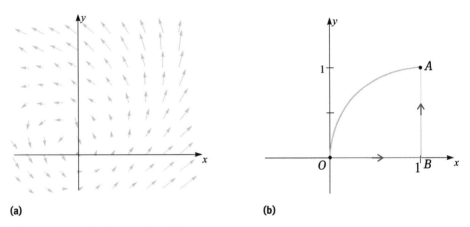

(a)   (b)

Figure 4.19: Green's theorem, Example 4.4.5.

We compute the line integral over the two line segments separately, i. e.,

$$\int_{\overrightarrow{BA}} (x^2 - y)dx - (x + \sin^2 y)dy = -\int_0^1 (1 + \sin^2 y)dy = \frac{1}{4}\sin 2 - \frac{3}{2} \quad \text{and}$$

$$\int_{\overrightarrow{OB}} (x^2 - y)dx - (x + \sin^2 y)dy = \int_0^1 x^2 dx = \frac{1}{3}.$$

Hence,

$$\int_C (x^2 - y)dx - (x + \sin^2 y)dy = \frac{1}{4}\sin 2 - \frac{3}{2} + \left(\frac{1}{3}\right) = \frac{1}{4}\sin 2 - \frac{7}{6}.$$

Finding areas by line integrals

Note that the area of a region D is given by $\iint_D 1 d\sigma$. Therefore, we could evaluate this using Green's theorem if we had $\frac{\partial g}{\partial x} - \frac{\partial f}{\partial y} = 1$. There are many choices of f and g that achieve this, and some examples are (in each case, L is the boundary curve of D with positive orientation)

when $f = 0$, $g = x$:

$$\iint_D 1 d\sigma = \iint_D \left(\frac{\partial g}{\partial x} - \frac{\partial f}{\partial y}\right) d\sigma = \oint_L f dx + g dy = \oint_L x dy, \tag{4.13}$$

when $f = -y$, $g = 0$:

$$\iint_D 1 d\sigma = \iint_D \left(\frac{\partial g}{\partial x} - \frac{\partial f}{\partial y}\right) d\sigma = \oint_L f dx + g dy = -\oint_L y dx, \tag{4.14}$$

when $f = -\frac{y}{2}$, $g = \frac{x}{2}$:

$$\iint_D 1 d\sigma = \iint_D \left(\frac{\partial g}{\partial x} - \frac{\partial f}{\partial y}\right) d\sigma = \oint_L f dx + g dy = \frac{1}{2} \oint_L x dy - y dx. \tag{4.15}$$

Example 4.4.6. Compute, using Green's theorem, the area of the region $D \subset \mathbb{R}^2$ that lies above the parabola $y = x^2$ and below $y = 4$.

Solution. Using the first of these formulas for finding areas and choosing a parameterization of the parabola as $x = t$, $y = t^2$, $-2 < t \le 2$, the area is

$$\oint_L x dy = \int_{-2}^{2} x \frac{dy}{dt} dt = \int_{-2}^{2} t \cdot 2t dt = \frac{32}{3}.$$

Note. As shown in Figure 4.20, the line integral should have included the flat top $y = 4$ of the region, but $dy = 0$ on this line segment, so it adds nothing to the area.

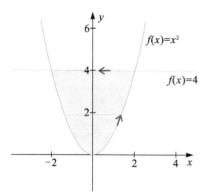

Figure 4.20: Green's theorem, Example 4.4.6.

Example 4.4.7. Find the area of the region D inside the ellipse L given parametrically by $x = a\cos t$, $y = b\sin t$, $0 \leq t \leq 2\pi$.

Solution. Using one of the equations (4.13)–(4.15) for the area A, we obtain

$$A(D) = \iint_D 1 d\sigma = \frac{1}{2}\oint_L xdy - ydx$$

$$= \frac{1}{2}\int_0^{2\pi} (a\cos t)(b\cos t)dt - (b\sin t)(-a\sin t)dt$$

$$= \frac{ab}{2}\int_0^{2\pi} dt = \pi ab.$$

Generalized Green's theorem

What happens if the region D is not simply connected? Let us investigate an example first.

Example 4.4.8. Investigate the line integral

$$\oint_C \frac{xdy - ydx}{x^2 + y^2}$$

for any circle C with center at the origin and radius $R > 0$ and oriented counterclockwise.

Solution. Note that $\mathbf{F} = \langle f, g \rangle$ with $f = -\frac{y}{x^2+y^2}$ and $g = \frac{x}{x^2+y^2}$. We compute

$$\frac{\partial f}{\partial y} = \left(-\frac{y}{x^2+y^2}\right)'_y = -\frac{(y)'_y(x^2+y^2) - (y)(x^2+y^2)'_y}{(x^2+y^2)^2} = \frac{y^2 - x^2}{(x^2+y^2)^2},$$

$$\frac{\partial g}{\partial x} = \left(\frac{x}{x^2+y^2}\right)'_x = -\frac{(x)'_x(x^2+y^2) - (x)(x^2+y^2)'_x}{(x^2+y^2)^2} = \frac{y^2 - x^2}{(x^2+y^2)^2}.$$

Therefore,

$$\frac{\partial f}{\partial y} = \frac{\partial g}{\partial x}.$$

Applying Green's theorem, the line integral would be 0! However, as seen in Example 4.4.1, the value is not 0! Why is this? The answer is that $\mathbf{F}$ is undefined at $(0,0)$, so $\mathbf{F}$ is not continuous on any region containing $(0,0)$ (in this case, the region is not simply connected).

Theorem 4.4.3 (Generalized Green's theorem). *Suppose the region $D \subset \mathbb{R}^2$ lies between two simple, closed, piecewise smooth curves L_1 and L_2, where L_1 is completely contained within L_2 (the curves can*

4.4 Green's theorem: circulation-curl form

intersect but not cross each other). Let $f(x,y)$ and $g(x,y)$ be defined on D and have continuous first partial derivatives on D. If L_2 and L_1 have positive orientation, then

$$\iint_D \left(\frac{\partial g}{\partial x} - \frac{\partial f}{\partial y}\right) dxdy = \int_{L_1} (fdx + gdy) + \int_{L_2} (fdx + gdy). \tag{4.16}$$

Proof. A proof can be developed by combining L_1 and L_2 into one positively oriented curve L that follows L_2, then follows a line segment S joining L_2 to L_1, continuing along L_1, and finally returns to L_2 over the line segment S but in the opposite direction. Applying Green's theorem to this curve L will give the generalized result. Figure 4.21 illustrates this idea. □

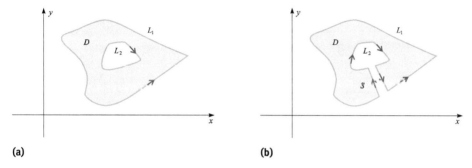

(a)   (b)

Figure 4.21: Generalized Green's theorem, the region is **not** simply connected.

Example 4.4.9. Show that

$$\oint_C \frac{xdy - ydx}{x^2 + y^2} = 2\pi$$

for every positively oriented simple closed curve C that encloses the origin. Note that $f = -\frac{y}{x^2+y^2}$ and $g = \frac{x}{x^2+y^2}$ are not defined at the origin. So, Green's theorem cannot be applied directly.

Solution. Since C is an arbitrary and unknown closed path, it is difficult to compute the integral directly. Hence, let us first consider a clockwise circle C' with its center at the origin and radius a, where a is chosen to be small enough that C' lies inside C, as shown in Figure 4.22. If D is the region enclosed between C and C' (D does not contain the origin), then using the generalized Green's theorem (4.16) with $f = -\frac{y}{x^2+y^2}$ and $g = \frac{x}{x^2+y^2}$ gives

$$\oint_C fdx + gdy + \oint_{C'} fdx + gdy = \iint_D \left(\frac{\partial g}{\partial x} - \frac{\partial f}{\partial y}\right) d\sigma$$

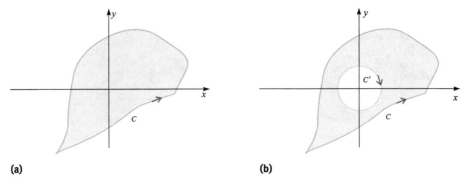

Figure 4.22: Generalized Green's theorem, the region is **not** simply connected, Example 4.4.9.

$$= \iint_D \left(\frac{y^2 - x^2}{(x^2 + y^2)^2} - \frac{y^2 - x^2}{(x^2 + y^2)^2} \right) d\sigma$$
$$= 0.$$

Therefore,

$$\oint_C f dx + g dy = -\oint_{C'} f dx + g dy$$
$$= \oint_{-C'} f dx + g dy,$$

where $-C'$ is the circle in the counterclockwise orientation. We now easily compute this last integral using the parameterization of $-C'$ given by $\mathbf{r}(t) = a\cos t \,\mathbf{i} + a\sin t \,\mathbf{j}$, $0 \le t \le 2\pi$. Thus,

$$\int_C f dx + g dy = \int_{-C'} f dx + g dy = \int_0^{2\pi} \langle f(\mathbf{r}(t)), g(\mathbf{r}(t)) \rangle \frac{d\mathbf{r}(t)}{dt} dt$$
$$= \int_0^{2\pi} \left\langle \frac{-a\sin t}{a^2 \cos^2 t + a^2 \sin^2 t}, \frac{a\cos t}{a^2 \cos^2 t + a^2 \sin^2 t} \right\rangle \cdot \langle -a\sin t, a\cos t \rangle dt$$
$$= \int_0^{2\pi} dt = 2\pi.$$

Example 4.4.10. Evaluate

$$\oint_L \frac{x dy - y dx}{x^2 + y^2},$$

where L is a piecewise smooth, simple closed curve that does not contain $(0, 0)$ on L or in its interior.

Solution. From the previous example the value is 2π when the origin is inside of L. If the origin is not inside of L or on L, then the answer is 0 by applying Green's theorem.

4.5 Green's theorem: flux-divergence form

4.5.1 Flux

Suppose instead we are interested in how much vector field is pointing outward of the given closed simple curve (in the normal direction). If the vector field models fluid flow, then the question is equivalent to finding the rate (mass per time) at which the fluid is flowing out of the region through the closed curve. This requires us to resolve the vector field to the outward normal vector direction at each point. Then we have

$$\text{flux integral} = \int_C (\mathbf{F} \cdot \mathbf{N}) ds. \tag{4.17}$$

But how do we find $\mathbf{N}$? Well, as shown in Figure 4.23(b) we can define the outward unit normal vector

$$\mathbf{N} = \mathbf{T} \times \mathbf{k}$$
$$= \frac{1}{|\mathbf{r}'(t)|} \frac{d\mathbf{r}}{dt} \times \mathbf{k} = \frac{1}{|\mathbf{r}'(t)|} \left\langle \frac{dx}{dt}, \frac{dy}{dt} \right\rangle \times \mathbf{k}$$
$$= \frac{1}{|\mathbf{r}'(t)|} \begin{vmatrix} \mathbf{i} & \mathbf{j} & \mathbf{k} \\ \frac{dx}{dt} & \frac{dy}{dt} & 0 \\ 0 & 0 & 1 \end{vmatrix}$$
$$= \frac{1}{|\mathbf{r}'(t)|} \left(\frac{dy}{dt} \mathbf{i} - \frac{dx}{dt} \mathbf{j} \right) = \frac{1}{|\mathbf{r}'(t)|} \left\langle \frac{dy}{dt}, -\frac{dx}{dt} \right\rangle.$$

So

$$\int_C (\mathbf{F} \cdot \mathbf{N}) ds = \int_C \langle f, g \rangle \cdot \frac{1}{|\mathbf{r}'(t)|} \left\langle \frac{dy}{dt}, -\frac{dx}{dt} \right\rangle ds = \int_C \langle f, g \rangle \cdot \left\langle \frac{dy}{dt}, -\frac{dx}{dt} \right\rangle dt.$$

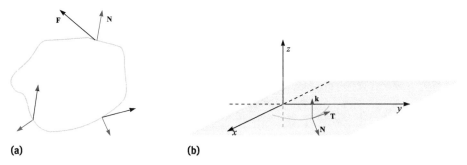

(a) (b)

Figure 4.23: Flux, outward normal vector.

Simplifying this, we obtain

$$\text{flux integral} \int_C (\mathbf{F} \cdot \mathbf{N})ds = \int_C f dy - g dx. \tag{4.18}$$

Example 4.5.1. Find the **flux** of the velocity flow $\mathbf{v}(x,y) = x\mathbf{i} + xy\mathbf{j}$ (cm/s) in two dimensions out of the circular region $C: x^2 + y^2 \leq 4$ for a fluid with constant density δ g/cm^2.

Solution. Choose the parametric form for C: $\mathbf{r}(t) = (2\cos t\, \mathbf{i} + 2\sin t\mathbf{j})$, $0 \leq t \leq 2\pi$. The flux is

$$\int_C \delta(\mathbf{v} \cdot \mathbf{N}) ds$$

$$= \delta \int_C x dy - xy dx = \delta \int_0^{2\pi} (2\cos t) d(2\sin t) - (4\cos t \sin t) d(2\cos t)$$

$$= 2\delta \int_0^{2\pi} (2\cos^2 t + 8\cos t \sin^2 t) dt$$

$$= 4\delta\pi \text{ g/s}.$$

4.5.2 Flux density – divergence

Similar to the density of circulation, we can define the density of flux, which is also called the divergence. The box shown in Figure 4.24 has length $2\Delta x$ and width $2\Delta y$.

Then, the flux crossing the
bottom is $\mathbf{F} \cdot (-\mathbf{j})2\Delta x = -g(x, y - \Delta y)2\Delta x$,
top is $(\mathbf{F} \cdot \mathbf{j})2\Delta x = g(x, y + \Delta y)2\Delta x$,

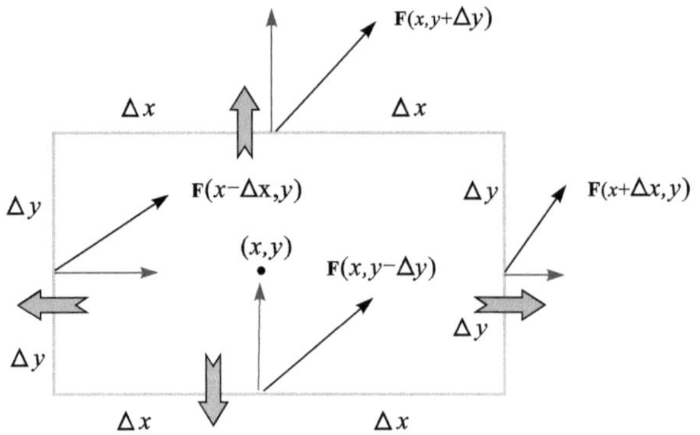

Figure 4.24: Density of flux, the divergence.

right is $(\mathbf{F} \cdot \mathbf{i})2\Delta y = f(x + \Delta x, y)2\Delta y$,
left is $\mathbf{F} \cdot (-\mathbf{i})2\Delta y = -f(x - \Delta x, y)2\Delta y$.
So the total flux is

$$g(x, y + \Delta y)2\Delta x - g(x, y - \Delta y)2\Delta x + f(x + \Delta x, y)2\Delta y - f(x - \Delta x, y)2\Delta y.$$

The density is, therefore,

$$\frac{g(x, y + \Delta y)2\Delta x - g(x, y - \Delta y)2\Delta x + f(x + \Delta x, y)2\Delta y - f(x - \Delta x, y)2\Delta y}{2\Delta x \times 2\Delta y},$$

which simplifies to

$$\frac{f(x + \Delta x, y) - f(x - \Delta x, y)}{2\Delta x} + \frac{g(x, y + \Delta y) - g(x - \Delta y)}{2\Delta y}.$$

Taking the limit gives

$$\frac{\partial f}{\partial x} + \frac{\partial g}{\partial y}.$$

The density is given a special name, the *divergence*, denoted by Div(**F**). So the divergence of a two-dimensional vector field $\mathbf{F} = \langle f, g \rangle$ is

$$\mathrm{Div}(\mathbf{F}) = \frac{\partial f}{\partial x} + \frac{\partial g}{\partial y},$$

or in vector form

$$\mathrm{Div}(\mathbf{F}) = \nabla \cdot \mathbf{F}.$$

4.5.3 The divergence-flux form of Green's theorem

With the ideas of the flux integral and the flux density (divergence), we can expect that the flux integral along a simple closed curve C is equal to the double integral of the flux density over the simply connected region enclosed by C. This is indeed true. We now state Green's theorem in the divergence-flux form.

Theorem 4.5.1 (Green's theorem: flux-divergence form). *Let L be a piecewise smooth simple closed curve in $\mathbb{R}^2$ which has **positive orientation** (the interior is on the left as you travel round L), and the region D inside of L is simply connected. Let $\vec{F} = f\vec{i} + g\vec{j}$ be a vector field for which f and g have continuous partial derivatives in a region containing C and D. Then*

$$\oint_L f(x,y)dy - g(x,y)dx = \iint_D \left(\frac{\partial f}{\partial x} + \frac{\partial g}{\partial y} \right) d\sigma, \qquad (4.19)$$

or, in vector form,

$$\oint_L (\mathbf{F} \cdot \mathbf{N})ds = \iint_D (\nabla \cdot \mathbf{F}) d\sigma. \qquad (4.20)$$

Proof. This is proved by using Green's theorem in the circulation-curl form. The flux integral

$$\oint_L (\mathbf{F} \cdot \mathbf{N})\,ds = \oint_L f\,dy - g\,dx \tag{4.21}$$

can be rearranged as

$$\oint_L (\mathbf{F} \cdot \mathbf{N})\,ds = \oint_L -g\,dx + f\,dy. \tag{4.22}$$

Compared with the circulation integral, this might be confusing because f and g are in different positions now. It may be helpful to memorize Green's theorem writing it as

$$\oint_L \clubsuit\,dx + \heartsuit\,dy = \iint_D \left(\frac{\partial \heartsuit}{\partial x} - \frac{\partial \clubsuit}{\partial y} \right) d\sigma. \tag{4.23}$$

Applying Green's theorem in the circulation-curl form to the flux integral, we obtain the flux-divergence form of Green's theorem

$$\oint_L f\,dy - g\,dx = \oint_L -g\,dx + f\,dy = \iint_D \left(\frac{\partial f}{\partial x} + \frac{\partial g}{\partial y} \right) d\sigma. \tag{4.24}$$

$\square$

Note. Recalling that the flux through C measures vector field "flowing" out of the closed curve, there must be some "source" in D. The term $\frac{\partial f}{\partial x} + \frac{\partial g}{\partial y}$ is the *divergence* of the vector field $\mathbf{F}$ in D. If at a point P, the divergence is positive, then it is a source. If at a point P, the divergence is negative, then it is a sink.

! **Example 4.5.2.** The graph of the vector field $\mathbf{F} = \langle x^2 - y^2, x - 3y \rangle$ and the curve $x^2 + y^2 = 1$ are shown in Figure 4.25(a).
1. Make a guess whether the flux along C is positive or negative.
2. Evaluate the two integrals in Green's theorem in flux-divergence form. Check to see if they are equal.

Solution.
1. From Figure 4.25(a), it looks like more vector field goes into the circle than comes out of the circle, so the flux might be negative.
2. We compute the flux integral

$$\oint_L (\mathbf{F} \cdot \mathbf{N})\,ds = \oint_L f\,dy - g\,dx$$
$$= \oint_L (x^2 - y^2)\,dy - (x - 3y)\,dx$$

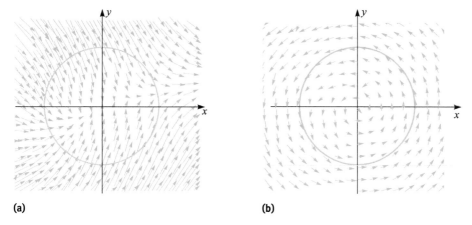

(a) (b)

Figure 4.25: Flux-divergence, Example 4.5.2 and Example 4.6.1.

$$= \int_0^{2\pi} (\cos^2 t - \sin^2 t) d(\sin t) - (\cos t - 3\sin t) d(\cos t)$$

$$= \int_0^{2\pi} (\cos^3 t - \cos t \sin^2 t + \cos t \sin t - 3\sin^2 t) dt = -3\pi.$$

We now compute the double integral

$$\iint_D \left(\frac{\partial f}{\partial x} + \frac{\partial g}{\partial y} \right) d\sigma = \iint_D \left(\frac{\partial (x^2 - y^2)}{\partial x} + \frac{\partial (x - 3y)}{\partial y} \right) d\sigma$$

$$= \iint_D (2x - 3) d\sigma = \iint_D 2x d\sigma - 3 \iint_D 1 d\sigma$$

$$= 0 - 3 \times \pi \times 1^2 = -3\pi.$$

Indeed, the two integrals have the same value.

4.6 Source-free vector fields

In Green's theorem in flux-divergence form, we have seen the term $\frac{\partial f}{\partial x} + \frac{\partial g}{\partial y}$, which measures the "source" of the vector field. Green's theorem in flux-divergence form in vector form is written as

$$\oint_L (\mathbf{F} \cdot \mathbf{N}) ds = \iint_D \nabla \cdot \mathbf{F} d\sigma.$$

If the divergence of a vector field is 0 at every point in a region D, then the field is called *source-free*. A point with positive divergence is called a *source*, and a point with negative divergence is called a *sink*. For a source-free vector field $\mathbf{F} = \langle f, g \rangle$, we define the stream function ψ, which satisfies

$$\frac{\partial \psi}{\partial y} = f \quad \text{and} \quad \frac{\partial \psi}{\partial x} = -g.$$

Since $f_x + g_y = 0$, or equivalently $f_x = -g_y$, we have

$$\frac{\partial^2 \psi}{\partial y \partial x} = \frac{\partial^2 \psi}{\partial x \partial y} \quad \text{or} \quad \psi_{yx} = \psi_{xy}.$$

Similar to a conservative field, for a source-free field, under suitable conditions, we have the following results:

$$\mathbf{F} = \langle f, g \rangle \text{ is source-free} \iff \frac{\partial f}{\partial x} + \frac{\partial g}{\partial y} = 0$$

$$\iff \int_C (\mathbf{F} \cdot \mathbf{N}) ds \text{ is path-independent}$$

$$\iff \oint_C (\mathbf{F} \cdot \mathbf{N}) ds = 0 \text{ for any simple closed curve}$$

$$\iff \text{there is a stream function } \psi \text{ such that}$$

$$\int_A^B (\mathbf{F} \cdot \mathbf{N}) ds = \psi(B) - \psi(A).$$

Example 4.6.1. Compute the divergence of the vector field $\mathbf{F} = \langle -y, x \rangle$. Does it have a stream function? If so, find one.

Solution. The graph of the vector field is shown in Figure 4.25(b). Since $f = -y$ and $g = x$,

$$\frac{\partial f}{\partial x} + \frac{\partial g}{\partial y} = \frac{\partial(-y)}{\partial x} + \frac{\partial(x)}{\partial y} = 0,$$

so it is a source-free field. Thus, there exists a stream function ψ. Since

$$\frac{\partial \psi}{\partial y} = f = -y, \quad \text{we have } \psi = \int (-y) dy = -\frac{y^2}{2} + C(x).$$

To find $C(x)$, we differentiate with respect to x to obtain

$$\frac{\partial \psi}{\partial x} = \left(-\frac{y^2}{2} + C(x)\right)'_x = C'(x), \quad \text{but } \frac{\partial \psi}{\partial x} = -g = -x.$$

So, $C'(x) = -x$. Thus, $C(x) = -\frac{x^2}{2} + C$. A family of stream functions for this vector field is

$$\psi(x, y) = -\frac{y^2}{2} - \frac{x^2}{2} + C.$$

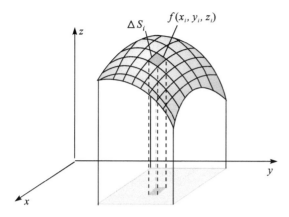

Figure 4.26: Surface integral.

4.7 Surface integral with respect to surface area

Suppose we have a thin metal shell that we can represent as a surface in $\mathbb{R}^3$. If the density of the shell is variable, how can we find its total mass? Sure, we can again employ the idea of elements, and then form an integral that we may be able to evaluate. To find, approximately, the mass of the metal shell, we divide the surface into small *patches* (subregions), as shown in Figure 4.26. Then add up the areas ΔS_i of each patch multiplied by the approximate density $f(x_i, y_i, z_i)$ of the patch. The approximation improves as the size of the patches decreases. This is exactly the process used in integration.

Definition 4.7.1. Let S be a bounded surface in space and let $f(x,y,z)$ be a bounded function defined on S. For any subdivision $\{S_k\}$ of S into n patches (small subregions), and arbitrarily choose a point $(x_k, y_k, z_k) \in S_k$ in each patch. Form the Riemann sum

$$\sum_{k=1}^{n} f(x_k, y_k, z_k) \Delta S_k,$$

where ΔS_k is the area of subregion S_k. If the limit of this Riemann sum exists as $n \to \infty$ and $\max |\Delta S_k| \to 0$ for all possible subdivisions and choices of points (x_k, y_k, z_k), then this value is the **surface integral** of $f(x,y,z)$ over the surface S, written

$$\iint_S f dS = \iint_S f(x,y,z) dS = \lim_{\max |\Delta S_k| \to 0,\ n \to \infty} \sum_{k=1}^{n} f(x_k, y_k, z_k) \Delta S_k. \tag{4.25}$$

Properties of the surface integral

The surface integral has the same basic properties as all integrals. Assuming that k, $l \in \mathbb{R}$ are two constants and that all the integrals exist:
1. $\iint_S 1 dS$ = the area of the surface S, $A(S)$,

2. $\iint_S (kf(x,y,z) + lg(x,y,z))dS = k\iint_S f(x,y,z)dS + l\iint_S g(x,y,z)dS$ (linearity),
3. $\iint_S f(x,y,z)dS + \iint_{S'} f(x,y,z)dS = \iint_{S+S'} f(x,y,z)dS$ (additivity),
4. if $f(x,y,z)$ is continuous, then there exists a point (a,b,c) such that
$$\iint_S f(x,y,z)dS = f(a,b,c) \cdot A(S).$$

Example 4.7.1. Suppose S is the surface of the unit ball. Evaluate the surface integral
$$\iint_S 2x\ln(1+y^2+z^2) - 3(x^2+y^2+z^2)dS.$$

Solution. By the properties of surface integrals,
$$\iint_S 2x\ln(1+y^2+z^2) - 3(x^2+y^2+z^2)dS$$
$$= 2\iint_S x\ln(1+y^2+z^2)dS - 3\iint_S (x^2+y^2+z^2)dS \quad \text{(linearity)}$$
$$= 2\times 0 - 3\iint_S (x^2+y^2+z^2)dS \quad \text{(symmetry)}$$
$$= 0 - 3\iint_S 1 dS \quad \text{(since all points on } S \text{ satisfy } x^2+y^2+z^2=1)$$
$$= -3 \times 4\pi 1^2 = -12\pi.$$

Surface integrals: surface is described by parametric equations

Note that in Chapter 3, for a parameterized surface $\mathbf{r}(u,v) = \langle x(u,v), y(u,v), z(u,v)\rangle$, we approximated ΔS_k by $|\mathbf{r}_u \times \mathbf{r}_v|\Delta u \Delta v$. Hence, equation (4.25) for the surface integral becomes
$$\iint_S f(x,y,z)dS = \lim_{\Delta S_k \to 0,\, n\to\infty} \sum f(x_k, y_k, z_k)\Delta S_k$$
$$= \lim_{\Delta S_k \to 0,\, n\to\infty} \sum f(x(u_k,v_k), y(u_k,v_k), z(u_k,v_k))|\mathbf{r}_u \times \mathbf{r}_v|\Delta u \Delta v,$$

and this is now a standard Riemann sum for a double integral, resulting in the final formula
$$\iint_S f(x,y,z)dS = \iint_{D_{uv}} f(x(u,v), y(u,v), z(u,v))|\mathbf{r}_u \times \mathbf{r}_v|dudv. \quad (4.26)$$

Example 4.7.2. Find $\iint_S \frac{1}{z}dS$, where S is the part of the sphere $x^2+y^2+z^2 = a^2$ that lies above the plane $z = h$ and h is a constant satisfying $0 < h \le a$.

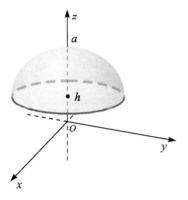

Figure 4.27: Surface integral, Example 4.7.2.

Solution. Figure 4.27 shows the surface S.
The surface S has a parametric description

$$\mathbf{r}(u,v) = \langle a\sin u \cos v, a\sin u \sin v, a\cos u\rangle,$$

where $0 \le v \le 2\pi$ and $0 \le u \le \cos^{-1}\frac{h}{a}$. So,

$$\mathbf{r}_u \times \mathbf{r}_v = \begin{vmatrix} \mathbf{i} & \mathbf{j} & \mathbf{k} \\ a\cos u \cos v & a\cos u \sin v & -a\sin u \\ -a\sin u \sin v & a\sin u \cos v & 0 \end{vmatrix}$$
$$= a^2 \sin^2 u \cos v \mathbf{i} + (a^2 \sin^2 u \sin v)\mathbf{j} + (a^2 \sin u \cos u)\mathbf{k},$$

and $|\mathbf{r}_u \times \mathbf{r}_v| = \sqrt{(a^2 \sin^2 u \cos v)^2 + (a^2 \sin^2 u \sin v)^2 + (a^2 \sin u \cos u)^2} = a^2 \sin u$. Thus,

$$\iint_S \frac{1}{z} dS = \iint_{D_{uv}} \frac{1}{a\cos u} a^2 \sin u\, du\, dv = a\int_0^{2\pi} dv \int_0^{\cos^{-1}\frac{h}{a}} \frac{\sin u}{\cos u} du$$
$$= 2\pi a \times (-\ln|\cos u|)\Big|_0^{\cos^{-1}\frac{h}{a}} = -2\pi a \ln\frac{h}{a} = 2\pi a \ln\frac{a}{h}.$$

Surface integrals: surface is given by an explicit equation $z = z(x,y)$

Also in Chapter 3, we have seen that if the surface is given by an explicit function $z = z(x,y)$, then the parametric description $\mathbf{r}(x,y) = \langle x, y, z(x,y)\rangle$ gives $dS = |\mathbf{r}_x \times \mathbf{r}_y| = \sqrt{1 + z_x^2 + z_y^2}\, dx\, dy$ and

$$\iint_S f(x,y,z)\, dS = \iint_{D_{xy}} f(x,y,z(x,y))\sqrt{1 + z_x^2 + z_y^2}\, dx\, dy. \qquad (4.27)$$

In fact, we can use all the ways that we have developed for the surface area element dS in Chapter 3 to evaluate a surface integral, so

$$\iint_S f(x,y,z)\,dS = \iint_{D_{yz}} f(x(y,z),y,z)\sqrt{1+x_y^2+x_z^2}\,dydz, \qquad (4.28)$$

$$\iint_S f(x,y,z)\,dS = \iint_{D_{xz}} f(x,y(x,z),z)\sqrt{1+y_x^2+y_z^2}\,dxdz. \qquad (4.29)$$

Example 4.7.3. Let S be the closed surface formed by S_1, the portion of the cone with equation $z = \sqrt{x^2+y^2}$ that lies below the plane $z = 1$, and S_2, the circular top of the cone given by $z = 1$, $x^2+y^2 \le 1$. Let f be defined on S by $f(x,y,z) = x^2+y^2$. Compute the area of S and evaluate $\iint_S f(x,y,z)\,dS$.

Solution. Figure 4.28 shows the cone and its circular top. The area of S does not need any integration, since standard formulas give the lateral surface area of the cone as $\pi\sqrt{2}$ and the area of the disk as π, so S has area $\pi(\sqrt{2}+1)$. We compute the integral of f by evaluating two surface integrals, i. e.,

$$\iint_S f(x,y,z)\,dS = \iint_{S_1} f(x,y,z)\,dS + \iint_{S_2} f(x,y,z)\,dS.$$

For the first integral, the projection of S_1 onto the xy-plane is $D: x^2+y^2 \le 1$, and since $z = \sqrt{x^2+y^2}$, the surface integral becomes

$$\iint_{S_1} f(x,y,z)\,dS = \iint_D (x^2+y^2)\sqrt{1+\left(\frac{\partial z}{\partial x}\right)^2+\left(\frac{\partial z}{\partial y}\right)^2}\,d\sigma$$

$$= \iint_D (x^2+y^2)\sqrt{1+\left(\frac{x}{\sqrt{x^2+y^2}}\right)^2+\left(\frac{y}{\sqrt{x^2+y^2}}\right)^2}\,d\sigma$$

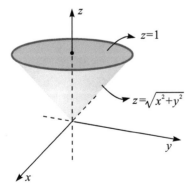

Figure 4.28: Surface integral, Example 4.7.3.

$$= \iint_D (x^2 + y^2)\sqrt{2}\,d\sigma.$$

Converting to polar coordinates over the region $D' : 0 \le r \le 1, 0 \le \theta \le 2\pi$ changes this to

$$\iint_{D'} (r^2)\sqrt{2}\,r\,dr\,d\theta = \sqrt{2}\int_0^{2\pi} d\theta \int_0^1 r^3\,dr = \frac{\sqrt{2}\pi}{2}.$$

For the second integral the surface S_2 is $z = 1$, so $z_x = z_y = 0$, and the domain D is the same as that of the first integral. Hence,

$$\iint_{S_2} f(x,y,z)\,dS = \iint_D (x^2 + y^2)\sqrt{1 + \left(\frac{\partial z}{\partial x}\right)^2 + \left(\frac{\partial z}{\partial y}\right)^2}\,d\sigma$$

$$= \iint_D (x^2 + y^2)\,d\sigma = \frac{\pi}{2}.$$

The value of this integral is obtained easily because it has exactly the same integrand as the first integral above except for a factor of $\sqrt{2}$. Hence, the complete integral of f is computed as

$$\iint_{S_1} (x^2 + y^2)\,dS + \iint_{S_2} (x^2 + y^2)\,dS = \frac{\sqrt{2}\pi}{2} + \frac{\pi}{2} = \frac{\sqrt{2}+1}{2}\pi.$$

4.8 Surface integrals of vector fields

4.8.1 Orientable surfaces

We have seen the line integral of a two-dimensional vector field, both for finding circulation and for finding flux. Now we discuss flux for a three-dimensional vector field. As was the case for flux in two dimensions, we first need to orient a surface so that we know which direction we are talking about. The surfaces we have encountered so far have two sides, as shown in Figure 4.29. Such surfaces are orientable. However, some surfaces are one-sided and are not orientable. For example, the *Moebius strip* is a one sided surface (Figure 4.30(a)). It is formed by taking a long and narrow strip of paper and joining the ends together after giving a half-twist to one end. Any point on the Moebius strip can be joined to any other point by a path that stays on the surface of the paper (the path does not go near the edge), showing that it really is one-sided. Consequently, on a Moebius strip it is not possible to define a unique unit normal vector, perpendicular to the surface, that changes continuously along any curve. For example suppose the unit normal $\mathbf{N}$ at any point is defined to be in the direction away from the

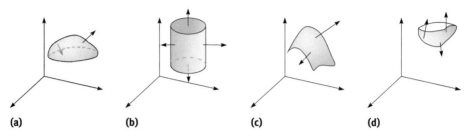

Figure 4.29: Orientable surfaces.

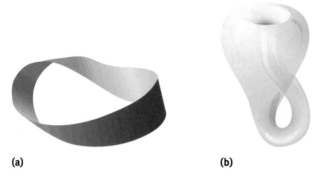

Figure 4.30: Nonorientable surfaces: Moebius strip and Klein bottle.

paper. Following a path from a point P with unit normal **N** around to the point P' on the other side of the paper from P will give a normal in exactly the opposite direction to **N**. This violates the continuity of the normal, since P and P' are essentially the same point. Another example for a nonorientable surface is the famous Klein bottle (Figure 4.30(b)). In the sequel, we only consider orientable surfaces which we can orient either upward or downward, outward or inward, leftward or rightward, and so forth. We now give the definition of an orientable surface.

Definition 4.8.1 (Orientable surface). Let S be a surface in $\mathbb{R}^3$. If at each point (x,y,z) of S we can assign a unit normal $\vec{N} = \vec{N}(x,y,z)$ that changes continuously along any curve on S, then we say that S is an orientable surface. Once $\vec{N}$ is defined for a surface S, the function $\vec{N}$ defines an **orientation** on S, and S is said to be **oriented**.

4.8.2 Flux integral $\iint_S (\mathbf{F} \cdot \mathbf{N})\, dS$

If $\mathbf{F}(x,y,z)$ is a vector field (think of it as the flow of a fluid at each point in space measured in mass/time) and **N** is the unit normal of an orientable surface, then at each point of the surface, $\mathbf{F} \cdot \mathbf{N}$ is the component of **F** perpendicular to S, as shown in Figure 4.31. Hence, we have the following definition of the flux.

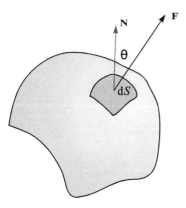

Figure 4.31: Flux integral.

Definition 4.8.2 (Flux of a three-dimensional vector field). If a surface S is oriented with unit normal vector $\vec{N}$, then the surface integral $\iint_S \vec{F} \cdot \vec{N} dS$ is called the flux of $\vec{F}$ across S (in the direction defined by $\vec{N}$); $\iint_S \vec{F} \cdot \vec{N} dS$ is also called the integral of the vector field $\vec{F}$ over S.

In particular, if $\mathbf{F}(x, y, z)$ is the velocity vector for the flow of a fluid through a region of space and the density function of the fluid at point (x, y, z) is $\delta(x, y, z) = 1$, then the flux element through a surface element ΔS in a given direction (or mass of fluid per unit time across ΔS) is $|\mathbf{F}| \cos\theta \Delta S$, where θ is the angle between $\mathbf{F}$ and $\mathbf{N}$ (the normal vector pointing in the given direction). Therefore, $|\mathbf{F}| \cos\theta \Delta S = |\mathbf{F}| \cos\theta |\mathbf{N}| \Delta S = (\mathbf{F} \cdot \mathbf{N}) \Delta S$, and the integration $\iint_S (\mathbf{F} \cdot \mathbf{N}) dS$ gives the total flux across S in the given direction. This is, indeed, a surface integral of the form already defined in equation (4.25), so it has the same properties (for example, linear and additive across two regions S_1 and S_2). In addition, we have

$$\iint_S (\mathbf{F} \cdot \mathbf{N}) dS = -\iint_{-S} (\mathbf{F} \cdot \mathbf{N}) dS, \tag{4.30}$$

where $-S$ denotes the negative orientation of the surface S.

If we denote $\mathbf{N} dS$ by $d\mathbf{S}$, then

$$\iint_S (\mathbf{F} \cdot \mathbf{N}) dS = \iint_S \mathbf{F} \cdot d\mathbf{S}. \tag{4.31}$$

If S is closed, we also adopt the notation $\oiint_S (\mathbf{F} \cdot \mathbf{N}) dS$, with a circle in the integral sign.

Example 4.8.1. Given the vector field $\mathbf{F} = (x^2 - \sin y^2 + z)\mathbf{i} - y\mathbf{j} + (z + z^2)\mathbf{k}$, find the flux out of the top and bottom faces of the cube

$$D = \{(x, y, z) | 0 \le x \le 2,\ 0 \le y \le 2,\ 0 \le z \le 2\}.$$

Solution. Since the flux is out of two faces, S_1, the top side of D, and S_2, the bottom side of D, and we orient S_1 upward and S_2 downward so that we will find outward flux leaving the cube, we choose $\mathbf{N}_1$ to be $\langle 0,0,1 \rangle$ and we choose $\mathbf{N}_2$ to be $\langle 0,0,-1 \rangle$. Then,

$$\iint_{S_1} (\mathbf{F} \cdot \mathbf{N}) dS = \iint_{S_1} \langle x^2 - \sin y^2 + z, -y, z + z^2 \rangle \cdot \langle 0,0,1 \rangle dS = \iint_{S_1} (z + z^2) dS$$

$$= \iint_{S_1} (2 + 2^2) dS \quad \text{(since the top side } z = 2\text{)}$$

$$= \iint_{S_1} (6) dS = 6 \times A(S_1)$$

$$= 24,$$

$$\iint_{S_2} (\mathbf{F} \cdot \mathbf{N}) dS = \iint_{S_2} \langle x^2 - \sin y^2 + z, -y, z + z^2 \rangle \cdot \langle 0,0,-1 \rangle dS = -\iint_{S_2} (z + z^2) dS$$

$$= \iint_{S_2} (0 + 0^2) dS \quad \text{(since the bottom side } z = 0\text{)}$$

$$= 0.$$

So, altogether, the desired flux is

$$\iint_{S_1+S_2} (\mathbf{F} \cdot \mathbf{N}) dS = 24 + 0 = 24.$$

Evaluating $\iint_S (\mathbf{F} \cdot \mathbf{N}) dS$ for a surface given parametrically

If a surface S is given parametrically by $\mathbf{r}(u,v) = \langle x(u,v), y(u,v), z(u,v) \rangle$, where $(u,v) \in D$, then two tangent vectors to the surface are $\mathbf{r}_u$ and $\mathbf{r}_v$. Hence, a unit normal vector is given by the unit vector in the direction of the cross product (vector product), i.e.,

$$\mathbf{N} = \frac{\mathbf{r}_u \times \mathbf{r}_v}{|\mathbf{r}_u \times \mathbf{r}_v|}.$$

But this $\mathbf{N}$ may or may not have the same direction as desired; hence,

$$\iint_S (\mathbf{F} \cdot \mathbf{N}) dS = \pm \iint_S \mathbf{F}(x(u,v), y(u,v), z(u,v)) \cdot \frac{\mathbf{r}_u \times \mathbf{r}_v}{|\mathbf{r}_u \times \mathbf{r}_v|} dS.$$

Using the surface integral evaluation formula, equation (4.26), this becomes

$$\iint_S (\mathbf{F} \cdot \mathbf{N}) dS = \pm \iint_D \mathbf{F} \cdot \frac{\mathbf{r}_u \times \mathbf{r}_v}{|\mathbf{r}_u \times \mathbf{r}_v|} |\mathbf{r}_u \times \mathbf{r}_v| du\, dv,$$

$$\iint_S (\mathbf{F} \cdot \mathbf{N}) dS = \pm \iint_D \mathbf{F} \cdot (\mathbf{r}_u \times \mathbf{r}_v) du\, dv. \tag{4.32}$$

4.8 Surface integrals of vector fields

Example 4.8.2. Find flux out of the lateral surface of the cylinder $x^2 + y^2 = 4$, $0 \le z \le 2$ for the vector field $\mathbf{F} = \langle x, y, x+z^2 \rangle$.

Solution. The cylinder has a parametric description $\mathbf{r}(u,v) = 2\cos u\,\mathbf{i} + 2\sin u\,\mathbf{j} + v\mathbf{k}$, $0 \le u \le 2\pi$ and $0 \le z \le 2$. Then

$$\mathbf{r}_u \times \mathbf{r}_v = \begin{vmatrix} \mathbf{i} & \mathbf{j} & \mathbf{k} \\ -2\sin u & 2\cos u & 0 \\ 0 & 0 & 1 \end{vmatrix} = 2\cos u\,\mathbf{i} + 2\sin u\,\mathbf{j}.$$

Note that this is an outward normal vector, so we take the positive sign and

$$\iint_S (\mathbf{F} \cdot \mathbf{N})\,dS = \iint_{D_{uv}} \langle 2\cos u, 2\sin u, 2\cos u + v^2 \rangle \cdot \langle 2\cos u, 2\sin u, 0 \rangle\,du\,dv$$

$$= \iint_{D_{uv}} (4\cos^2 u + 4\sin^2 u)\,du\,dv = 4 \iint_{D_{uv}} du\,dv = 16\pi.$$

Evaluating $\iint_S (\mathbf{F} \cdot \mathbf{N})\,dS$ for a surface $z = z(x,y)$

A point $(x,y,z(x,y))$ on the surface S given explicitly by $z = z(x,y)$ satisfies $F = z - z(x,y) = 0$, so ∇F is a normal vector to this surface. Since $\nabla F = \langle F_x, F_y, F_z \rangle = \langle -z_x, -z_y, 1 \rangle$, a unit normal is

$$\mathbf{N} = \frac{1}{\sqrt{1 + z_x^2 + z_y^2}} \langle -z_x, -z_y, 1 \rangle.$$

This normal vector may or may not have the same direction as desired. Hence, for a vector field $\vec{F} = \langle f, g, h \rangle$ defined on S, the surface integral becomes

$$\iint_S (\mathbf{F} \cdot \mathbf{N})\,dS = \pm \iint_S \langle f, g, h \rangle \cdot \langle -z_x, -z_y, 1 \rangle \frac{1}{\sqrt{1 + z_x^2 + z_y^2}}\,dS.$$

Using the surface integral evaluation formula, equation (4.27), this becomes

$$\iint_S (\mathbf{F} \cdot \mathbf{N})\,dS = \pm \iint_{D_{xy}} \langle f, g, h \rangle \cdot \langle -z_x, -z_y, 1 \rangle \frac{1}{\sqrt{1 + z_x^2 + z_y^2}} \sqrt{1 + z_x^2 + z_y^2}\,d\sigma$$

$$= \pm \iint_{D_{xy}} \langle f, g, h \rangle \cdot \langle -z_x, -z_y, 1 \rangle\,d\sigma$$

$$= \pm \iint_{D_{xy}} (-fz_x - gz_y + h)\,dx\,dy,$$

where D_{xy} is the projection of S onto the xy-plane. So, we have

$$\iint_S (\mathbf{F} \cdot \mathbf{N})\,dS = \pm \iint_{D_{xy}} (-fz_x - gz_y + h)\,dx\,dy. \tag{4.33}$$

Similar formulas hold for surfaces $S : x = x(y,z)$ and $S : y = y(x,z)$. If D_{yz} and D_{xz} are projections of S onto the yz-plane and xz-plane, respectively, then

$$\iint_S (\mathbf{F} \cdot \mathbf{N})\,dS = \pm \iint_{D_{yz}} (f - gx_y - hx_z)\,dydz, \qquad (4.34)$$

$$\iint_S (\mathbf{F} \cdot \mathbf{N})\,dS = \pm \iint_{D_{xz}} (-fy_x + g - hy_z)\,dxdz. \qquad (4.35)$$

! **Example 4.8.3.** Evaluate $\iint_S \mathbf{F} \cdot d\mathbf{S}$, where $\vec{F}(x,y,z) = y\,\vec{i} + x\,\vec{j} + z\,\vec{k}$ and S is the boundary of the solid region R enclosed by the paraboloid $z = 1 - x^2 - y^2$ and the plane $z = 0$. Assume S is oriented inward.

Solution. The boundary S consists of a parabolic top surface S_1 with $z = 1 - x^2 - y^2$ and a circular flat bottom surface S_2 with $z = 0$; S is a closed surface where the outward normal for S_1 is oriented upward and is $\langle -z_x, -z_y, 1 \rangle = \langle 2x, 2y, 1 \rangle$, so the inward normal is $\langle -2x, -2y, -1 \rangle$. The normal to S_2 can be taken as $\langle 0, 0, 1 \rangle$. Both S_1 and S_2 have the same projection in the xy-plane, namely, the disk $D : x^2 + y^2 \le 1$. Equation (4.33) gives

$$\iint_S (\mathbf{F} \cdot \mathbf{N})\,dS = \iint_{S_1} (\mathbf{F} \cdot \mathbf{N})\,dS + \iint_{S_2} (\mathbf{F} \cdot \mathbf{N})\,dS$$

$$= \iint_D \langle y, x, 1 - x^2 - y^2 \rangle \cdot \langle -2x, -2y, -1 \rangle\,d\sigma + \iint_D \langle y, x, 0 \rangle \cdot \langle 0, 0, 1 \rangle\,d\sigma$$

$$= \iint_D -2xy - 2xy - (1 - x^2 - y^2)\,dxdy + \iint_D y \cdot 0 + x \cdot 0 + 0 \cdot (1)\,dxdy$$

$$= \iint_D (-4xy - 1 + x^2 + y^2)\,dxdy + 0$$

$$= \int_0^{2\pi}\!\!\int_0^1 (-1 + r^2)r\,drd\theta \quad \text{(changing to polar coordinates)}$$

$$= \int_0^{2\pi} d\theta \int_0^1 (-r + r^3)\,dr = -\frac{\pi}{2}.$$

Other forms of $\iint_S (\mathbf{F} \cdot \mathbf{N})\,dS$

Recall that the unit normal vector $\mathbf{N}$ has the form $\langle \cos\alpha, \cos\beta, \cos\gamma \rangle$, where α, β, and γ are direction angles, and the projection element of dS onto coordinate planes has the following relations:

$$\cos\alpha\,dS = dydz, \quad \cos\beta\,dS = dxdz, \quad \text{and} \quad \cos\gamma\,dS = dxdy.$$

Then,

$$\iint_S (\mathbf{F} \cdot \mathbf{N})dS = \iint_S \langle f, g, h \rangle \cdot \langle \cos\alpha, \cos\beta, \cos\gamma \rangle dS$$

$$= \iint_S f\cos\alpha\, dS + g\cos\beta\, dS + h\cos\gamma\, dS$$

$$= \iint_S f\,dydz + g\,dxdz + h\,dxdy. \tag{4.36}$$

This can also be derived as follows.

Equation (4.33) for evaluating surface integrals as double integrals shows that when the field is $\mathbf{F} = \langle f, 0, 0 \rangle$, then

$$\iint_S (\mathbf{F} \cdot \mathbf{N})dS = \pm \iint_{D_{yz}} (f - 0x_y - 0x_z)\,dydz = \pm \iint_{D_{yz}} f\,dydz.$$

Similarly, using the alternative forms of the equation (4.33), we can show that if $\mathbf{F} = \langle 0, g, 0 \rangle$, then

$$\iint_S (\mathbf{F} \cdot \mathbf{N})dS = \pm \iint_{D_{xz}} (-0y_x + g - 0y_z)\,dxdz = \pm \iint_{D_{xz}} g\,dxdz,$$

and if $\mathbf{F} = \langle 0, 0, h \rangle$,

$$\iint_S (\mathbf{F} \cdot \mathbf{N})dS = \pm \iint_{D_{xy}} (-0z_x - 0z_y + h)\,dxdy = \pm \iint_{D_{xy}} h\,dxdy.$$

Putting the above three equations together and writing $\mathbf{F} = \langle f, g, h \rangle = \langle 0, 0, h \rangle + \langle 0, g, 0 \rangle + \langle f, 0, 0 \rangle$, we obtain

$$\iint_S (\mathbf{F} \cdot \mathbf{N})dS = \pm \iint_{D_{yz}} f\,dydz \pm \iint_{D_{xz}} g\,dxdz \pm \iint_{D_{xy}} h\,dxdy$$

or, as it is usually written,

$$\iint_S (\mathbf{F} \cdot \mathbf{N})dS = \iint_S f\,dxdy + g\,dxdz + h\,dydz.$$

Note. Again, one must be aware that the normal vectors must be consistent with the desired direction.

Example 4.8.4. Find the flux integral

$$\iint_S x\,dy\,dz + (y-z)\,dx\,dz + x\,dx\,dy$$

when S is part of the plane $x + 2y + z = 3$ in the first octant with the unit normal **N** of S pointing to the side of the surface away from the origin.

Solution. If we let $\mathbf{F}(x, y, z) = \langle x,\ y - z,\ x \rangle$, then we rewrite the integral as

$$\iint_S x\,dy\,dz + (y-z)\,dx\,dz + x\,dx\,dy = \iint_S (\mathbf{F} \cdot \mathbf{N})\,dS.$$

Since $z = 3 - x - 2y$ on S, the normal vector pointing away from the origin is $\langle -z_x, -z_y, 1 \rangle = \langle 1, 2, 1 \rangle$. Then,

$$\iint_S (\mathbf{F} \cdot \mathbf{N})\,dS = \iint_{D_{xy}} \langle x, y - z, x \rangle \cdot \langle 1, 2, 1 \rangle\,d\sigma = \iint_{D_{xy}} x + 2(y - z) + x\,d\sigma$$

$$= \iint_{D_{xy}} x + 2(y - 3 + x + 2y) + x\,d\sigma$$

$$= \iint_{D_{xy}} 4x + 6y - 6\,d\sigma$$

$$= \int_0^{\frac{3}{2}} \left(\int_0^{3-2y} (4x + 6y - 6)\,dx \right) dy$$

$$= \int_0^{\frac{3}{2}} (-4y^2 + 6y)\,dy$$

$$= \frac{9}{4}.$$

4.9 Divergence theorem

4.9.1 Divergence of a three-dimensional vector field

As seen with two-dimensional vector fields, the divergence which measures as "source" of a vector field **F** is defined to be $\nabla \cdot \mathbf{F}$. This definition can be extended to three-dimensional vector fields as well. For example, if $\mathbf{F} = \langle f, g, h \rangle$, where f, g, and h are three functions of three variables, then the divergence of **F** is

$$\text{divergence of } \mathbf{F} = \text{Div } \mathbf{F} = \nabla \cdot \mathbf{F} = \frac{\partial f}{\partial x} + \frac{\partial g}{\partial y} + \frac{\partial h}{\partial z}.$$

Physical interpretation of the divergence ∇ · F

Construct a box with center (x, y, z) of volume $2\Delta x \times 2\Delta y \times 2\Delta z$ with sides parallel to the coordinate planes with one corner at $(x + \Delta x, y + \Delta y, z + \Delta z)$ and the diagonally opposite corner at $(x - \Delta x, y - \Delta y, z - \Delta z)$, as in Figure 4.32. The values of Δx, Δy, and Δz are small changes in x, y, and z, respectively. For a continuous differentiable vector field

$$\mathbf{F}(x, y, z) = \langle f(x, y, z), g(x, y, z), h(x, y, z) \rangle,$$

we calculate the flux per unit volume out of this box, the density of the flux, and then let $\Delta x, \Delta y, \Delta z \to 0$ to show that the rate of change of the "quantity" of $\mathbf{F}$ at (x, y, z) is $\nabla \cdot \mathbf{F}$.

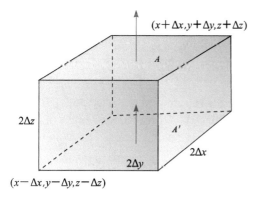

Figure 4.32: Divergence, the density of flux, box model.

The component of $\mathbf{F}$ in the z-direction is $h(x, y, z)$. Hence, the flow out of the face A (Figure 4.32) of the box is approximately $h(x, y, z + \Delta z)4\Delta x \Delta y$, i.e., the flow per unit time at the center of the face multiplied by the area of the face. Similarly, the flow into face A' of the box is approximately $h(x, y, z - \Delta z)4\Delta x \Delta y$. Hence, the change in the z-direction per unit volume is approximately

$$\frac{h(x, y, z + \Delta z)4\Delta x \Delta y - h(x, y, z - \Delta z)4\Delta x \Delta y}{8\Delta x \Delta y \Delta z}$$
$$= \frac{h(x, y, z + \Delta z) - h(x, y, z - \Delta z)}{2\Delta z},$$

and in the limiting case as $\Delta z \to 0$, the flux change per unit volume in the z-direction is

$$\lim_{\Delta z \to 0} \left[\frac{h(x, y, z + \Delta z) - h(x, y, z - \Delta z)}{2\Delta z} \right] = \frac{\partial h(x, y, z)}{\partial z}.$$

Similarly, the changes in the x- and y-directions are $\frac{\partial f}{\partial x}$ and $\frac{\partial g}{\partial y}$, so the total rate of change per unit volume is

$$\frac{\partial f}{\partial x} + \frac{\partial g}{\partial y} + \frac{\partial h}{\partial z} = \nabla \cdot \mathbf{F} = \text{Div } \mathbf{F}.$$

Note. If, for example, $\mathbf{F} = \langle f, g, h \rangle$ is fluid flow at (x, y, z), then the flow may be in any direction. However, we know that the flow $\mathbf{F}$ is equivalent to the flow of its three components f, g, and h in the direction of the three coordinate axes. This allows us to use the box method above to compute the total flux.

4.9.2 Divergence theorem

Recall the divergence-flux form of Green's theorem,

$$\oint_C (\mathbf{F} \cdot \mathbf{N}) ds = \iint_D (\nabla \cdot \mathbf{F}) d\sigma = \iint_D \left(\frac{\partial f}{\partial x} + \frac{\partial g}{\partial y} \right) d\sigma,$$

which states that the integral of the divergence over a simply connected region D gives the total flux out of the boundary C of the region D. The three-dimensional version of Green's theorem in three-dimensional space is the following divergence theorem.

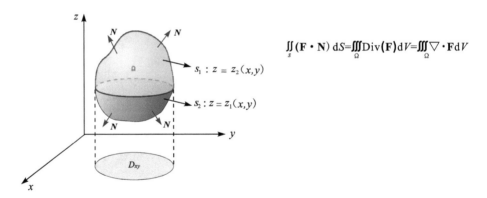

Figure 4.33: Divergence theorem.

---

i   **Theorem 4.9.1** (The divergence theorem). *Let S be a closed surface in $\mathbb{R}^3$ oriented outward, enclosing a simply connected region Ω. Let the vector field $\mathbf{F} = \langle f(x,y,z), g(x,y,z), h(x,y,z) \rangle$ be defined and have continuous partial derivatives on a region containing Ω and S. Then*

$$\iint_S (\mathbf{F} \cdot \mathbf{N}) dS = \iiint_\Omega \text{Div } \mathbf{F} dV = \iiint_\Omega \nabla \cdot \mathbf{F} dV.$$

Note. As noted previously, the quantity $\iint_S (\mathbf{F} \cdot \mathbf{N}) dS$ measures the flux of the vector field across the surface S. If, for example, $\mathbf{F}$ measures a fluid flow (mass/unit time in the direction of $\mathbf{F}$) at each point (x, y, z), then $\iint_S (\mathbf{F} \cdot \mathbf{N}) dS$ measures the amount per unit time of fluid crossing the surface S in the direction of $\mathbf{N}$. In this case $\iiint_\Omega \nabla \cdot \mathbf{F} dV$ measures the rate of change of fluid mass in the region Ω. This is because $\nabla \cdot \mathbf{F}$ measures the flux per unit volume at the point, or in other words, the rate (per unit volume) at which the fluid quantity is changing at the point (x, y, z).

Verifying the divergence theorem

Example 4.9.1. Compute the flux of $F(x, y, z) = z\mathbf{i} + y\mathbf{j} + x\mathbf{k}$ out of $S : x^2 + y^2 + z^2 = 1$ using the two integrals in the divergence theorem.

Solution. We use the parameterization $\mathbf{r}(u, v) = \langle \sin u \cos v, \sin u \sin v, \cos u \rangle$, $0 \leq u \leq \pi$, $0 \leq v \leq 2\pi$. Then as computed before

$$\mathbf{r}_u \times \mathbf{r}_v = (\sin^2 u \cos v)\mathbf{i} + (\sin^2 u \sin v)\mathbf{j} + (\sin u \cos u)\mathbf{k},$$

$$\iint_S (\mathbf{F} \cdot \mathbf{N}) dS = \iint_{D_{uv}} \langle f, g, h \rangle \cdot (\mathbf{r}_u \times \mathbf{r}_v) du dv$$

$$= \iint_{D_{uv}} \langle \cos u, \sin u \sin v, \sin u \cos v \rangle$$

$$\cdot \langle \sin^2 u \cos v, \sin^2 u \sin v, \sin u \cos u \rangle du dv$$

$$= \iint_{D_{uv}} (\sin^2 u \cos u \cos v + \sin^3 u \sin^2 v + \sin^2 u \cos u \cos v) du dv$$

$$= \iint_{D_{uv}} \sin^3 u \sin^2 v \, du dv$$

$$= \int_0^\pi \sin^3 u \, du \int_0^{2\pi} \sin^2 v \, dv = -\int_0^\pi (1 - \cos^2 u) d(\cos u) \int_0^{2\pi} \frac{1 - \cos 2v}{2} dv$$

$$= \left(-\cos u + \frac{1}{3} \cos^3 u \Big|_0^\pi \right) \times \pi = \frac{4\pi}{3}.$$

We now evaluate the triple integral, where Ω (the unit ball) is the interior of S. We have

$$\iiint_\Omega \nabla \cdot \mathbf{F} dV = \iiint_\Omega \left(\frac{\partial}{\partial x}(z) + \frac{\partial}{\partial y}(y) + \frac{\partial}{\partial z}(x) \right) dV$$

$$= \iiint_\Omega 1 dV = \frac{4\pi 1^3}{3} = \frac{4\pi}{3}.$$

Note. The last integral did not require a computation because it is the volume of the sphere of radius 1.

Example 4.9.2. Evaluate the integral

$$\iint_S (xy\,\vec{i} + yz\,\vec{j} + xz\,\vec{k}) \cdot d\vec{S},$$

where S is the cube bounded by the coordinate planes, by $x = 1$, $y = 1$, and $z = 1$, orientated outward.

Solution. Using the divergence theorem, with B denoting the inside of the box, we have

$$\iint_S (xy\,\vec{i} + yz\,\vec{j} + xz\,\vec{k}) \cdot d\vec{S}$$

$$= \iiint_B \nabla \cdot (xy\,\vec{i} + yz\,\vec{j} + xz\,\vec{k})\,dV = \iiint_B (y + z + x)\,dV$$

$$= \int_0^1 \int_0^1 \int_0^1 (y + z + x)\,dx\,dy\,dz = \int_0^1 \int_0^1 \left(y + z + \frac{1}{2}\right)dy\,dz$$

$$= \int_0^1 (z + 1)\,dz = \frac{3}{2}.$$

Example 4.9.3. Find the flux across S, the top hemisphere $z = \sqrt{4 - x^2 - y^2}$, oriented outward, if $\mathbf{F} = \langle x^3 + 2y\sin z, x^4 + y^3, e^{-y^2-x^2} + z^3 \rangle$.

Solution. It would be hard to evaluate $\iint_S \mathbf{F} \cdot d\mathbf{S}$ directly. Instead, we add S': $z = 0$, $x^2 + y^2 \leq 4$ with orientation downward. So $S + S'$ is closed and oriented outward. Then,

$$\iint_S \mathbf{F} \cdot d\mathbf{S} + \iint_{S'} \mathbf{F} \cdot d\mathbf{S} = \iint_{S+S'} \mathbf{F} \cdot d\mathbf{S} = \iiint_\Omega \nabla \cdot \mathbf{F}\,dV$$

$$= \iiint_\Omega 3x^2 + 3y^2 + 3z^2\,dV$$

$$= 3 \int_0^{2\pi} d\theta \int_0^{\pi/2} d\phi \int_0^2 \rho^4 \sin\phi\,d\rho$$

$$= 3 \int_0^{2\pi} d\theta \int_0^{\pi/2} \sin\phi\,d\phi \int_0^2 \rho^4\,d\rho = \frac{192\pi}{5}.$$

Now we need to subtract $\iint_{S'} \mathbf{F} \cdot d\mathbf{S}$ in order to get the desired flux.

$$\iint_S \mathbf{F} \cdot d\mathbf{S} = \frac{192\pi}{5} - \iint_{S'} \mathbf{F} \cdot d\mathbf{S}$$

$$= \frac{192\pi}{5} - \iint_{S'} \langle x^3 + 2y\sin z, x^4 + y^3, e^{-y^2-x^2} + z^3 \rangle \cdot \langle 0, 0, -1 \rangle dS$$

$$= \frac{192\pi}{5} + \iint_{S'} e^{-y^2-x^2} dS$$

$$= \frac{192\pi}{5} + \iint_{x^2+y^2 \le 4} e^{-y^2-x^2} d\sigma = \frac{192\pi}{5} + \int_0^{2\pi} d\theta \int_0^2 e^{-r^2} r \, dr$$

$$= \frac{192\pi}{5} - 2\pi \left(\frac{1}{2} e^{-4} - \frac{1}{2} \right).$$

Example 4.9.4. An electric charge q at the origin creates an electric field $\mathbf{F}$ at $\mathbf{r}$ (the position vector from the origin) given by

$$\mathbf{F} = \frac{q}{4\pi \epsilon_0} \frac{\mathbf{r}}{r^3},$$

where $r = |\mathbf{r}|$ and ϵ_0 is a constant. Compute $\nabla \cdot \mathbf{F}$ and find the flux across the surface S of the sphere B : $x^2 + y^2 + z^2 \le b^2$. Find the flux across any closed surface S_1 containing the charge.

Solution. A vector field of the form

$$k\frac{\mathbf{r}}{r^3} = k \left\langle \frac{x}{r^3}, \frac{y}{r^3}, \frac{z}{r^3} \right\rangle,$$

where k is a constant, is the *inverse square law* (such as electric charge and gravity). The divergence is zero, because $r = (x^2 + y^2 + z^2)^{\frac{1}{2}}$ and $r^2 = x^2 + y^2 + z^2$. So $2r\frac{\partial r}{\partial x} = 2x$, and $\frac{\partial r}{\partial x} = \frac{x}{r}$. Thus,

$$\frac{\partial}{\partial x}\left(\frac{x}{r^3}\right) = \frac{r^3 - x3r^2 \frac{\partial r}{\partial x}}{r^6} = \frac{1}{r^3} - \frac{3x^2}{r^5}.$$

By symmetry, we have

$$\nabla \cdot k\frac{\mathbf{r}}{r^3} = k\nabla \cdot \left\langle \frac{x}{r^3}, \frac{y}{r^3}, \frac{z}{r^3} \right\rangle$$

$$= k\left(\frac{1}{r^3} - \frac{3x^2}{r^5}\right) + k\left(\frac{1}{r^3} - \frac{3y^2}{r^5}\right) + k\left(\frac{1}{r^3} - \frac{3z^2}{r^5}\right)$$

$$= \frac{3k}{r^3} - 3k\left(\frac{x^2 + y^2 + z^2}{r^5}\right),$$

$$\nabla \cdot \mathbf{F} = \frac{3k}{r^3} - \frac{3k}{r^3} = 0.$$

Using the divergence theorem when $\mathbf{F}$ is an inverse square field would give $\iint_S \mathbf{F} \cdot d\mathbf{S} = \iiint_\Omega \nabla \cdot \mathbf{F} dV = 0$, but this is a wrong answer, because the vector field $\mathbf{F}$ is not defined at the origin (in fact it approaches ∞ as $x, y, z \to 0$).

Hence, we must integrate the flux integral directly without using the divergence theorem. The outward unit normal to the sphere $x^2 + y^2 + z^2 = b^2$ at $(x, y, z) \in S$ is $\frac{1}{b}(x, y, z)$, so

$$\iint_S \mathbf{F} \cdot d\mathbf{S} = \iint_S (\mathbf{F} \cdot \mathbf{N}) dS$$

$$= \frac{q}{4\pi \epsilon_0} \iint_S \left(\frac{x\mathbf{i} + y\mathbf{j} + z\mathbf{k}}{(\sqrt{x^2 + y^2 + z^2})^3} \right) \cdot \left(\frac{x\mathbf{i} + y\mathbf{j} + z\mathbf{k}}{b} \right) dS$$

$$= \frac{q}{4\pi \epsilon_0} \iint_S \frac{x^2 + y^2 + z^2}{b^4} dS$$

$$= \frac{q}{4\pi \epsilon_0} \iint_S \frac{1}{b^2} dS$$

$$= \frac{q}{4\pi b^2 \epsilon_0} \cdot 4\pi b^2 \quad \text{(since } 4\pi b^2 \text{ is the area of } S\text{)}$$

$$= \frac{q}{\epsilon_0}.$$

Hence, the flux through the sphere of any radius is $\frac{q}{\epsilon_0}$.

In fact, the flux through any closed orientable surface S_1 containing the charge is the same value $\frac{q}{\epsilon_0}$. To see this, let the radius b of the sphere S be sufficiently large so that S totally encloses S_1. If $\mathbf{F} = \frac{q}{4\pi\epsilon_0} \frac{\mathbf{r}}{r^3}$, then the region R between the two surfaces satisfies the divergence theorem because this region no longer contains the origin. Hence,

$$\iint_{S+S_1} (\mathbf{F} \cdot \mathbf{N}) dS = \iiint_R \nabla \cdot \mathbf{F} dV = 0 \Longrightarrow$$

$$\iint_S (\mathbf{F} \cdot \mathbf{N}) dS + \iint_{S_1} (\mathbf{F} \cdot \mathbf{N}) dS = 0,$$

$$-\iint_{S_1} (\mathbf{F} \cdot \mathbf{N}) dS = \iint_S (\mathbf{F} \cdot \mathbf{N}) dS = \frac{q}{\epsilon_0}.$$

But the unit normal $\mathbf{N}$ on S_1 will point out of the region R, which means it points into the region containing the charge. Changing the direction of $\mathbf{N}$ to point outwards from the charge q shows that

$$\iint_{S_1} (\mathbf{F} \cdot \mathbf{N}) dS = \frac{q}{\epsilon_0}.$$

Proof of the divergence theorem under special conditions

Suppose the solid Ω is bounded by S, which consists of two surfaces, the lower one S_1: $z = z_1(x,y)$ and the upper one $S_2 : z = z_2(x,y)$. The projection of S onto the xy-plane is D_{xy}, as shown in Figure 4.33. Since

$$\iint_S (\mathbf{F} \cdot \mathbf{N}) dS = \iint_S (f\mathbf{i} + g\mathbf{j} + h\mathbf{k}) \cdot \mathbf{N} dS = \iint_S f\mathbf{i} \cdot \mathbf{N} dS + \iint_S g\mathbf{j} \cdot \mathbf{N} dS + \iint_S h\mathbf{k} \cdot \mathbf{N} dS$$

and

$$\iiint_\Omega \nabla \cdot \mathbf{F} dV = \iiint_\Omega \frac{\partial f}{\partial x} + \frac{\partial g}{\partial y} + \frac{\partial h}{\partial z} dV,$$

we start by showing that

$$\iiint_\Omega \frac{\partial h}{\partial z} dV = \iint_S h\mathbf{k} \cdot \mathbf{N} dS. \tag{4.37}$$

But this follows from

$$\iiint_\Omega \frac{\partial h}{\partial z} dV = \iint_{D_{xy}} \left(\int_{z_1(x,y)}^{z_2(x,y)} \frac{\partial h}{\partial z} dz \right) dxdy$$

$$= \iint_{D_{xy}} (h(x,y,z_2(x,y)) - h(x,y,z_1(x,y))) dxdy$$

and

$$\iint_S h\mathbf{k} \cdot \mathbf{N} dS = \iint_{S_1} h\mathbf{k} \cdot \mathbf{N} dS + \iint_{S_2} h\mathbf{k} \cdot \mathbf{N} dS$$

$$= -\iint_{D_{xy}} h\mathbf{k} \cdot \langle -z_x, -z_y, 1 \rangle dxdy + \iint_{D_{xy}} h\mathbf{k} \cdot \langle -z_x, -z_y, 1 \rangle dxdy$$

$$= -\iint_{D_{xy}} h(x,y,z_1(x,y)) dxdy + \iint_{D_{xy}} h(x,y,z_2(x,y)) dxdy$$

$$= \iint_{D_{xy}} h(x,y,z_2(x,y)) - h(x,y,z_1(x,y)) dxdy.$$

Similarly, we also have

$$\iiint_\Omega \frac{\partial g}{\partial y} dV = \iint_S g\mathbf{j} \cdot \mathbf{N} dS \quad \text{and} \quad \iiint_\Omega \frac{\partial f}{\partial x} dV = \iint_S f\mathbf{i} \cdot \mathbf{N} dS.$$

Adding them up gives the divergence theorem.

4.10 Stokes theorem

4.10.1 The curl of a three-dimensional vector field

Recall that when we were trying to find the circulation along a plane curve C over a vector field $\mathbf{F} = \langle f, g \rangle$, we saw the term $\frac{\partial g}{\partial x} - \frac{\partial f}{\partial y}$ in Green's theorem in curl-circulation form, and we noted that

$$\frac{\partial g}{\partial x} - \frac{\partial f}{\partial y}$$

is the **k**-component of the curl, a vector relating to the rotational effect of a vector field.

If we want to find circulation along a simple closed curve over a vector field in $\mathbb{R}^3$, what can we expect for the **i**- or **j**-components of the curl? What do the **i**- or **j**-components look like? In general, we define the curl of the vector field $\mathbf{F} = \langle f, g, h \rangle$ as

$$\operatorname{curl} \mathbf{F} = \left(\frac{\partial h}{\partial y} - \frac{\partial g}{\partial z} \right) \mathbf{i} + \left(\frac{\partial f}{\partial z} - \frac{\partial h}{\partial x} \right) \mathbf{j} + \left(\frac{\partial g}{\partial x} - \frac{\partial f}{\partial y} \right) \mathbf{k}.$$

This is better written as

$$\operatorname{curl} \mathbf{F} = \begin{vmatrix} \mathbf{i} & \mathbf{j} & \mathbf{k} \\ \frac{\partial}{\partial x} & \frac{\partial}{\partial y} & \frac{\partial}{\partial z} \\ f & g & h \end{vmatrix}.$$

We recall the notation $\nabla = \langle \frac{\partial}{\partial x}, \frac{\partial}{\partial y}, \frac{\partial}{\partial z} \rangle$, so we have the even simpler notation

$$\operatorname{curl} \mathbf{F} = \nabla \times \mathbf{F}.$$

If the curl of a vector field is always $\mathbf{0}$, then the field is *irrotational*. An irrotational vector field $\mathbf{F}$ might be conservative, that is, there exists a potential function $\varphi(x, y, z)$ such that $\mathbf{F} = \nabla \varphi$.

Example 4.10.1. First compute the curl of the vector field given by

$$\mathbf{F}(x, y, z) = \langle 2xy - z,\ x^2 + 4y,\ -x + 2z \rangle.$$

Is the field irrotational? Attempt to find a potential function φ for the vector field $\mathbf{F}$.

Solution. Note that $f = 2xy - z$, $g = x^2 + 4y$, and $h = -x + 2z$. We first compute curl $\mathbf{F}$. We have

$$\operatorname{curl} \mathbf{F} = \nabla \times \mathbf{F} = \begin{vmatrix} \mathbf{i} & \mathbf{j} & \mathbf{k} \\ \frac{\partial}{\partial x} & \frac{\partial}{\partial y} & \frac{\partial}{\partial z} \\ 2xy - z & x^2 + 4y & -x + 2z \end{vmatrix}$$

$$= \left(\frac{\partial(-x+2z)}{\partial y} - \frac{\partial(x^2+4y)}{\partial z}\right)\mathbf{i} + \left(\frac{\partial(2xy-z)}{\partial z} - \frac{\partial(-x+2z)}{\partial x}\right)\mathbf{j}$$
$$+ \left(\frac{\partial(x^2+4y)}{\partial x} - \frac{\partial(2xy-z)}{\partial y}\right)\mathbf{k}$$
$$= (0-0)\mathbf{i} + (-1-(-1))\mathbf{j} + (2x - 2x)\mathbf{k} = 0\mathbf{i} + 0\mathbf{j} + 0\mathbf{k}.$$

This field is irrotational. Now, we attempt to find a potential function φ. We know that φ satisfies

$$\frac{\partial \varphi}{\partial x} = 2xy - z, \quad \frac{\partial \varphi}{\partial y} = x^2 + 4y, \quad \text{and} \quad \frac{\partial \varphi}{\partial z} = -x + 2z.$$

Step 1: We integrate the first component of $\mathbf{F}$ with respect to x to find φ (incompletely), i.e.,

$$\varphi(x,y,z) = \int (2xy - z)dx,$$
$$\varphi(x,y,z) = x^2 y - xz + C(y,z)$$

Step 2: We differentiate this φ with respect to y giving the second component of $\mathbf{F}$, and use this to deduce more information about φ, i.e.,

$$\frac{\partial \varphi}{\partial y} = x^2 + \frac{\partial C(y,z)}{\partial y} \quad \text{(but this must be equal to } x^2 + 4y)$$
$$\implies \frac{\partial C(y,z)}{\partial y} = 4y$$
$$\implies C(y,z) = 2y^2 + C(z)$$
$$\implies \varphi(x,y,z) = x^2 y - xz + 2y^2 + C(z).$$

Step 3: We differentiate φ again with respect to z giving the third component of $\mathbf{F}$, and we use this to determine $\varphi(x,y,z)$ (up to a constant), i.e.,

$$\frac{\partial \varphi}{\partial z} = -x + \frac{dC(z)}{dz} \quad \text{(but this must be equal to } -x + 2z)$$
$$\implies \frac{dC(z)}{dz} = 2z$$
$$\implies C(z) = z^2 + C.$$

Hence, we have

$$\varphi(x,y,z) = x^2 y - xz + 2y^2 + z^2 + C,$$

where C is an arbitrary constant. The simplest answer is when $C = 0$: $\varphi(x,y,z) = x^2 y - xz + 2y^2 + z^2$.

4.10.2 Stokes theorem

Recall Green's theorem in curl-circulation form. If C is a simple and closed plane curve which is the boundary of a simply connected region D, we have

$$\oint_C \mathbf{F} \cdot \mathbf{T} ds = \iint_D \frac{\partial g}{\partial x} - \frac{\partial f}{\partial y} dxdy.$$

Now, in three-dimensional space, if C is a simple closed space curve, which is the boundary of a smooth orientable surface, is there any similar result? The answer is Stokes theorem, stated below.

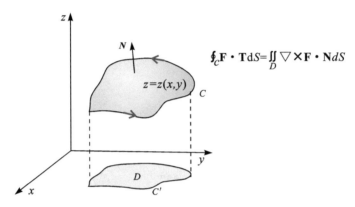

Figure 4.34: Stokes theorem.

Theorem 4.10.1 (Stokes theorem). *Suppose that S is a bounded simple orientable smooth surface with unit normal $\mathbf{N}$ and boundary curve C that is oriented with unit tangent $\mathbf{T}$ as described in the preceding sections (a corkscrew following the direction of $\mathbf{N}$ turns in the same direction as the positive direction on C). Let $\mathbf{F}(x,y,z)$ be a continuously differentiable vector-valued function defined on S. Then*

$$\oint_C \mathbf{F} \cdot \mathbf{T} ds = \iint_S (\nabla \times \mathbf{F}) \cdot \mathbf{N} dS = \iint_S (\mathrm{Curl}\, \mathbf{F} \cdot \mathbf{N}) dS,$$

or in another form,

$$\oint_C f dx + g dy + h dz = \iint_S \left(\frac{\partial h}{\partial y} - \frac{\partial g}{\partial z} \right) dydz + \left(\frac{\partial f}{\partial z} - \frac{\partial h}{\partial x} \right) dzdx + \left(\frac{\partial g}{\partial x} - \frac{\partial f}{\partial y} \right) dxdy.$$

Note. Note that

$$\oint_C \mathbf{F} \cdot \mathbf{T} ds = \iint_S \left(\frac{\partial h}{\partial y} - \frac{\partial g}{\partial z} \right) dydz + \left(\frac{\partial f}{\partial z} - \frac{\partial h}{\partial x} \right) dzdx + \left(\frac{\partial g}{\partial x} - \frac{\partial f}{\partial y} \right) dxdy,$$

which can be written in different forms, i. e.,

$$\oint_C \mathbf{F} \cdot \mathbf{T} ds = \iint_S \left(\frac{\partial h}{\partial y} - \frac{\partial g}{\partial z} \right) dydz + \left(\frac{\partial f}{\partial z} - \frac{\partial h}{\partial x} \right) dzdx + \left(\frac{\partial g}{\partial x} - \frac{\partial f}{\partial y} \right) dxdy$$

$$= \iint_S \begin{vmatrix} dydz & dzdx & dxdy \\ \frac{\partial}{\partial x} & \frac{\partial}{\partial y} & \frac{\partial}{\partial z} \\ f & g & h \end{vmatrix}$$

$$= \iint_S \begin{vmatrix} \cos \alpha & \cos \beta & \cos \gamma \\ \frac{\partial}{\partial x} & \frac{\partial}{\partial y} & \frac{\partial}{\partial z} \\ f & g & h \end{vmatrix} dS \quad \text{(note that } \mathbf{N} = \langle \cos \alpha, \cos \beta, \cos \gamma \rangle\text{)}$$

$$= \iint_S (\nabla \times \mathbf{F} \cdot \mathbf{N}) dS.$$

Note. In the two-dimensional case, the surface S is a planar region D, and the unit normal vector $\mathbf{N}$ is the basis vector $\mathbf{k} = \langle 0, 0, 1 \rangle$, so Stokes theorem becomes

$$\oint_C \mathbf{F} \cdot \mathbf{T} ds = \iint_D \nabla \times \mathbf{F} \cdot \mathbf{k} \, dxdy \tag{4.38}$$

$$= \iint_D \begin{vmatrix} 0 & 0 & 1 \\ \frac{\partial}{\partial x} & \frac{\partial}{\partial y} & \frac{\partial}{\partial z} \\ f & g & h \end{vmatrix} dxdy = \iint_D \frac{\partial g}{\partial x} - \frac{\partial f}{\partial y} dxdy, \tag{4.39}$$

which is exactly Green's theorem in circulation-curl form. Thus, Stokes theorem is an extension of Green's theorem in three-dimensional space.

Note. There is more than one smooth surface with the same boundary C; however, the integral $\iint_S \nabla \times \mathbf{F} \cdot \mathbf{N} dS$ gives the same value for each of these smooth surfaces satisfying the conditions of the theorem.

Verifying Stokes theorem

Example 4.10.2. Let the vector field $\mathbf{F}$ be $\langle 2z - y, x, y \rangle$, and let the surface S be the upper hemisphere $z = \sqrt{4 - x^2 - y^2}$ oriented outward. The boundary curve C of S is $x^2 + y^2 = 4$ in the xy-plane oriented counterclockwise. Evaluate:
1. $\iint_S \nabla \times \mathbf{F} \cdot \mathbf{N} dS.$
2. $\iint_{S_1} \nabla \times \mathbf{F} \cdot \mathbf{N} dS$, where S_1 is the disk with boundary C and with $\mathbf{N}$ pointing upward.
3. $\oint_C \mathbf{F} \cdot \mathbf{T} ds.$

Solution. The surface and its orientation is shown in Figure 4.35. We first compute the curl of $\mathbf{F}$, which is

$$\text{curl } \mathbf{F} = \begin{vmatrix} \mathbf{i} & \mathbf{j} & \mathbf{k} \\ \frac{\partial}{\partial x} & \frac{\partial}{\partial y} & \frac{\partial}{\partial z} \\ 2z - y & x & y \end{vmatrix} = \mathbf{i} + 2\mathbf{j} + 2\mathbf{k}.$$

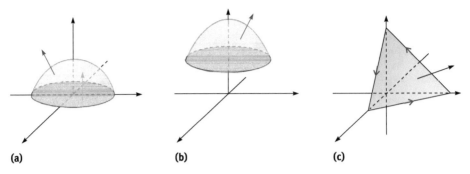

Figure 4.35: Stokes theorem, Example 4.10.2, Example 4.10.4, and Example 4.10.5.

1. Since $z_x = \dfrac{-x}{\sqrt{4-x^2-y^2}}$ and $z_y = \dfrac{-y}{\sqrt{4-x^2-y^2}}$, we have

$$\iint_S \nabla \times \mathbf{F} \cdot \mathbf{N} dS = \iint_{D_{xy}} \langle 1, 2, 2 \rangle \cdot \langle -z_x, -z_y, 1 \rangle dxdy$$

$$= \iint_{D_{xy}} \dfrac{x}{\sqrt{4-x^2-y^2}} + 2 \dfrac{y}{\sqrt{4-x^2-y^2}} + 2 dxdy$$

$$= \iint_{D_{xy}} 2 dxdy = 2 \times \pi 2^2 = 8\pi.$$

2. Since $\mathbf{N}$ can be taken as $\langle 0, 0, 1 \rangle$, we have

$$\iint_S \nabla \times \mathbf{F} \cdot \mathbf{N} dS = \iint_{D_{xy}} \langle 1, 2, 2 \rangle \cdot \langle 0, 0, 1 \rangle dxdy = \iint_{D_{xy}} 2 dxdy = 8\pi.$$

3. We parameterize $C : \mathbf{r}(t) = \langle 2\cos t, 2\sin t, 0 \rangle$, $\mathbf{r}'(t) = \langle -2\sin t, 2\cos t, 0 \rangle$. Then

$$\oint_C \mathbf{F} \cdot \mathbf{T} ds = \int \langle 2z - y, x, y \rangle \cdot \langle -2\sin t, 2\cos t, 0 \rangle dt$$

$$= \int_0^{2\pi} \langle 2 \times 0 - 2\sin t, 2\cos t, 2\sin t \rangle \cdot \langle -2\sin t, 2\cos t, 0 \rangle dt$$

$$= \int_0^{2\pi} (4\sin^2 t + 4\cos^2 t) dt = 8\pi.$$

> **Example 4.10.3.** Evaluate $\iint_S (\nabla \times \vec{F}) \cdot d\vec{S} = \iint_S (\nabla \times \vec{F}) \cdot \vec{N} dS$, when $\vec{F} = \langle xz, yz, xy \rangle$ and S is the part of the sphere $x^2 + y^2 + z^2 = 4$ inside the cylinder $x^2 + y^2 = 1$ with $z \geq 0$.

4.10 Stokes theorem

Solution. By Stokes theorem,

$$\iint_S (\nabla \times \vec{F}) \cdot \vec{N} \, dS = \oint_C \vec{F} \cdot \vec{T} \, ds,$$

where C is the boundary curve $x^2+y^2 = 1$ and z is given by $z^2 = 4-(x^2+y^2) = 3 \Rightarrow z = \sqrt{3}$. We can represent C in vector form with positive orientation as

$$\vec{r}(t) = \cos t \, \vec{i} + \sin t \, \vec{j} + \sqrt{3} \, \vec{k}, \quad 0 \le t \le 2\pi.$$

Hence, on C: $\vec{F} = \langle xz, yz, xy \rangle = \langle \sqrt{3}\cos t, \sqrt{3}\sin t, \cos t \sin t \rangle$,

$$\oint_C \vec{F} \cdot \vec{T} \, ds = \int_0^{2\pi} \vec{F} \cdot \frac{d\vec{r}}{dt} \, dt$$

$$= \int_0^{2\pi} \langle \sqrt{3}\cos t, \sqrt{3}\sin t, \cos t \sin t \rangle \cdot \langle -\sin t, \cos t, 0 \rangle \, dt$$

$$= \int_0^{2\pi} (-\sqrt{3}\cos t \sin t + \sqrt{3}\sin t \cos t) \, dt$$

$$= 0.$$

Therefore, $\iint_S (\nabla \times \vec{F}) \cdot d\vec{S} = 0$.

Example 4.10.4. If $F = \langle z^2 y, -3xy, e^{-x^2} y^3 \rangle$ and S is part of the surface $z = 5 - x^2 - y^2$ above $z = 4$, oriented upward, find $\iint_S \text{curl } F \cdot dS$.

Solution. The curl of **F** is

$$\text{curl } \mathbf{F} = \begin{vmatrix} \mathbf{i} & \mathbf{j} & \mathbf{k} \\ \frac{\partial}{\partial x} & \frac{\partial}{\partial y} & \frac{\partial}{\partial z} \\ z^2 y & -3xy & e^{-x^2} y^3 \end{vmatrix} = \begin{pmatrix} 3y^2 e^{-x^2} \\ 2xe^{-x^2} y^3 + 2zy \\ -z^2 - 3y \end{pmatrix}.$$

To evaluate $\iint_S \text{curl } \mathbf{F} \cdot d\mathbf{S}$ directly would be hard. We now use Stokes theorem. We first find the boundary curve $z = 4$ and $x^2 + y^2 < 1$, and orient it counterclockwise as viewed from above. We parameterize the curve by $\mathbf{r}(t) = \langle \cos t, \sin t, 4 \rangle$, $0 \le t \le 2\pi$. Then,

$$\iint_S \text{curl } \mathbf{F} \cdot d\mathbf{S} = \oint_C \mathbf{F} \cdot d\mathbf{r} = \oint_C \langle z^2 y, -3xy, e^{-x^2} y^3 \rangle \cdot d\langle \cos t, \sin t, 4 \rangle$$

$$= \int_0^{2\pi} \langle 4^2 \sin t, -3\cos t \sin t, e^{-\cos^2 t} (\sin t)^3 \rangle \cdot \langle -\sin t, \cos t, 0 \rangle \, dt$$

$$= \int_0^{2\pi} (-16\sin^2 t - 3\sin t \cos^2 t) dt$$

$$= -8 \int_0^{2\pi} (1 - \cos 2t) dt = -16\pi.$$

Or, we choose an alternative surface with the boundary. This alternative surface could be the disk $x^2 + y^2 \leq 1$ and $z = 4$. The outward unit normal vector is $\langle 0, 0, 1 \rangle$. Thus,

$$\iint_S \text{curl } \mathbf{F} \cdot d\mathbf{S} = \iint_S \text{curl } \mathbf{F} \cdot \langle 0, 0, 1 \rangle dS$$

$$= \iint_S \langle 3y^2 e^{-x^2}, 2xe^{-x^2} y^3 + 2zy, -z^2 - 3y \rangle \cdot \langle 0, 0, 1 \rangle dS$$

$$= \iint_S -z^2 - 3y \, dS = \iint_S -16 - 3y \, dS = \iint_D -16 - 3y \, d\sigma$$

$$= \int_0^{2\pi} d\theta \int_0^1 (-16 - 3r\sin\theta) r \, dr$$

$$= \int_0^{2\pi} -8 - \sin\theta \, d\theta = -16\pi.$$

Example 4.10.5. Compute $\int_C \vec{F} \cdot d\vec{r}$, where $\vec{F} = xz\vec{i} + xy\vec{j} + 3xz\vec{k}$ and C is the triangular closed curve with vertices followed in the order $(1, 0, 0)$, $(0, 2, 0)$, $(0, 0, 2)$, $(1, 0, 0)$.

Solution. If S is the triangular part of the plane $2x + y + z = 2$ defined by the three vertices (in the first octant and with boundary curve C), then Stokes theorem gives

$$\oint_C \vec{F} \cdot d\vec{r} = \oint_C \vec{F} \cdot \vec{T} ds = \iint_S (\nabla \times \vec{F}) \cdot \vec{N} dS,$$

and $\nabla \times \vec{F} = \langle 0, x - 3z, y \rangle$. Hence,

$$\iint_S (\nabla \times \vec{F}) \cdot \vec{N} dS = \iint_S \langle 0, x - 3z, y \rangle \cdot \vec{N} dS.$$

Evaluating this using equation (4.33), we find $\iint_D (-fz_x - gz_y + h) dx dy$, using $z = z(x, y) = 2 - 2x - y$, $f = 0$, $g = x - 3z$, and $h = y$, where the projection of D onto the xy-plane is given by the triangle with vertices $(0, 0, 0)$, $(1, 0, 0)$, $(0, 2, 0)$, and $\langle -z_x, -z_y, 1 \rangle = \langle 2, 1, 1 \rangle$. This is the correct orientation for the plane. Hence, the integral becomes

$$\oint_C \vec{F} \cdot d\vec{r} = \iint_S (\nabla \times \vec{F}) \cdot \vec{N} dS = \iint_D ((x - 3z) + y) d\sigma$$

$$= \iint_D (x+y - 3(2-2x-y))d\sigma$$

$$= \int_0^1 \int_0^{2-2x} (7x + 4y - 6) dy\, dx$$

$$= \int_0^1 (-6x^2 + 10x - 4) dx$$

$$= -1.$$

Example 4.10.6. Show, using Stokes theorem, $\oint_C \vec{F} \cdot \vec{T} ds = 0$ for any gradient vector field $\vec{F}$ with continuous partial derivatives and simple orientable smooth surface S with unit normal $\vec{N}$ and boundary curve C in $\mathbb{R}^3$.

Solution. We know that if $\mathbf{F} = \nabla \varphi$ for some function $\varphi(x,y,z)$ (a potential function of $\mathbf{F}$), then $\nabla \times \nabla \varphi = \vec{0}$ because

$$\nabla \times \nabla \varphi = \begin{vmatrix} \mathbf{i} & \mathbf{j} & \mathbf{k} \\ \frac{\partial}{\partial x} & \frac{\partial}{\partial y} & \frac{\partial}{\partial z} \\ \frac{\partial \varphi}{\partial x} & \frac{\partial \varphi}{\partial y} & \frac{\partial \varphi}{\partial z} \end{vmatrix}$$

$$= \left(\frac{\partial^2 \varphi}{\partial y \partial z} - \frac{\partial^2 \varphi}{\partial z \partial y} \right) \mathbf{i} + \left(\frac{\partial^2 \varphi}{\partial z \partial x} - \frac{\partial^2 \varphi}{\partial x \partial z} \right) \mathbf{j} + \left(\frac{\partial^2 \varphi}{\partial x \partial y} - \frac{\partial^2 \varphi}{\partial y \partial x} \right) \mathbf{k}$$

$$= \mathbf{0}.$$

Hence, by Stokes theorem,

$$\oint_C \mathbf{F} \cdot \mathbf{T} ds = \iint_S (\nabla \times \mathbf{F}) \cdot \mathbf{N} dS = \iint_S (\nabla \times \nabla \varphi) \cdot \mathbf{N} dS = \iint_S 0\, dS = 0.$$

By Stokes theorem, we can conclude that if f, g, and h have continuous partial derivatives, then

$$\mathbf{F} = \langle f, g, h \rangle \text{ is conservative} \iff \oint_C \mathbf{F} \cdot \mathbf{T} ds = 0 \iff \text{curl } \mathbf{F} = \mathbf{0}.$$

for any simple closed curve C

Note that the proof of the fundamental theorem of line integrals can be extended to three-dimensional vector fields in ways similar to the results we have obtained for two-dimensional vector fields. For a three-dimensional vector field $\mathbf{F} = \langle f, g, h \rangle$, where f, g, and h all have continuous partial derivatives and D is a simply connected region in $\mathbb{R}^3$ bounded by the simple curve C, we have

$\mathbf{F} = \langle f, g, h \rangle$ is conservative

$\iff$ there is a function φ such that $\mathbf{F} = \nabla \varphi$, or $d\varphi = f dx + g dy + h dz$

$$\iff \int_C \mathbf{F} \cdot \mathbf{T} ds = \int_C f dx + g dy + h dz \text{ is path-independent}$$

$$\iff \oint_C \mathbf{F} \cdot \mathbf{T} ds = 0 \text{ for any simple closed curve}$$

$$\iff \text{curl } \mathbf{F} = \mathbf{0}. \text{ This means } \varphi_{yz} = \varphi_{zy}, \varphi_{xz} = \varphi_{zx}, \text{ and } \varphi_{xy} = \varphi_{yx}.$$

To find a potential function for a conservative field, we can follow the method used in Example 4.10.1.

Example 4.10.7. Evaluate $\int_C y dx + x dy + 2z dz$, where

$$C : \mathbf{r}(t) = \frac{t(t-1)}{e^{\sqrt{t}}} \mathbf{i} + \sin\left(\frac{\pi}{2} t^2\right) \mathbf{j} + \frac{t}{t^2 + 1} \mathbf{k}, \quad 0 \le t \le 1.$$

Solution. Since

$$\nabla \times \mathbf{F} = \begin{vmatrix} \mathbf{i} & \mathbf{j} & \mathbf{k} \\ \frac{\partial}{\partial x} & \frac{\partial}{\partial y} & \frac{\partial}{\partial z} \\ y & x & 2z \end{vmatrix} = \mathbf{0},$$

this field is conservative and, therefore, it is path-independent. The two endpoints are $(0,0,0)$ and $(0,1,\frac{1}{2})$. We could have found a potential function, but we simply choose a simple route between the endpoints. If we choose the line segment $\mathbf{r}_1(t) = t\langle 0, 1, \frac{1}{2}\rangle$, for $0 \le t \le 1$, then we have $x = 0, y = t$, and $z = \frac{t}{2}$ and

$$\int_C y dx + x dy + 2z dz = \int_0^1 0 + 0 + \frac{t}{2} dt = \frac{1}{4}.$$

Proof of Stokes theorem under special conditions

See Figure 4.34. Suppose the surface $S : z = z(x,y)$ is smooth and C is its boundary with compatible orientations. The projection of S onto the xy-plane is D, and the projection of C onto the xy-plane is C'. Furthermore, we assume $z_{xy} = z_{yx}$. Note that $dz = z_x dx + z_y dy$. Then,

$$\oint_C \mathbf{F} \cdot \mathbf{T} ds = \oint_C f dx + g dy + h dz = \oint_{C'} f dx + g dy + h(z_x dx + z_y dy)$$

$$= \oint_{C'} (f + h z_x) dx + (g + h z_y) dy$$

$$= \iint_D \left(\frac{\partial}{\partial x}(g + h z_y) - \frac{\partial}{\partial y}(f + h z_x) \right) d\sigma$$

$$= \iint_D \begin{pmatrix} (g_x + g_z z_x + (h_x + h_z z_x) z_y + h z_{yx}) \\ -(f_y + f_z z_y + (h_y + h_z z_y) z_x + h z_{xy}) \end{pmatrix} d\sigma$$

$$= \iint_D (g_x - f_y) + z_x(g_z - h_y) + z_y(h_x - f_z) d\sigma$$

$$= \iint_D \langle h_y - g_z, f_z - h_x, g_x - f_y \rangle \cdot \langle -z_x, -z_y, 1 \rangle d\sigma$$

$$= \iint_S \nabla \times \mathbf{F} \cdot \mathbf{N} dS.$$

This completes the proof.

Interpretation of curl

Now we can shed more light on the meaning of the curl vector using Stokes theorem. Let P_0 be a point on the surface S and let S_{P_0} be a very small patch of S containing P_0. Let $A(S_{P_0})$ be the area of the small patch. Then, under the continuity assumption, we have

$$\oint_C \mathbf{F} \cdot d\mathbf{r} = \iint_{S_{P_0}} (\text{curl } \mathbf{F} \cdot \mathbf{N}) dS \approx (\text{curl } \mathbf{F}(P_0) \cdot \mathbf{N}(P_0)) \cdot A(S_{P_0}),$$

$$(\text{curl } \mathbf{F}(P_0) \cdot \mathbf{N}(P_0)) \approx \frac{\oint_C \mathbf{F} \cdot d\mathbf{r}}{A(S_{P_0})}.$$

When taking limit as the small patch contracts to P_0, we see that curl $\mathbf{F}(P_0) \cdot \mathbf{N}(P_0)$ is the circulation density at the point P_0. Thus, the integration of curl $\mathbf{F} \cdot \mathbf{N}$ generates the total circulation along the boundary curve C. Also, one sees that the greatest circulation occurs when curl $\mathbf{F}$ is parallel to $\mathbf{N}$, in which case we have the greatest curling effect.

4.11 Review

Main concepts discussed in this chapter are listed below.
1. Line integral of $f(x, y)$ or $f(x, y, z)$ along a curve C with respect to arc length:

$$\int_C f(x, y) ds \quad \text{or} \quad \int_C f(x, y, z) ds.$$

2. Some equivalent notations for the line integral of a vector field $\mathbf{F} = \langle f, g \rangle$ along a curve C:

$$\int_C \mathbf{F} \cdot \mathbf{T} ds = \int_C \mathbf{F} \cdot d\mathbf{r} = \int_C f dx + g dy,$$

$$\int_C \mathbf{F} \cdot \mathbf{N}\,ds = \int_C \mathbf{F} \cdot d\mathbf{s} = \int_C f\,dy - g\,dx.$$

3. The fundamental theorem of line integrals:

$$\int_C \nabla\varphi \cdot d\mathbf{r} = \varphi(B) - \varphi(A), \quad \text{where points } A \text{ and } B \text{ are two endpoints of } C.$$

4. Circulation and flux integral for $\mathbf{F} = \langle f, g \rangle$:

$$\int_C \mathbf{F} \cdot d\mathbf{r} = \int_C f\,dx + g\,dy \quad \text{circulation integral,}$$

$$\int_C \mathbf{F} \cdot d\mathbf{s} = \int_C f\,dy - g\,dx \quad \text{flux integral.}$$

5. Green's theorem:

$$\oint_C f\,dx + g\,dy = \iint_D \left(\frac{\partial g}{\partial x} - \frac{\partial f}{\partial y}\right) d\sigma \quad \text{circulation-curl form,}$$

$$\oint_C f\,dy - g\,dx = \iint_D \left(\frac{\partial f}{\partial x} + \frac{\partial g}{\partial y}\right) d\sigma \quad \text{flux-divergence form.}$$

6. Under suitable conditions,

vector field $\mathbf{F} = \langle f, g \rangle$ is conservative

$\iff$ there is a function $\varphi(x, y)$ such that $\mathbf{F} = \nabla\varphi$

$\iff$ there is a function $\varphi(x, y)$ such that $d\varphi(x, y) = f\,dx + g\,dy$

$\iff \int_C \mathbf{F} \cdot d\mathbf{r}$ is path-independent

$\iff \oint_C \mathbf{F} \cdot d\mathbf{r} = 0$

$\iff \varphi_{xy} = \varphi_{yx} \left(\dfrac{\partial g}{\partial x} = \dfrac{\partial f}{\partial y}\right).$

7. Under suitable conditions

vector field $\mathbf{F} = \langle f, g \rangle$ is source-free

$\iff$ there is a function $\psi(x, y)$ such that $d\psi(x, y) = f\,dy - g\,dx$

$\iff \int_C \mathbf{F} \cdot d\mathbf{s}$ is path-independent

$\iff \oint_C \mathbf{F} \cdot d\mathbf{s} = 0$

$\iff \psi_{xy} = \psi_{yx} \left(\dfrac{\partial g}{\partial x} + \dfrac{\partial f}{\partial y} = 0\right).$

8. Surface integral with respect to surface area:

$$\iint_S f(\mathbf{r}(u,v))dS = \iint_{D_{uv}} f(\mathbf{r}(u,v))|\mathbf{r}_u \times \mathbf{r}_v|dudv \quad \text{for surface } \mathbf{r} = \mathbf{r}(u,v),$$

$$\iint_S f(x,y,z)dS = \iint_{D_{xy}} f(x,y,z)\sqrt{1+z_x^2+z_y^2}dxdy \quad \text{for surface } z = f(x,y),$$

$$\iint_S f(x,y,z)dS = \iint_{D_{xy}} f(x,y,z)\frac{|\nabla F|}{|F_z|}dxdy \quad \text{for surface } F(x,y,z) = 0.$$

Similar results hold for surfaces that can be projected to coordinate planes other than the xy-plane.

9. Divergence of a vector field $\mathbf{F} = \langle f, g, h \rangle$:

$$\text{Div}(\mathbf{F}) = \frac{\partial f}{\partial x} + \frac{\partial g}{\partial y} + \frac{\partial h}{\partial z} = \nabla \cdot \mathbf{F}.$$

10. Some equivalent notations:

$$\iint_S (\mathbf{F} \cdot \mathbf{N})dS = \iint_S \mathbf{F} \cdot d\mathbf{S} = \iint_S fdydz + gdzdx + hdxdy.$$

11. Flux of a vector field $\mathbf{F} = \langle f, g, h \rangle$ crossing a given surface S in a given direction:

$$\iint_S \mathbf{F} \cdot d\mathbf{S} = \pm \iint_{D_{xy}} (-fz_x - gz_y + h)dxdy.$$

Similar results hold for a surface that can be projected onto coordinate planes other than the xy-plane.

12. The divergence theorem: the outward flux crossing a closed surface S is

$$\iint_S fdydz + gdzdx + hdxdy = \iiint_\Omega \left(\frac{\partial f}{\partial x} + \frac{\partial g}{\partial y} + \frac{\partial h}{\partial z}\right)dV,$$

$$\iint_S \mathbf{F} \cdot d\mathbf{S} = \iiint_\Omega (\nabla \cdot \mathbf{F})dV.$$

13. The curl of a vector field $\mathbf{F} = \langle f, g, h \rangle$:

$$\text{curl } \mathbf{F} = \left(\frac{\partial h}{\partial y} - \frac{\partial g}{\partial z}\right)\mathbf{i} + \left(\frac{\partial f}{\partial z} - \frac{\partial h}{\partial x}\right)\mathbf{j} + \left(\frac{\partial g}{\partial x} - \frac{\partial f}{\partial y}\right)\mathbf{k}$$

$$= \begin{vmatrix} \mathbf{i} & \mathbf{j} & \mathbf{k} \\ \frac{\partial}{\partial x} & \frac{\partial}{\partial y} & \frac{\partial}{\partial z} \\ f & g & h \end{vmatrix} = \nabla \times \mathbf{F}.$$

14. Stokes theorem:

$$\oint_C f dx + g dy + h dz = \iint_S \left(\frac{\partial h}{\partial y} - \frac{\partial g}{\partial z} \right) dy dz + \left(\frac{\partial f}{\partial z} - \frac{\partial h}{\partial x} \right) dz dx + \left(\frac{\partial g}{\partial x} - \frac{\partial f}{\partial y} \right) dx dy,$$

$$\oint_C \mathbf{F} \cdot d\mathbf{r} = \iint_S (\nabla \times \mathbf{F} \cdot \mathbf{N}) dS = \iint_S \text{curl } \mathbf{F} \cdot d\mathbf{S}.$$

15. Under suitable conditions, for an irrotational vector field:

$$\text{curl } \mathbf{F} = \mathbf{0} \iff \text{there is a function } \varphi \text{ such that } \mathbf{F} = \nabla \varphi$$
$$\iff d\varphi = f dx + g dy + h dz$$
$$\iff \int_C \mathbf{F} \cdot d\mathbf{r} \text{ is path-independent.}$$

4.12 Exercises

4.12.1 Line integrals

1. Evaluate each of the following line integrals for the given curve C:
   (1) $\int_C \sqrt{2y} ds$, $C : x = a(t - \sin t), y = a(1 - \cos t), 0 \le t \le 2\pi$,
   (2) $\int_C (x + y) ds$, C consists of three line segments with vertices $(0,0)$, $(1,0)$, and $(0,1)$,
   (3) $\int_C \cos \sqrt{x^2 + y^2} ds$, C is the boundary of the region in the first quadrant bounded by $x = y$, $y = \sqrt{R^2 - x^2}$, and $y = 0$,
   (4) $\int_C \sqrt{2y^2 + z^2} ds$, $C : \begin{cases} x^2+y^2+z^2=a^2, \\ x-y=0, \end{cases}$
   (5) $\int_C (x^2 + y^2 + z^2) ds$, $C : x = e^t \cos t, y = e^t \sin t$, and $z = e^t$, $0 \le t \le 2\pi$.
2. Evaluate each of the following line integrals of a vector field:
   (1) $\int_C (x^2 + 2xy) dx + (y^2 - 2xy) dy$, C is the arc of the parabola $y = x^2$ from $(-1, 1)$ to $(1, 1)$,
   (2) $\int_C (y^2 - z^2) dx + 2yz dy - x^2 dz$, $C : x = t, y = t^2, z = t^3, 0 \le t \le 1$,
   (3) $\oint_C x dy$, C is the triangular path consisting of the line segments from $(0, 0)$ to $(2, 0)$, from $(2, 0)$ to $(0, 3)$, and from $(0, 3)$ to $(0, 0)$,
   (4) $\int_C \mathbf{F} \cdot \mathbf{T} ds$, where $\mathbf{F} = \langle -y, x \rangle$ and C is the unit circle with counterclockwise orientation,
   (5) $\int_C \mathbf{F} \cdot d\mathbf{r}$, where $\mathbf{F} = \langle y^2 \cos z, x^2, zy \rangle$ and $C: \mathbf{r}(t) = \langle 2\cos t, 2\sin t, 3t \rangle, 0 \le t \le \pi$.
3. Evaluate each of the following line integrals:
   (1) $\int_C \nabla(x^2 + y^2) \cdot d\mathbf{r}$, where C is the curve $\mathbf{r}(t) = \langle \frac{1}{1+t^2}, \cos(t\pi) \rangle, 0 \le t \le 1$,
   (2) $\int_C \nabla f \cdot d\mathbf{r}$, where $f(x, y, z) = e^{xy} + z^2$ and $C: \mathbf{r}(t) = \langle \ln(1 + t^2), t, \frac{8}{\pi} \arctan t \rangle$, $0 \le t \le 1$.
4. Use Green's theorem to evaluate each of the following line integrals (assume the boundary of each region is positively orientated):

(1) $\oint_C (x+y)^2 dx + (x^2-y^2)dy$, C is the triangle with vertices $A(1,1)$, $B(3,3)$, and $C(3,5)$,

(2) $\oint_C xy^2 dx - x^2 y dy$, C is the circle $x^2+y^2 = R^2$,

(3) $\int_C (y+2xy)dx + (x^2+2x+y^2)dy$, C is the top-half-arc of the circle $x^2+y^2 = 4x$ from $(4,0)$ to $(0,0)$,

(4) $\oint_C \mathbf{F}\cdot d\mathbf{r}$, where $\mathbf{F}(x,y) = \langle e^x(1-\cos y), e^x(y-\sin y)\rangle$ and C is the boundary of the region enclosed by $x=0$, $x=\pi$, $y=0$, and $y = \sin x$,

(5) $\oint_C \nabla(e^x + \sin(yx^2))\cdot d\mathbf{r}$, where C is any smooth simple closed curve in the xy-plane,

(6) $\int_C \mathbf{F}\cdot \mathbf{T} ds$, where $\mathbf{F}(x,y) = \langle e^{x^4} + y^2, xy + \sin(\ln y)\rangle$ and C is the boundary of the quadrilateral with vertices $(1,1)$, $(1,2)$, $(2,3)$, and $(2,1)$.

5. Determine whether each of the following vector fields is conservative; if so, find a potential function:
   (1) $\mathbf{F} = x\mathbf{i} - y\mathbf{j}$,
   (2) $\mathbf{F} = \langle \tan y, x\sec^2 y\rangle$,
   (3) $\mathbf{F} = \langle 1 - ye^{-x}, e^{-x}\rangle$,
   (4) $\mathbf{F} = \langle y + 2xy, x^2 + x + y^2\rangle$,
   (5) $\mathbf{F} = -y\mathbf{i} + x\mathbf{j}$,
   (6) $\mathbf{F} = \langle e^x \cos y, -e^x \sin y\rangle$,
   (7) $\mathbf{F} = (x^2 + 2xy - y^2)\mathbf{i} + (x^2 - 2xy - y^2)\mathbf{j}$

6. Use a line integral to find the area of the region enclosed by the curve $x = a\cos^3 t$ and $y = a\sin^3 t$.

7. Use Green's theorem to prove that the centroid of a plane region D in the xy-plane has coordinates $(\bar{x}, \bar{y})$ given by

$$\bar{x} = \frac{1}{2A(D)}\oint_C x^2 dy \quad \text{and} \quad \bar{y} = -\frac{1}{2A(D)}\oint_C y^2 dx,$$

where $A(D)$ is the area of the region D. Hence, find the coordinates of the centroid of the semicircle $y = \sqrt{a^2 - x^2}$.

8. Evaluate the outward flux of each of the following vector fields across the given curve C:
   (1) $\mathbf{F} = xy^2\mathbf{i} + xy\mathbf{j}$, and C is the boundary of the annulus $1 \le x^2 + y^2 \le 4$,
   (2) $\mathbf{F} = \langle -y, x\rangle$, and C is the circle with center the origin and radius a.

9. Consider the vector field $\mathbf{F} = \frac{x}{x^2+y^2}\mathbf{i} + \frac{y}{x^2+y^2}\mathbf{j}$.
   (1) Show that $\text{Div}(\mathbf{F}) = 0$.
   (2) Show that the outward flux across any circle centered at $(0,0)$ with radius a is 2π.
   (3) Does this example contradict Green's theorem in flux-divergence form? Explain.

4.12.2 Surface integrals

1. Evaluate each of the following surface integrals:
   (1) $\iint_S (x^2 + y^2) dS$, S is the sphere $x^2 + y^2 + z^2 = R^2$,

(2) $\iint_S xyz\, dS$, S is the part of the plane $x + y + z = 1$ that lies in the first octant,

(3) $\iint_S (xy + yz + zx)\, dS$, S is the part of the cone $z = \sqrt{x^2 + y^2}$ that lies inside the cylinder $x^2 + y^2 = 2x$.

2. Evaluate each surface integral $\iint_S \mathbf{F} \cdot d\mathbf{S}$ for each of the following vector fields $\mathbf{F}$ and oriented surfaces S:

   (1) $\mathbf{F} = \langle 0, 0, xyz \rangle$, S is the part of the cylinder $x^2 + z^2 = R^2$ in the first and fifth octants and between the two planes $y = 0$ and $y = h$, with outward orientation,

   (2) $\mathbf{F} = (y - z)\mathbf{i} + (z - x)\mathbf{j} + (x - y)\mathbf{k}$, S is the surface of the region E bounded by the cone $z = \sqrt{x^2 + y^2}$ and the plane $z = 1$, with outward orientation.

3. Evaluate each of the following surface integrals $\iint_S P\,dydz + Q\,dxdz + R\,dxdy$:

   (1) $\iint_S xy\,dydz + yz\,dxdz + zx\,dxdy$, where S is the surface of the solid bounded by $z = 0, y = 0, z = 0$, and $x + y + z = 1$, oriented outward,

   (2) $\iint_S (e^{-x^2y} + x)\,dydz + (2e^{-x^2y} + y)\,dxdz + (e^{-x^2y} + z)\,dxdy$, where S is the part of the plane $x - y + z = 1$ that is in the fourth octant, oriented outward,

   (3) $\iint_S (x^2 - y)\,dxdz + \sin(xy)\,dxdy$, where S is the part of the cylinder $x^2 + y^2 = 1$ that is cut by $z = 0$ and $z = 2$, oriented outward.

4. Compute the divergence of each of the following vector fields:

   (1) $\mathbf{F} = (x^2 + \sin y^2)\mathbf{i} + (y^2 - x)\mathbf{j}$, (2) $\mathbf{F} = \langle x + x^3 + yz^2, e^{-x^2} + \ln(y^2 + 1), z + xy \rangle$,

   (3) $\mathbf{F} = (x^3 + yz)\mathbf{i} - xz\mathbf{j} + yz\mathbf{k}$, (4) $\mathbf{F} = \langle x - \frac{1}{1+xy^2}, \tan^{-1} z + y, z^2 + 3x \rangle$.

5. Use the divergence theorem to find $\iint_S x^3\,dydz + y^3\,dzdx + z^3\,dxdy$, where S is the top half of the sphere $x^2 + y^2 + z^2 = a^2$, with outward orientation.

6. Evaluate the integral

$$\iint_S \frac{x\,dydz + y\,dzdx + z\,dxdy}{\sqrt{(x^2 + y^2 + z^2)^3}},$$

where S is the ellipsoid $x^2 + 2y^2 + 5z^2 = 1$, with outward orientation.

7. We define the solid W by $x^2 + y^2 + z^2 \leq 1$, and

$$\mathbf{F} = \left\langle x^3 + 3x + \frac{1}{z^2 + y^2 + 1}, y^3 + xy, z^3 - xz + \sin(xy) \right\rangle$$

is a vector field.

   (1) Compute the divergence of $\mathbf{F}$.
   (2) Find the flux out of W (that is, evaluate $\iint_S \mathbf{F} \cdot \mathbf{N}\,dS$).

8. To evaluate $\iint_\Sigma xyz\,dxdy$, where $\Sigma : x^2 + y^2 + z^2 = 1, (x \geq 0, y \geq 0)$, oriented outward, two students provided the following solutions:

   Solution 1
   The integration surface is symmetric about the xy-plane, with half the plane above the xy-plane and the other half below the xy-plane. The integrand xyz, if keeping

x and y fixed, is an odd function with respect to the variable z. Therefore

$$\iint_\Sigma xyz dxdy = \iint_{\Sigma, z\geq 0} xyz dxdy + \iint_{\Sigma, z\leq 0} xyz dxdy$$

$$= \iint_{\Sigma, z\geq 0} xy\sqrt{1-x^2-y^2} dxdy + \iint_{\Sigma, z\leq 0} xy(-\sqrt{1-x^2-y^2}) dxdy = 0.$$

Solution 2
The second student writes

$$\iint_\Sigma xyz dxdy = \iint_\Sigma xy\sqrt{1-x^2-y^2} dxdy$$

$$= \int_0^{\frac{\pi}{2}} d\theta \int_0^1 (r\cos\theta)(r\sin\theta)\sqrt{1-r^2} rdr$$

$$= \int_0^{\frac{\pi}{2}} (\cos\theta \sin\theta) d\theta \int_0^1 r^3 \sqrt{1-r^2} dr = \frac{1}{15}.$$

Is one of the solutions correct, and if so, which one?
Evaluate the integral directly first as a flux integral, and then evaluate it by using the divergence theorem. Hint: Add some surfaces so that one has a closed surface.

9. Compute the curl for each of the following vector fields:
   (1) $\mathbf{F} = (e^x \cos y)\mathbf{i} - (e^x \sin y)\mathbf{j}$,   (2) $\mathbf{F} = \langle x+y, x^2+2z, 2y-xz \rangle$,
   (3) $\mathbf{F} = \langle z, 1, x \rangle$,   (4) $\mathbf{F} = \frac{\langle x,y,z \rangle}{\sqrt{x^2+y^2+z^2}}$.

10. Given a vector field $\mathbf{F} = (2xy - z^2)\mathbf{i} + (x^2 + 2z)\mathbf{j} + (2y - 2xz)\mathbf{k}$.
    (1) Show that $\mathbf{F}$ is conservative.
    (2) Find a potential function for this vector field.
    (3) Evaluate $\int_C \mathbf{F} \cdot d\mathbf{r}$, where C is the curve $\mathbf{r}(t) = \langle 1+t^3, \frac{9}{1+t^2}, \sin(\frac{\pi}{4}t) \rangle$ for $0 \leq t \leq 2$.

11. Given $\mathbf{F} = \langle y + \sin x, z^2 + \cos y, x^3 \rangle$.
    (1) Find Curl($\mathbf{F}$).
    (2) Evaluate $\int_C (y + \sin x) dx + (z^2 + \cos y) dy + x^3 dz$, where C is the curve $\mathbf{r}(t) = \langle \sin t, \cos t, \sin 2t \rangle$, $0 \leq t \leq 2\pi$.
    (3) Evaluate $\int_C \mathbf{F} \cdot d\mathbf{r}$, where C is the curve $\mathbf{r}(t) = \langle \cos t, \sin t + 2\cos t, \sin t \rangle$, for $0 \leq t \leq 2\pi$, and is orientated counterclockwise as viewed from above.

12. If $\mathbf{F}(x, y, z) = (x + y^2)\mathbf{i} + (y + z^2)\mathbf{j} + (z + x^2)\mathbf{k}$, find $\int_C \mathbf{F} \cdot d\mathbf{r}$, where C is the triangle with vertices $(1, 0, 0)$, $(0, 1, 0)$, and $(0, 0, 1)$ and is orientated counterclockwise as viewed from above.

13. Let S be the part of the spherical surface $x^2 + y^2 + z^2 = 2$ lying in $z > 1$. Orient S upwards and let C be its bounding circle lying in the plane $z = 1$ with compatible orientation.

(a) Parameterize C and use the parameterization to evaluate the line integral

$$I = \oint_C xz\,dx + x\,dy + 4\,dz.$$

(b) Compute the curl of the vector field $\mathbf{F} = \langle xz, x, 4 \rangle$.
(c) Evaluate the flux integral $\iint_S (\nabla \times \mathbf{F} \cdot \mathbf{N})\,dS$ directly.
(d) Which theorem can be applied to evaluate (c) by using I directly?

14. Let $\mathbf{F} = \langle 2x - y + z, x + y + z, 2y - 3z \rangle$.
   (a) Find $\nabla \cdot \mathbf{F}$, the divergence of $\mathbf{F}$.
   (b) Find $\nabla \times \mathbf{F}$, the curl of $\mathbf{F}$.
   (c) Evaluate $\iint_S \mathbf{F} \cdot d\mathbf{S}$, where S:

   $$\{(x,y,z) \mid |2x - y + z| + |x + y + z| + |2y - 3z| = 1\}$$

   with outward orientation, using the divergence theorem.
   (d) Evaluate $\oint_C \mathbf{F} \cdot d\mathbf{r}$, where $C : \begin{cases} x^2 + 2y^2 = 4, \\ z = 4, \end{cases}$ oriented counterclockwise as viewed from above, using Stokes theorem.

5 Introduction to ordinary differential equations

As we have seen before, equations are used as mathematical models built to solve practical problems. Algebra is sufficient to solve static problems. However, in many cases, natural phenomena involve quantities that are changing and can only be described by equations that describe these changes. Those changes usually are described by derivatives of some functions. An equation relating an unknown function and one or more of its derivatives is called a *differential equation*. In this chapter, we develop methods for finding exact solutions for certain types of differential equations. Also, we introduce some other approaches for finding approximate solutions by numerical or graphical methods.

5.1 Introduction

We first investigate several examples. We assume the population $P(t)$ is a function of time, t, subject to constant birth and death rates. Then the rate of change of P with respect to time t can be modeled as

$$\frac{dP}{dt} = kP, \quad \text{where } k = \text{birth rate} - \text{death rate} = \text{some constant}.$$

This equation involves the unknown function P and its first derivative P'. It is a *first-order differential equation* because it involves a first-order derivative. One can check that $P = Ce^{kt}$, where C is an arbitrary constant, is a solution to this equation. If k is positive, it is an exponential growth model, and if k is negative, it is an exponential decay model. This is the case for the population of a family of bacteria growing or disappearing over a short period of time. Also, many radioactive materials satisfy this law.

Newton's law of cooling says that the rate of change of the temperature $T(t)$ of a body is proportional to the difference between T and the temperature A of the surrounding medium. If we know $T(0) = 50\,°C$, then we have

$$\frac{dT}{dt} = -k(T - A),\, T(0) = 50\,°C,$$

where k is a positive constant. Note that if $T > A$, then $\frac{dT}{dt} < 0$, so the temperature is a decreasing function of t and the body is cooling, but if $T < A$, then $\frac{dT}{dt} > 0$, so that T is increasing. The condition $T(0) = 50\,°C$ is called an *initial condition*.

In a spring-mass model, the mass moves back and forth about an equilibrium point. If F is the force exerted by the spring on the mass and F_r is the resistance force, then by Newton's second law we have

$$F - F_r = -mx'',$$

where x is the displacement from the equilibrium point, and the derivative is with respect to time. By Hooke's law, we have $F = -kx$, where k is some constant. The negative sign in the right-hand side indicates that the resultant force and the displacement are in opposite directions. If F_r, the resistance force, is proportional to the velocity of the mass, then $F_r = lx'$ for some constant l, so we have

$$-kx - lx' = -mx'' \quad \text{or} \quad mx'' - lx' - kx = 0.$$

This equation involves an unknown function $x(t)$ and its first- and second-order derivatives, so it is a *second-order differential equation*.

The examples above are all *ordinary differential equations* (ODEs), since the unknown functions only depend on a single variable. The following differential equations are all examples of ODEs:

(a) $\dfrac{d^3y}{dx^3} + 2\dfrac{dy}{dx} - 3y = \dfrac{1}{x}$, (b) $y^{(4)} + y^{(2)}y = y'$,

(c) $\left(\dfrac{d^2x}{dt^2}\right)^3 + 2\left(\dfrac{dx}{dt}\right)^4 = x$, (d) $\dot{x} = 2\sqrt{x} + t^2$.

The *order* of an ODE is the highest derivative that is found in the ODE. Thus, in the above four ODEs, (a) is of order 3, (b) is of order 4, (c) is of order 2, and (d) has order 1.

The *degree* of an ODE is the highest power of the highest derivative in that ODE. Thus, in the above four ODEs, (a), (b), and (d) all have degree 1 while (c) has degree 3.

Scientists and economists also use *partial differential equations* (PDEs) to solve problems. For example, the heat equation is

$$\frac{\partial u}{\partial t} - \alpha\left(\frac{\partial^2 u}{\partial x^2} + \frac{\partial^2 u}{\partial y^2} + \frac{\partial^2 u}{\partial z^2}\right) = 0,$$

where $u(x, y, z, t)$ is the temperature of a body and α is the thermal diffusivity. The wave equation

$$\frac{\partial^2 u}{\partial t^2} = c^2 \frac{\partial^2 u}{\partial x^2},$$

the harmonic equation

$$\frac{\partial^2 u}{\partial t^2} + \frac{\partial^2 u}{\partial x^2} = 0,$$

and the famous Black–Scholes model for option pricing

$$\frac{\partial V}{\partial t} + \frac{1}{2}\sigma^2 S^2 \frac{\partial^2 V}{\partial S^2} = rV - rS\frac{\partial V}{\partial S}$$

are all examples of PDEs.

Both ODEs and PDEs have enormously important applications, as seen above. In this text, we only consider some ODEs.

5.2 First-order ODEs

5.2.1 General and particular solutions and direction fields

We will discuss a *first-order differential equation* that can be written in the form

$$\frac{dy}{dx} = f(x,y).$$

Example 5.2.1. Solve the differential equation $\frac{dy}{dx} = 2x$.

Solution. Integrating both sides gives all solutions, i.e.,

$$y(x) = \int 2x\,dx,$$
$$y = x^2 + C.$$

This is a *general solution* of the differential equation $\frac{dy}{dx} = 2x$ since it gives every possible solution to the equation. If, furthermore, we know $y(1) = 2$, then we will be able to determine the constant $C = 1$ to obtain the *particular solution* $y = x^2 + 1$.

The graphs of a general solution are a family of curves, called *solution curves* or *integral curves*. Figure 5.1 shows some solution curves for $C = 0, \pm 1$, and 3.

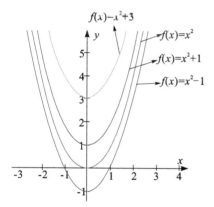

Figure 5.1: Example 5.2.1, some solution curves to $y' = 2x$.

In general, a general solution to a differential equation

$$\frac{dy}{dx} = f(x,y)$$

involves an arbitrary constant, and the graphs are a family of curves.

The problem

$$\begin{cases} \frac{dy}{dx} = f(x,y), \\ y(a) = b, \end{cases} \text{ alternatively written as } \frac{dy}{dx} = f(x,y), \ y(a) = b,$$

is called an *initial value problem*. The solution to an initial value problem is called a *particular solution*.

There is a nice theorem about the existence of a particular solution to an initial value problem. We give, without proof, the theorem that guarantees the existence of such solutions for a first-order ODE of the form $\frac{dy}{dx} = f(x,y)$.

Theorem 5.2.1 (Existence and uniqueness of solutions). *Suppose that the function $f(x,y)$ and its partial derivatives $f_x(x,y)$ and $f_y(x,y)$ are continuous on some region R in the xy-plane that contains the point (a,b) in its interior. Then the initial value problem*

$$\frac{dy}{dx} = f(x,y), \quad y(a) = b$$

has one and only one solution that is defined on an open interval I containing the point a.

When given a first-order ODE in the form

$$\frac{dy}{dx} = f(x,y) \quad \text{or} \quad F(x,y,y') = 0,$$

we may not be able to see its solution at first sight, for example,

$$\frac{dy}{dx} = x^2 + y^2 \quad \text{or} \quad \frac{dy}{dx} = x^3 - 2xy.$$

However, we do know, at each point, the derivative of the unknown function $y(x)$. The derivative is the slope of the tangent line at that point, so we can sketch a small line segment to indicate its tangent at some points. For example, for the differential equation $\frac{dy}{dx} = x^2 + y^2$, we can compute the derivative at points $(0,0)$, $(1,1)$, and $(2,1)$ to obtain $y'(0) = 0$, $y'(1) = 2$, and $y'(2) = 5$. Thus, we can sketch a diagram as in Figure 5.2(a).

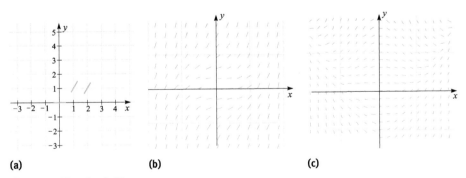

(a) (b) (c)

Figure 5.2: Direction fields.

This type of diagram is called a *direction field* or *slope field* of the differential equation. With the help of a computer algebra system, we have the direction fields for the above two differential equations as shown in Figure 5.2(b) and (c). If we know an additional condition, $y(x_0) = y_0$, then we are able to sketch a solution curve that passes through the point (x_0, y_0). Several particular solution curves for $y' = x^3 - 2xy$ are shown in Figure 5.3.

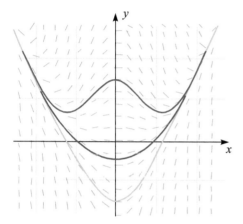

Figure 5.3: Direction field and solution curves.

5.2.2 Separable differential equations

For some first-order ODEs of the form

$$\frac{dy}{dx} = f(x, y),$$

if we can factor $f(x, y) = h(x)g(y)$, then we may separate the variables x and y to obtain

$$\frac{dy}{dx} = f(x, y) = h(x)g(y),$$

$$\frac{dy}{g(y)} = h(x)dx.$$

Then, we can integrate both sides separately, the left side as a function of y and the right side as a function of x, i.e.,

$$\int \frac{1}{g(y)} dy = \int h(x) dx.$$

In this way, we may be able to find an exact solution.

Example 5.2.2. Solve the exponential growth/decay model equation

$$\frac{dP}{dt} = kP, \quad \text{where } k \text{ is a nonzero constant.}$$

When $k < 0$, find the time when $P = P_0/2$, where $P_0 = P(0)$.

Solution. We separate the variables to obtain

$$\frac{1}{P}dP = kdt.$$

We integrate

$$\int \frac{1}{P}dP = \int kdt$$

to get

$$\ln|P| = kt + C_1, \quad \text{where } C_1 \text{ is an arbitrary constant.}$$

Simplifying this we obtain $P = \pm e^{kt+C_1} = \pm e^{C_1}e^{kt}$. Note that $\pm e^{C_1}$ is also an arbitrary constant except 0, but $y = 0$ is a solution, so we can write this as

$$P = Ce^{kt}, \quad \text{where } C \text{ is an arbitrary constant.}$$

When $t = 0$, $P = P_0$, this means $C = P_0$ and $P = P_0 e^{kt}$. Set $P = P_0/2$. We have

$$P_0/2 = P_0 e^{kt},$$
$$\ln \frac{1}{2} = kt,$$
$$t = \frac{-\ln 2}{k}, \quad \text{for } k < 0.$$

Note. The radioactive half-life for a given radioisotope is given by the above formula.

Example 5.2.3. Suppose a curve $y = y(x)$ has a derivative $2x(y^2 + 1)$ at each point (x, y).
1. Find any such curve.
2. Furthermore, if the curve passes through the point $(0, 1)$, then find this particular curve.

Solution.
1. The curve $y = y(x)$ satisfies the differential equation

$$\frac{dy}{dx} = 2x(y^2 + 1).$$

We separate the variables to obtain

$$\frac{1}{y^2+1} dy = 2x dx.$$

Then, we integrate

$$\int \frac{1}{y^2+1} dy = \int 2x dx$$

to obtain $\tan^{-1} y = x^2 + C$. Thus, we have $y = \tan(x^2 + C)$.

2. If, furthermore, we know $y(0) = 1$, then $1 = \tan C$ and $C = \frac{\pi}{4}$. So we get the particular curve

$$y = \tan\left(x^2 + \frac{\pi}{4}\right).$$

5.2.3 Substitution methods

A *homogeneous* first-order differential equation $\frac{dy}{dx} = f(x, y)$ is one where $f(x, y)$ can be rewritten as a function of $\frac{y}{x}$, i. e.,

$$\frac{dy}{dx} = F\left(\frac{y}{x}\right). \tag{5.1}$$

This typically happens when $f(x, y)$ is made up of polynomial terms in x and y where the exponents of x and y add to the same value for all such terms.

If $\frac{dy}{dx} = F(\frac{y}{x})$ and we make the substitution

$$v = \frac{y}{x} \quad \text{so that } y = vx \text{ and } \frac{dy}{dx} = v + x\frac{dv}{dx},$$

then the differential equation (5.1) is transformed into a separable equation with independent variable x and dependent variable v, i. e.,

$$v + x\frac{dv}{dx} = F(v),$$

$$x\frac{dv}{dx} = F(v) - v.$$

So we may solve the differential equation by using separation of variables.

Example 5.2.4. Solve the differential equation

$$2xy\frac{dy}{dx} = 4x^2 + 3y^2.$$

Solution. This is a homogeneous equation (the degree of each term is the same – two in this case). We rewrite it as

$$\frac{dy}{dx} = \frac{4x^2 + 3y^2}{2xy} = 2\left(\frac{x}{y}\right) + \frac{3}{2}\left(\frac{y}{x}\right).$$

Then, the substitution $y = vx$, $\frac{dy}{dx} = v + x\frac{dv}{dx}$, transforms this to

$$v + x\frac{dv}{dx} = \frac{2}{v} + \frac{3}{2}v,$$

$$x\frac{dv}{dx} = \frac{4 + v^2}{2v},$$

and, hence,

$$\int \frac{2v}{v^2 + 4}\,dv = \int \frac{1}{x}\,dx,$$

$$\ln(v^2 + 4) = \ln|x| + C_1.$$

Thus,

$$v^2 + 4 = |x|e^{C_1},$$

$$\frac{y^2}{x^2} + 4 = Cx, \quad \text{where } C = \pm e^{C_1},$$

$$y^2 + 4x^2 = Cx^3,$$

where $C > 0$ if $x > 0$, and $C < 0$ if $x < 0$.

Example 5.2.5. Solve the differential equation

$$\frac{dy}{dx} = (x + y + 3)^2.$$

Solution. This is not separable or homogeneous. Let us try to simplify it by making the substitution $v = x + y + 3$, so $y = v - x - 3$. Differentiating this gives

$$\frac{dy}{dx} = \frac{dv}{dx} - 1.$$

So the transformed equation is

$$\frac{dv}{dx} = 1 + v^2.$$

This is a separable equation and

$$\int \frac{dv}{1 + v^2} = \int dx,$$

$$\tan^{-1} v = x + C,$$
$$v = \tan(x + C).$$

So
$$y(x) = \tan(x + C) - x - 3.$$

5.2.4 Exact differential equations

Suppose that an ODE has the form
$$f(x,y) + g(x,y)\frac{dy}{dx} = 0 \quad \text{or} \quad \frac{dy}{dx} = -\frac{f(x,y)}{g(x,y)},$$

or, equivalently, its differential form
$$f(x,y)dx + g(x,y)dy = 0. \tag{5.2}$$

If there exists a function $\varphi(x,y)$ such that
$$\frac{\partial \varphi(x,y)}{\partial x} = f(x,y) \quad \text{and} \quad \frac{\partial \varphi(x,y)}{\partial y} = g(x,y),$$

then
$$f(x,y)dx + g(x,y)dy = \frac{\partial \varphi(x,y)}{\partial x}dx + \frac{\partial \varphi(x,y)}{\partial y}dy = d\varphi(x,y) = 0.$$

This means that $\varphi(x,y) = C$, which is a general solution of the differential equation $f(x,y)dx + g(x,y)dy = 0$.

Equations of this type are called *exact differential equations*. In Chapter 4, we showed the following result holds.

Theorem 5.2.2 (Criterion for exactness). *Suppose that the functions $f(x,y)$ and $g(x,y)$ are continuous and have continuous first-order derivatives in the simply connected region D. Then the ODE*
$$f(x,y)dx + g(x,y)dy = 0$$
is exact in D if and only if at each point of D
$$\frac{\partial f}{\partial y} = \frac{\partial g}{\partial x}.$$

Example 5.2.6. Solve the differential equation $\frac{dy}{dx} = \frac{y^2 - 2x + 3}{y - 2xy}$.

Solution. We rearrange the terms to obtain

$$(y^2 - 2x + 3)dx + (2xy - y)dy = 0.$$

Since

$$\frac{\partial(y^2 - 2x + 3)}{\partial y} = 2y = \frac{\partial(2xy - y)}{\partial x},$$

this is an exact differential equation. Assume $d\varphi(x,y) = (y^2 - 2x + 3)dx + (2xy - y)dy$. Then

$$\frac{\partial \varphi}{\partial x} = y^2 - 2x + 3 \quad \text{and} \quad \varphi(x,y) = \int (y^2 - 2x + 3)dx = y^2 x - x^2 + 3x + h(y).$$

To find $h(y)$, we differentiate $\varphi(x,y)$ with respect to y to obtain

$$\frac{\partial \varphi}{\partial y} = 2xy + 0 + h'(y).$$

But $\frac{\partial \varphi}{\partial y} = 2xy - y$, so $h'(y) = -y$ and such an $h(y) = -\frac{y^2}{2}$. Therefore,

$$\varphi(x,y) = y^2 x - x^2 + 3x - \frac{y^2}{2},$$

and

$$\varphi(x,y) = C \quad \text{or} \quad y^2 x - x^2 + 3x - \frac{y^2}{2} = C$$

is a general solution to the original differential equation. Figure 5.4 shows the direction field and several solution curves.

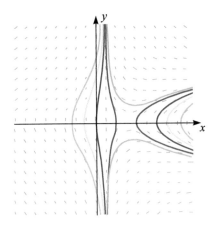

Figure 5.4: Direction field and solution curves, Example 5.2.6.

5.2.5 First-order linear differential equations

The first-order linear differential equation has the form

$$a(x)y' + b(x)y = c(x) \quad \text{or} \quad a(x)\frac{dy}{dx} + b(x)y = c(x). \tag{5.3}$$

The term "linear" refers to the "y" terms that appear as y and $\frac{dy}{dx}$, but not are raised to any power or combined in some other function, whereas $a(x)$, $b(x)$, and $c(x)$ are allowed to be nonlinear functions. Thus, in the differential equations

(a) $x^2 y' + 2y = \sin x$, (b) $y'^2 + x + 2y = 0$, (c) $yy' + x = 0$, (d) $y' + \sqrt{y} = x$,

only (a) is linear. If $a(x) = 0$, it would not be a differential equation, so we assume $a(x) \neq 0$ and dividing by $a(x)$ gives

$$\frac{dy}{dx} + \frac{b(x)}{a(x)} y = \frac{c(x)}{a(x)}. \tag{5.4}$$

Let $P(x) = \frac{b(x)}{a(x)}$ and $Q(x) = \frac{c(x)}{a(x)}$. Then equation (5.4) becomes

$$\frac{dy}{dx} + P(x)y = Q(x). \tag{5.5}$$

There is a nice technique for solving equation (5.5). Suppose there exists a function $\rho(x)$ such that multiplying both sides of the equation $\frac{dy}{dx} + P(x)y = Q(x)$ by $\rho(x)$ transforms the left-hand side into the derivative of the product $\rho(x) \times y$. Such a function $\rho(x)$ is called an *integrating factor*. Then,

$$\rho(x)\frac{dy}{dx} + P(x)\rho(x)y = \rho(x)Q(x),$$

$$\frac{d}{dx}(\rho(x)y) = \rho(x)Q(x). \tag{5.6}$$

We integrate both sides to obtain

$$\rho(x)y = \int \rho(x)Q(x)dx.$$

Therefore, we get a general solution

$$y = \frac{1}{\rho(x)} \int \rho(x)Q(x)dx. \tag{5.7}$$

Now the question left is, how do we find $\rho(x)$? Applying the product rule on the left-hand side of equation (5.6) gives

$$\rho(x)\frac{dy}{dx} + y\frac{d\rho(x)}{dx} = \rho(x)Q(x).$$

Comparing this with the original equation multiplied by $\rho(x)$ shows that

$$\frac{d\rho(x)}{dx} = P(x)\rho(x).$$

This is a separable equation, and solving it gives the integrating factor $\rho(x) = e^{\int P(x)dx}$. Substituting $\rho(x)$ in equation (5.7), we obtain a general solution to equation (5.5), i.e.,

$$y(x) = e^{-\int P(x)dx}\left(\int e^{\int P(x)dx} Q(x)dx + C\right). \tag{5.8}$$

Note. One can easily check that $y(x) = Ce^{-\int P(x)dx}$ is a general solution of the first-order linear homogeneous equation (right-hand side function $Q(x) = 0$)

$$\frac{dy}{dx} + P(x)y = 0.$$

Note that the function $y^* = e^{-\int P(x)dx}(\int e^{\int P(x)dx} Q(x)dx)$ is a particular solution to $\frac{dy}{dx} + P(x)y = Q(x)$. Thus, a general solution to $\frac{dy}{dx} + P(x)y = Q(x)$ can be written as

| a general solution of $\frac{dy}{dx} + P(x)y = 0$ | + | a particular solution of $\frac{dy}{dx} + P(x)y = Q(x)$. |

Example 5.2.7. Solve the first-order linear ODE $\frac{dy}{dx} = x^3 - 2xy$.

Solution. Since $P(x) = 2x$ and $Q(x) = x^3$, by equation (5.8), a general solution is

$$y = e^{-\int 2xdx}\left(\int e^{\int 2xdx} x^3 dx + C\right) = e^{-x^2}\left(\int e^{x^2} x^3 dx + C\right)$$

$$= e^{-x^2}\left(\frac{1}{2}\int x^2 e^{x^2} d(x^2) + C\right)$$

$$= e^{-x^2}\left(\frac{1}{2}x^2 e^{x^2} - \frac{1}{2}e^{x^2} + C\right)$$

$$= \frac{x^2}{2} - \frac{1}{2} + Ce^{-x^2}.$$

Figure 5.3 shows the direction field and several solution curves to this ODE.

Example 5.2.8. Solve the initial value problem

$$\frac{dy}{dx} - y = 2e^{-x/3}, \quad y(0) = -1.$$

Solution. This is a first-order linear differential equation with $P(x) = -1$ and $Q(x) = 2e^{-x/3}$, so a general solution is

$$y(x) = e^{-\int P(x)dx}\left(\int Q(x)e^{\int P(x)dx}dx + C\right)$$

$$= e^{-\int(-1)dx}\left(\int (2e^{-x/3}e^{\int(-1)dx})dx + C\right)$$

$$= e^{x}\left(-\frac{3}{2}e^{-\frac{4}{3}x} + C\right).$$

Substitution of $x = 0$ and $y = -1$ shows that $C = \frac{1}{2}$, so the desired particular solution is

$$y(x) = \frac{1}{2}e^{x} - \frac{3}{2}e^{-x/3}.$$

Example 5.2.9. Solve the differential equation $ydx + (y^3 - x)dy = 0$ (assume that $y > 0$).

Solution. If we rewrite the equation as

$$\frac{dy}{dx} + \frac{y}{y^3 - x} = 0,$$

it is not linear, homogeneous, separable, or exact as we have discussed so far. However, if we rewrite it as

$$\frac{dx}{dy} + \frac{y^3 - x}{y} = 0,$$

then

$$\frac{dx}{dy} - \frac{1}{y}x = -y^2.$$

It is now linear in x, as a function of y with $P = -\frac{1}{y}$ and $Q = -y^2$. The general solution is given by

$$x = e^{\int \frac{1}{y}dy}\left(\int y^2 e^{-\int \frac{1}{y}dy}dy + C_1\right) = e^{\ln y}\left(-\int y^2 e^{\ln y}dy + C_1\right)$$

$$= y\left(-\int y^2 \times \frac{1}{y}dy + C_1\right)$$

$$= y\left(-\frac{y^2}{2} + C_1\right).$$

A general solution is, therefore, $2x = -y^3 + Cy$ (we replaced the constant $2C_1$ by C).

Bernoulli equations

A first-order differential equation of the form

$$\frac{dy}{dx} + P(x)y = Q(x)y^n \tag{5.9}$$

is called a *Bernoulli equation*.

Remark. This type of equation was named after Jacob Bernoulli, who was one of the many prominent mathematicians in the Bernoulli family.

If either $n = 0$ or $n = 1$, then the equation is linear. Otherwise, dividing both sides by y^n and making the substitution

$$v = y^{1-n}$$

transforms it into the linear equation

$$\frac{dv}{dx} + (1-n)P(x)v = (1-n)Q(x).$$

Rather than memorizing the form of this transformed equation, it is more efficient to make the substitution explicitly, after dividing both sides by y^n, as in the following example.

Example 5.2.10. Solve the ODE

$$2xy' = 3y + 4x^2 y^3.$$

Solution. Divide the ODE by $2x$ to obtain

$$\frac{dy}{dx} - \frac{3}{2x}y = 2xy^3.$$

We see that this is a Bernoulli equation with $P(x) = -\frac{3}{2x}$ and $Q(x) = 2x$ with $n = 3$. We divide the equation by y^3 to obtain

$$y^{-3}\frac{dy}{dx} - \frac{3}{2x}y^{-2} = 2x.$$

Note that the first term $y^{-3}\frac{dy}{dx}$ is exactly $-\frac{1}{2}\frac{d(y^{-2})}{dx}$. Hence, we let $v = y^{-2}$, and the above equation becomes linear, i.e.,

$$-\frac{1}{2}\frac{d(y^{-2})}{dx} - \frac{3}{2x}y^{-2} = 2x,$$

$$\frac{dv}{dx} - \frac{3 \times (-2)}{2x}v = 2 \times (-2)x,$$

$$\frac{dv}{dx} + \frac{3}{x}v = -4x.$$

A general solution is

$$v = e^{-\int \frac{3}{x}dx}\left(\int -4xe^{\int \frac{3}{x}dx}dx + C\right)$$

$$= x^{-3}\left(\int -4x \times x^3 dx + C\right)$$

$$= x^{-3}\left(-\frac{4}{5}x^5 + C\right),$$

$$v = -\frac{4}{5}x^2 + \frac{C}{x^3}.$$

Since $v = y^{-2}$, a general solution to the original differential equation is

$$y^{-2} = -\frac{4}{5}x^2 + \frac{C}{x^3}.$$

5.3 Second-order ODEs

5.3.1 Reducible second-order equations

Differential equations of higher order appear in many applications in science and engineering. For example, the well-known simple harmonic motion equation

$$\frac{d^2x}{dt} = -\omega^2 x$$

is a second-order differential equation. A *second-order differential equation* involves the second derivative of an unknown function $y(x)$. Thus, it has the general form

$$F(x, y, y', y'') = 0. \tag{5.10}$$

A general solution to a second-order ODE involves two arbitrary constants. Thus, to find a particular solution, we need two additional conditions, say, $y(x_0) = a$ and $y'(x_0) = b$.

Some types of second-order ODEs can be reduced to a first-order equation and then solve the first-order equation using methods from previous sections. These are called reducible second-order ODEs. This is often the case if either the dependent variable y or the independent variable x is missing from a second-order ODE.

1. The dependent variable y and its derivative y' are missing

If y and y' are both missing, then the equation has the form $y'' = f(x)$. Thus, integrating the equation twice gives a general solution,

$$y' = \int f(x)dx + C_1,$$

$$y = \int\left[\int f(x)dx + C_1\right]dx + C_2.$$

Note. This is equivalent to solving two first-order ODEs $\frac{dP}{dx} = f(x)$ and $\frac{dy}{dx} = P$.

Example 5.3.1. Solve the differential equation

$$y'' = x + 1.$$

Solution. Integrating once gives

$$y' = \int (x+1)dx = \frac{x^2}{2} + x + C_1.$$

Integrating again gives

$$y = \int\left(\frac{x^2}{2} + x + C_1\right)dx = \frac{x^3}{6} + \frac{x^2}{2} + C_1 x + C_2.$$

2. The dependent variable y is missing

If y is missing, then equation (5.10) takes the form

$$F(x, y', y'') = 0. \tag{5.11}$$

The substitution

$$y' = p, \quad y'' = \frac{dp}{dx}$$

results in a first-order differential equation in p and x, i.e.,

$$F(x, p, p') = 0.$$

If we can find a general solution $p(x, C_1)$ involving an arbitrary constant C_1, then we can write the solution of the original equation as

$$y(x) = \int y'(x)dx = \int p(x, C_1)dx + C_2.$$

This gives us a solution of equation (5.11) that involves two arbitrary constants C_1 and C_2, as is to be expected in the case of a second-order differential equation.

Example 5.3.2. Solve the equation $xy'' + 2y' = 6x$, in which the dependent variable y is missing. Determine the particular solution if $y(1) = 2$ and $y'(1) = 1$.

Solution. Let $y' = p$, so that $y'' = \frac{dp}{dx}$. The substitution defined above gives the first-order equation

$$x\frac{dp}{dx} + 2p = 6x, \quad \text{that is,} \quad \frac{dp}{dx} + \frac{2}{x}p = 6.$$

This is linear (in p). So, solving it by the method given by equation (5.8) gives

$$p(x) = 2x + \frac{C_1}{x^2}.$$

This means $\frac{dy}{dx} = 2x + \frac{C_1}{x^2}$. A final integration with respect to x yields a general solution of the original equation $xy'' + 2y' = 6x$, i.e.,

$$y(x) = \int p(x)dx = \int \left(2x + \frac{C_1}{x^2}\right)dx$$

$$= x^2 - \frac{C_1}{x} + C_2.$$

Since $y'(1) = 1$, this means $p(1) = 1 = 2 + \frac{C_1}{1}$, so $C_1 = -1$. Since $y(1) = 2$, we have

$$2 = 1^2 - \frac{-1}{1} + C_2, \quad \text{so } C_2 = 0.$$

Thus, the particular solution is

$$y(x) = x^2 + \frac{1}{x}.$$

3. The independent variable x is missing

If x is missing, then equation (5.10) takes the form

$$F(y, y', y'') = 0. \tag{5.12}$$

The substitution

$$p = y', \quad y'' = \frac{dp}{dx} = \frac{dp}{dy}\frac{dy}{dx} = p\frac{dp}{dy}$$

results in a first-order differential equation in terms of p as a function of y, i.e.,

$$F\left(y, p, p\frac{dp}{dy}\right) = 0.$$

If we can solve this equation for a general solution $p(y, C_1)$ involving an arbitrary constant C_1, then (assuming that $y' \neq 0$) we can find a solution of the original equation, with x as a function of y, by solving the first-order ODE $\frac{dy}{dx} = p(y, C_1)$ as follows:

$$p(y, C_1) = \frac{dy}{dx}, \quad \text{so that } dx = \frac{1}{p(y, C_1)}dy,$$

$$x = \int \frac{1}{p(y, C_1)} dy + C_2.$$

This leads to the implicit solution $y = y(x)$ of equation (5.12).

Example 5.3.3. Solve the initial value problem $yy'' = (y')^2$ in which the independent variable x is missing, with the initial conditions $y(0) = 2$ and $y'(0) = 1$.

Solution. We substitute

$$y' = p \quad \text{and} \quad y'' = \frac{dp}{dx} = \frac{dp}{dy}\frac{dy}{dx} = p\frac{dp}{dy}.$$

The original equation becomes

$$yp\frac{dp}{dy} = p^2.$$

One solution is $p = 0 \Longrightarrow y = k$ is a constant. Otherwise divide by p and use separation of variables. Then we have

$$\int \frac{dp}{p} = \int \frac{dy}{y},$$
$$\ln |p| = \ln |y| + C,$$
$$e^{\ln |p|} = e^{\ln |y|} e^C,$$
$$|p| = |y|e^C,$$
$$p = C_1 y, \quad \text{where } C_1 = \pm e^C,$$
$$\frac{dy}{dx} = C_1 y.$$

The initial condition $y'(0) = 1$ when $y(0) = 2$ gives $C_1 = \frac{1}{2}$. Hence, integrating again we obtain

$$dx = \frac{2}{y} dy,$$
$$x = 2\ln |y| + C_2.$$

Simplifying the expression gives

$$y^2 = e^{x - C_2} = e^{-C_2} e^x = A e^x,$$

where $A = e^{-C_2}$ is an arbitrary constant. Substituting the initial condition $y(0) = 2$ gives $A = 4$. Thus, the particular solution satisfying the initial conditions is

$$y = 2\sqrt{e^x}.$$

Note that $y(0) = 2$, so we take y as being positive.

5.3.2 Second-order linear differential equations

Early in this chapter, we derived the following second-order ODE for a mass-spring system if the resistance is proportional to the mass's velocity:

$$mx'' - lx' + kx = 0.$$

If, furthermore, there is an external force $f(t)$ acting on the mass, we will have

$$mx'' - lx' + kx = f(t).$$

In general, the so-called nonhomogeneous second-order linear differential equations have the form

$$a(x)\frac{d^2y}{dx^2} + b(x)\frac{dy}{dx} + c(x)y = f(x). \tag{5.13}$$

The term "linear" applies to y, y', and y'', and it means that they appear in separate terms of the ODE without an exponent (other than one) and are not part of another function (such as $\sqrt{1+y}$). The functions $a(x)$, $b(x)$, $c(x)$, and $f(x)$ are allowed to be nonlinear. We also assume that $a(x) \neq 0$.

If in addition $f(x) = 0$ for all x in the above equation, then the differential equation is called a *homogeneous* linear equation:

$$a(x)\frac{d^2y}{dx^2} + b(x)\frac{dy}{dx} + c(x)y = 0. \tag{5.14}$$

The ODE

$$e^x y'' + (\cos x)y' + (1 + \sqrt{x})y = x$$

is linear and nonhomogeneous. By contrast, the equations

$$y'' = yy' \quad \text{and} \quad y'' - 3(y')^2 + 4y^3 = 0$$

are not linear because they contain products and powers of y or its derivative.

The second-order ODE

$$x^2 y'' + 2xy' + 3y = \cos x$$

is linear and nonhomogeneous, whereas the following is linear and homogeneous:

$$x^2 y'' + 2xy' + 3y = 0.$$

Homogeneous second-order linear differential equations

We now explore the solutions to equation (5.14).

Theorem 5.3.1 (Principle of superposition for homogeneous equations). *Let $y_1(x)$ and $y_2(x)$ be two solutions of the homogeneous linear equation (5.14), $a(x)y'' + b(x)y' + c(x)y = 0$, defined on the interval I. If C_1 and C_2 are constants, then the linear combination*

$$y = C_1 y_1(x) + C_2 y_2(x)$$

is also a solution of this homogeneous ODE.

Proof. Since y_1 and y_2 are solutions of equation (5.14), we have

$$a(x)y_1'' + b(x)y_1' + c(x)y_1 = 0 \quad \text{and} \quad a(x)y_2'' + b(x)y_2' + c(x)y_2 = 0.$$

Substituting $y = C_1 y_1 + C_2 y_2$ into equation (5.14), we have

$$a(x)(C_1 y_1 + C_2 y_2)'' + b(x)(C_1 y_1 + C_2 y_2)' + c(x)(C_1 y_1 + C_2 y_2)$$
$$= a(x)C_1 y_1'' + a(x)C_2 y_2'' + b(x)C_1 y_1' + b(x)C_2 y_2' + c(x)C_1 y_1 + c(x)C_2 y_2$$
$$= C_1[a(x)y_1'' + b(x)y_1' + c(x)y_1] + C_2[a(x)y_2'' + b(x)y_2' + c(x)y_2]$$
$$= 0.$$

So, $y = C_1 y_1 + C_2 y_2$ is a solution of equation (5.14). □

Thus, if we can find two particular solutions to equation (5.14), and they are linearly independent, then the linear combination of the two particular solutions gives a general solution to equation (5.14). The definition of two linearly independent functions is given below.

Definition 5.3.1 (Linear independence of two functions). Two functions defined on an open interval I are said to be linearly independent on I provided that neither is a constant multiple of the other (alternatively, neither of the two functions $\frac{f}{g}$ or $\frac{g}{f}$ is a constant-valued function on I).

For example, e^x and e^{2x} are two linearly independent functions, while e^{2x} and $2e^{2x}$ are linearly dependent. By the superposition theorem, we have the following theorem.

Theorem 5.3.2 (General solution of second-order homogeneous linear ODEs). *Let y_1 and y_2 be two linearly independent solutions of the homogeneous linear differential equation*

$$a(x)y'' + b(x)y' + c(x)y = 0$$

where $a(x)(\neq 0)$, $b(x)$, and $c(x)$ are continuous on some interval I. Then, a general solution is

$$y(x) = C_1 y_1(x) + C_2 y_2(x),$$

where C_1 and C_2 are two arbitrary constants.

Note. This theorem could be generalized to a linear homogeneous differential equation of order n. Then we have

$$y^{(n)} + p_1(x)y^{(n-1)} + p_2(x)y^{(n-2)} + \cdots + p_{n-1}(x)y' + p_n(x)y = 0. \tag{5.15}$$

That is, if $y_1, y_2, \ldots, y_n$ are n linearly independent solutions of this equation, then a general solution is

$$y(x) = c_1 y_1 + c_2 y_2 + \cdots + c_n y_n, \quad \text{where } c_1, c_2, \ldots, c_n \text{ are } n \text{ arbitrary constants.}$$

Example 5.3.4. For the differential equation

$$y'' - 4y = 0,$$

we can verify that $y_1(x) = e^{2x}$ and $y_2(x) = e^{-2x}$ are two solutions, and $\frac{y_1}{y_2} = e^{4x}$ is not a constant. Therefore, y_1 and y_2 are two linearly independent solutions. So, a general solution is $y(x) = C_1 e^{2x} + C_2 e^{-2x}$.

Homogeneous second-order linear differential equations with constant coefficients
As an illustration of the general theory for solutions to a second-order linear ODE, we now discuss the general second-order homogeneous linear differential equation

$$ay'' + by' + cy = 0 \tag{5.16}$$

with constant coefficients $a(\neq 0)$, b, and c. We first look for a single solution of this equation, and begin with the observation that exponential functions often play a role in such solutions, such as the solution of $y'' - 4y = 0$ in the previous example. Note that if r is a constant, then

$$(e^{rx})' = re^{rx} \quad \text{and} \quad (e^{rx})'' = r^2 e^{rx}.$$

Hence, if substituting $y = e^{rx}$ into equation (5.16), we find

$$ar^2 e^{rx} + br e^{rx} + c e^{rx} = 0.$$

However, e^{rx} is never zero, so we can divide this out of the equation. We conclude that $y(x) = e^{rx}$ will satisfy the homogeneous linear differential equation (5.16) with constant coefficients precisely when r is a root of the algebraic equation

$$ar^2 + br + c = 0. \tag{5.17}$$

This quadratic equation is called the *characteristic equation* or *auxiliary equation* of equation (5.16).

It is easy to see that when the characteristic equation (5.17) has two distinct (unequal) roots r_1 and r_2, then these give two solutions $y_1 = e^{r_1 x}$ and $y_2 = e^{r_2 x}$ that are linearly independent (since $e^{r_1 x}/e^{r_2 x}$ is not a constant). This gives the following result.

Theorem 5.3.3 (Homogeneous linear ODEs – distinct real roots). *If r_1 and r_2 are real and distinct roots of the characteristic equation $ar^2 + br + c = 0$ of the ODE $ay'' + by' + cy = 0$, then*

$$y(x) = C_1 e^{r_1 x} + C_2 e^{r_2 x},$$

where C_1 and C_2 are arbitrary constants, is a general solution of $ay'' + by' + cy = 0$.

Example 5.3.5. Find a general solution of

$$2y'' - 7y' + 3y = 0.$$

Solution. The solutions of the characteristic equation

$$2r^2 - 7r + 3 = 0$$

are $r_1 = \frac{1}{2}$ and $r_2 = 3$. So, a general solution is

$$y(x) = C_1 e^{\frac{1}{2}x} + C_2 e^{3x}.$$

If the characteristic equation (5.17) has a root of multiplicity 2, then $r_1 = r_2$. In this case, we only get one particular solution. We cannot say that

$$y = C_1 e^{r_1 x} + C_2 e^{r_2 x}$$

is a general solution since $e^{r_1 x} = e^{r_2 x}$, and they are not linearly independent. There is, in fact, only one arbitrary constant. To find another solution, we use the *method of variation of parameter*. We assume a particular solution has the form

$$y^* = C(x) e^{r_1 x}.$$

Then, we will look for such a $C(x)$. We plug y^* into the equation to obtain

$$a(y^*)'' + b(y^*)' + cy^* = 0,$$
$$a(C(x)e^{r_1 x})'' + b(C(x)e^{r_1 x})' + c(C(x)e^{r_1 x}) = 0,$$
$$a(C''(x)e^{r_1 x} + 2r_1 C'(x)e^{r_1 x} + r_1^2 C(x)e^{r_1 x}) + b(C'(x)e^{r_1 x} + C(x)r_1 e^{r_1 x}) + c(C(x)e^{r_1 x}) = 0.$$

Since $e^{r_1 x} \neq 0$, we can simplify this to

$$a[C''(x) + 2r_1 C'(x) + r_1^2 C(x)] + b[C'(x) + r_1 C(x)] + cC(x) = 0,$$
$$aC''(x) + (2ar_1 + b)C'(x) + (ar_1^2 + br_1 + c)C(x) = 0.$$

Since $ar_1^2 + br_1 + c = 0$ and $2ar_1 + b = 0$ ($r_1 = r_2$ is a repeated root), we have

$$aC''(x) = 0.$$

Therefore, such a $C(x)$ does exist; we can choose a simple one, say, $C(x) = x$ (choosing any $C(x) = kx + l$ also works). Then, we have another particular solution $y = xe^{r_1 x}$ which is linearly independent of $y = e^{r_1 x}$. So, we have the following theorem.

Theorem 5.3.4 (Homogeneous linear ODEs – repeated roots). *If the characteristic equation $ar^2 + br + c = 0$ of the ODE $ay'' + by' + cy = 0$ has only one root r (a double root, or repeated root), then*

$$y(x) = (C_1 + C_2 x)e^{rx},$$

where C_1 and C_2 are arbitrary constants, is a general solution of $ay'' + by' + cy = 0$.

Example 5.3.6. Solve the equation $9y'' + 12y' + 4y = 0$.

Solution. The auxiliary equation $9r^2 + 12r + 4 = 0$ can be factored as

$$(3r + 2)^2 = 0.$$

The only root is $r = -\frac{2}{3}$. Thus, a general solution is

$$y = C_1 e^{-2x/3} + C_2 x e^{-2x/3}.$$

Example 5.3.7. Solve the initial value problem

$$\begin{cases} y'' + 2y' + y = 0, \\ y(0) = 5, \ y'(0) = -3. \end{cases}$$

Solution. We note first that the characteristic equation $r^2 + 2r + 1 = 0$ has repeated roots $r_1 = r_2 = -1$. Hence, a general solution for the ODE is

$$y(x) = C_1 e^{-x} + C_2 x e^{-x}.$$

In order to use the initial conditions, we differentiate to find y', i.e.,

$$y'(x) = -C_1 e^{-x} + C_2 e^{-x} - C_2 x e^{-x}.$$

So, the initial conditions substituted into the equations for $y(x)$ and $y'(x)$ give

$$y(0) = C_1 = 5,$$
$$y'(0) = -C_1 + C_2 = -3,$$

which imply $C_1 = 5$ and $C_2 = 2$. Therefore, the desired solution is

$$y(x) = 5e^{-x} + 2xe^{-x}.$$

The third case is when the discriminant of the auxiliary equation, $b^2 - 4ac$, is less than 0, then the auxiliary equation has two complex roots, and they are conjugate pairs of the form $r_{1,2} = \alpha \pm \beta i$. The theory still implies that $e^{(\alpha+\beta i)x}$ and $e^{(\alpha-\beta i)x}$ are particular solutions of the linear ODE, but we would not expect to have complex numbers in the solution of a real problem. The next theorem shows that we can, in fact, find real solutions via these complex solutions.

Theorem 5.3.5 (Homogeneous linear ODEs – complex conjugate roots). *If $r_1 = \alpha + \beta i$ and $r_2 = \alpha - \beta i$ are complex conjugate roots of the characteristic equation $ar^2 + br + c = 0$ of the ODE $ay'' + by' + c = 0$, then*

$$y = e^{\alpha x}(C_1 \cos \beta x + C_2 \sin \beta x),$$

where C_1 and C_2 are arbitrary constants, is a general solution of $ay'' + by' + cy = 0$.

Proof. For the proof, Euler's theorem is required, which states that for any θ, we have $e^{i\theta} = \cos \theta + i \sin \theta$. The theory of solutions developed above applies, so we can write a general solution as

$$\begin{aligned} y &= Ae^{r_1 x} + Be^{r_2 x} = Ae^{(\alpha+\beta i)x} + Be^{(\alpha-\beta i)x} \\ &= Ae^{\alpha x}(\cos \beta x + i \sin \beta x) + Be^{\alpha x}(\cos \beta x - i \sin \beta x) \\ &= e^{\alpha x}[(A+B)\cos \beta x + i(A-B)\sin \beta x] \\ &= e^{\alpha x}(C_1 \cos \beta x + C_2 \sin \beta x), \end{aligned}$$

where $C_1 = A + B$ and $C_2 = i(A - B)$. The solutions are, therefore, real when C_1 and C_2 are both real. □

Note. In fact, without using Euler's theorem, one can still derive a general solution by proving that $e^{\alpha x} \cos \beta x$ and $e^{\alpha x} \sin \beta x$ are two linearly independent solutions.

Example 5.3.8. Solve $x''(t) = -\omega^2 x$.

Solution. Since the characteristic equation, $r^2 + \omega^2 = 0$, has roots $r_1 = \omega i$ and $r_2 = -\omega i$, it follows that a general solution is

$$\begin{aligned} x(t) &= C_1 e^{0t} \cos \omega t + C_2 e^{0t} \sin \omega t \\ &= \sqrt{C_1^2 + C_2^2} \sin(\omega t + B). \end{aligned}$$

Note that if we denote $A = \sqrt{C_1^2 + C_2^2}$, then this general solution could also be written as

$$x(t) = A \sin(\omega t + B).$$

This is the general solution for a simple harmonic motion where A is the amplitude and ω is the angular velocity. The period is $T = \frac{2\pi}{\omega}$.

Example 5.3.9. Find a general solution of the differential equation $y'' - 2y' + 5y = 0$.

Solution. The characteristic equation is

$$r^2 - 2r + 5 = 0$$

with roots $r_{1,2} = 1 \pm 2i$. A general solution is, therefore,

$$y = e^x(C_1 \cos 2x + C_2 \sin 2x).$$

Summary

A general solution of $ay'' + by' + cy = 0$ has one of the following forms:

Roots of $ar^2 + br + c = 0$	General solution
r_1, r_2 real and distinct	$y = C_1 e^{r_1 x} + C_2 e^{r_2 x}$
$r_1 = r_2 = r$	$y = C_1 e^{rx} + C_2 x e^{rx}$
$r_1, r_2 = \alpha \pm \beta i$	$y = e^{\alpha x}(C_1 \cos \beta x + C_2 \sin \beta x)$

For higher-order homogeneous linear differential equations of the form

$$y^{(n)} + p_1 y^{(n-1)} + p_2 y^{(n-2)} + \cdots + p_n y = 0, \quad \text{where all } p_i \text{ are constants,}$$

the results are similar to those that we have obtained for second-order linear ODEs. The principal difference is that there will be n roots of the auxiliary equation when the order of the ODE is n. A general solution is a linear combination of n independent solutions, and each root (or conjugate pair of roots) of the auxiliary equation

$$r^n + p_1 r^{n-1} + \cdots + p_n = 0$$

corresponds to one particular solution (or a pair of particular solutions), as shown in the above table.

If there is a root r with multiplicity 3, or higher, then this root will create a term in the general solution with three arbitrary constants, C_1, C_2, and C_3, i. e., $C_1 e^{rx} + C_2 x e^{rx} + C_3 x^2 e^{rx}$. Roots of still higher multiplicity will extend this result in an analogous way.

Example 5.3.10. Find a general solution of the equation $y^{(4)} + 2y''' + 3y'' = 0$.

Solution. The auxiliary equation is $r^4 + 2r^3 + 3r^2 = 0$. Solutions are $r = 0$ (repeated) and $r = -1 \pm \sqrt{2}i$, so a general solution is

$$y = C_1 e^{0x} + C_2 x e^{0x} + e^{-1x}(C_3 \cos \sqrt{2}x + C_4 \sin \sqrt{2}x)$$
$$= C_1 + C_2 x + C_3 e^{-x} \cos \sqrt{2}x + C_4 e^{-x} \sin \sqrt{2}x.$$

Nonhomogeneous second-order linear differential equations

We now discuss nonhomogeneous second-order linear differential equations of the form

$$a(x)y'' + b(x)y' + c(x)y = f(x). \qquad (5.18)$$

The associated homogeneous equation

$$a(x)y'' + b(x)y' + c(x)y = 0, \qquad (5.19)$$

where the right-hand side function $f(x)$ is replaced by zero, is called the *complimentary equation*. A general solution of this equation is called a *complimentary function*. In cases where equation (5.18) models a physical system, the nonhomogeneous term $f(x)$ frequently corresponds to some external influence on the system being modeled.

There is a nice connection between the solutions of equation (5.18) and equation (5.19).

Theorem 5.3.6 (General solution of nonhomogeneous linear ODEs). *A general solution of a nonhomogeneous differential equation*

$$a(x)\frac{d^2y}{dx^2} + b(x)\frac{dy}{dx} + c(x)y = f(x)$$

can be written as

$$y(x) = y_c(x) + y_p(x), \qquad (5.20)$$

where $y_c(x)$ is a complementary function (a general solution of the associated homogeneous equation (5.19)), and $y_p(x)$ is a particular solution of equation (5.18).

Proof. We first show that $y(x) = y_p(x) + y_c(x)$ is a solution of equation (5.18). Substituting into that equation, we obtain

$$a(x)(y_p(x) + y_c(x))'' + b(x)(y_p(x) + y_c(x))' + c(x)(y_p(x) + y_c(x))$$
$$= [a(x)y_p''(x) + b(x)y_p'(x) + c(x)y_p(x)] + [a(x)y_c'' + b(x)y_c'(x) + c(x)y_c(x)]$$
$$= f(x) + 0$$
$$= f(x).$$

Now we show that any solution of equation (5.18) must be of the form of equation (5.20). If y^* is any particular of equation (5.18), then

$$a(x)y^{*''} + b(x)y^{*'} + c(x)y^* = f(x).$$

But we also have

$$a(x)y_p'' + b(x)y_p' + c(x)y_p = f(x).$$

So
$$a(x)(y^* - y_p)'' + b(x)(y^* - y_p)' + c(x)(y^* - y_p) = 0.$$

This means that $y^* - y_p$ must be a solution of the complementary equation $a(x)y'' + b(x)y' + c(x)y = 0$. So, $y^* - y_p = y_c$ for some suitable constants in y_c. This means
$$y^* = y_c + y_p. \qquad \square$$

We now apply the theory to nonhomogeneous second-order linear differential equations with constant coefficients.

Nonhomogeneous second-order linear differential equations with constant coefficients

We have derived theorems for finding general solutions to
$$ay'' + by' + cy = 0,$$

which have the form $C_1 y_1 + C_2 y_2$, where y_1 and y_2 are two linearly independent solutions. If we can find a particular solution y_p to the nonhomogeneous second-order linear differential equation with constant coefficients of the form
$$ay'' + by' + cy = f(x), \qquad (5.21)$$

then, according to Theorem 5.3.6, we obtain a general solution to equation (5.21),
$$y(x) = C_1 y_1 + C_2 y_2 + y_p.$$

In general, it is very hard to find a particular solution y_p for a nonhomogeneous equation. In the following, we only discuss the cases where the right-hand side function $f(x)$ of equation (5.21) is a linear combination of products of the form $P_m(x)e^{\lambda x}$, where λ is a real or complex constant and $P_m(x)$ is a polynomial of degree m. We use the *method of undetermined coefficients*, in which we choose for y_p the most likely function, such as a polynomial multiplied by an exponential, $Q(x)e^x$, and then determine the unknown coefficients by substituting $y_p(x)$ into the ODE.

Example 5.3.11. Solve the differential equation $y'' + y' - 2y = 2x + 1$.

Solution. The roots of the auxiliary equation $r^2 + r - 2 = 0$ are $r = 1$ and $r = -2$. Hence, a complementary function is
$$y_c = C_1 e^x + C_2 e^{-2x}.$$

It seems likely that a polynomial will give a particular solution, because the right-hand side of the differential equation is a polynomial. Since the right-hand side, $2x + 1$, is a polynomial of degree 1, we try $y_p = Ax + B$ of degree 1. Substituting into the given differential equation we have

$$(Ax + B)'' + (Ax + B)' - 2(Ax + B) = 2x + 1,$$
$$A - 2Ax - 2B = 2x + 1.$$

However, the polynomial on the left-hand side equals the polynomial on the right-hand side exactly when their coefficients are equal. Thus,

$$-2A = 2 \quad \text{and} \quad A - 2B = 1.$$

This gives $A = -1$ and $B = -1$. So, $y_p = -x - 1$ is a particular solution. Therefore, a general solution is

$$y = y_c + y_p = C_1 e^x + C_2 e^{-2x} - x - 1.$$

Example 5.3.12. Find a particular solution for each of the following differential equations:

(a) $3y'' - 2y' - y = e^{2x}$, (b) $y'' - 2y' - 3y = e^{3x}$.

Solution. For (a), we try $y_p = Ae^{2x}$. Then

$$3(Ae^{2x})'' - 2(Ae^{2x})' - (Ae^{2x}) = e^{2x},$$
$$12Ae^{2x} - 4Ae^{2x} - Ae^{2x} = e^{2x},$$
$$A = \frac{1}{7}.$$

So a particular solution is $y_p = \frac{1}{7}e^{2x}$.

For (b), if we try $y_p = Ae^{3x}$, it will not work. Can you see why? Instead, we try $y_p = Axe^{3x}$. Then

$$(Axe^{3x})'' - 2(Axe^{3x})' - 3(Axe^{3x}) = e^{3x},$$
$$A(3xe^{3x} + e^{3x})' - 2A(3xe^{3x} + e^{3x}) - 3Axe^{3x} = e^{3x},$$
$$A(3e^{3x} + 9xe^{3x} + 3e^{3x}) - 2A(3xe^{3x} + e^{3x}) - 3Axe^{3x} = e^{3x},$$
$$A(6 + 9x) - 2A(1 + 3x) - 3xA = 1,$$
$$A = \frac{1}{4}.$$

A particular solution is, therefore, $y_p = \frac{1}{4}xe^{3x}$.

Example 5.3.13. Solve the differential equation

$$y'' - 5y' + 6y = xe^{2x}, \quad y(0) = 1, \quad y'(0) = 2.$$

Solution. The auxiliary equation $r^2 - 5r + 6 = 0$ has roots $r_1 = 2$ and $r_2 = 3$, so a complementary function is $y_c = C_1 e^{2x} + C_2 e^{3x}$. For a particular solution, shall we try $y_p = (Ax + B)e^{2x}$, because it is similar to xe^{2x}? The answer is "No" since the complementary function already has the term Ce^{2x}. So, instead, we try $y_p = x(Ax + B)e^{2x}$, for which the derivatives are

$$y'_p = (Ax^2 + Bx)' e^{2x} + (Ax^2 + Bx)(e^{2x})'$$
$$= (2Ax + B)e^{2x} + 2(Ax^2 + Bx)e^{2x},$$
$$y''_p = 2Ae^{2x} + 2(2Ax + B)e^{2x} + (4Ax + 2B)e^{2x} + 4(Ax^2 + Bx)e^{2x}$$
$$= 2e^{2x}(A + 2B + 4Ax + 2Bx + 2Ax^2).$$

Substituting into the differential equation, $y'' - 5y' + 6y = xe^{2x}$, gives

$$2e^{2x}(A + 2B + 4Ax + 2Bx + 2Ax^2) - 5e^{2x}(B + 2Ax + 2Bx + 2Ax^2) + 6e^{2x}(Bx + Ax^2) = xe^{2x}.$$

Dividing by e^{2x} and collecting coefficients yields

$$2(A + 2B + 4Ax + 2Bx + 2Ax^2) - 5(B + 2Ax + 2Bx + 2Ax^2) + 6(Bx + Ax^2) = x,$$
$$2A - B - 2Ax = x.$$

Equating coefficients gives

$$A = -\frac{1}{2} \quad \text{and} \quad B = -1,$$

so a particular solution is

$$y_p = x\left(-\frac{1}{2}x - 1\right)e^{2x}.$$

Thus, we obtain a general solution

$$y = C_1 e^{2x} + C_2 e^{3x} - x\left(\frac{1}{2}x + 1\right)e^{2x}.$$

Since $y' = 2C_2 e^{2x} + 3C_2 e^{3x} - 2x(\frac{x}{2} + 1)e^{2x} - (x + 1)e^{2x}$, under the condition $y(0) = 1$ and $y'(0) = 2$, we have

$$1 = C_1 + C_2,$$
$$2 = 2C_1 + 3C_2 - 1.$$

So, $C_1 = 0$ and $C_2 = 1$, and the particular solution is

$$y = e^{3x} - x\left(\frac{x}{2} + 1\right)e^{2x}.$$

A general approach to the method of undetermined coefficients

In general, we can make a reasonable guess to get a particular solution of equation (5.21) (or a higher-order extension of this equation), i. e.,

$$a\frac{d^2y}{dx^2} + b\frac{dy}{dx} + cy = f(x),$$

when $f(x)$ is of the form $P_m(x)e^{\lambda x}$, where $P_m(x)$ is a degree m polynomial, and λ is a constant. Our choice for a particular solution takes the form $y_p(x) = Q(x)e^{\lambda x}$, where $Q(x)$ is a polynomial. We substitute $y = Q(x)e^{\lambda x}$ into the ODE above to obtain

$$a[Q''(x)e^{\lambda x} + 2\lambda Q'(x)e^{\lambda x} + \lambda^2 Q(x)e^{\lambda x}] + b[Q'(x)e^{\lambda x} + \lambda Q(x)e^{\lambda x}] + cQ(x)e^{\lambda x} = P_m(x)e^{\lambda x}.$$

We cancel out the factor $e^{\lambda x}$ from both sides of the equation, resulting in

$$aQ''(x) + (2a\lambda + b)Q'(x) + (a\lambda^2 + b\lambda + c)Q(x) = P_m(x). \qquad (5.22)$$

We now consider the following cases.

Case 1: $a\lambda^2 + b\lambda + c \neq 0$. That is, λ is **not** a root of the characteristic equation of the associated homogeneous equation. Thus, we deduce that $Q(x)$ needs to be of the same degree, m, as the polynomial $P_m(x)$.

Case 2: $a\lambda^2 + b\lambda + c = 0$ but $2a\lambda + b \neq 0$. That is, λ is a root of the characteristic equation of multiplicity 1. Then equation (5.22) becomes

$$aQ''(x) + (2a\lambda + b)Q'(x) = P_m(x).$$

This tells us that $Q(x)$ should be chosen to be of degree $m+1$. That is, in order to use the method of undetermined coefficients, we choose $Q(x)$ to have degree one more than the degree of $P_m(x)$.

Case 3: $a\lambda^2 + b\lambda + c = 0$ and $2a\lambda + b = 0$. That is, λ is a root of the characteristic equation of multiplicity 2. Then, equation (5.22) becomes

$$aQ''(x) = P_m(x).$$

This tells us that $Q(x)$ should be of degree $m + 2$. That is, in order to use the method of undetermined coefficients, we choose $Q(x)$ to have degree two more than the degree of $P_m(x)$.

We summarize the results. If the right-hand side of the ODE is $f(x) = P_m(x)e^{\lambda x}$, then we initially choose $y_p(x) = Q(x)e^{\lambda x}$, where $Q(x)$ is a polynomial of degree m. We modify y_p by multiplying it by x if λ is a root of the auxiliary equation, and by x^2 if λ is a repeated root of the auxiliary equation. We determine the undetermined coefficients by substituting $y = y_p$ into the differential equation.

Example 5.3.14. Find a particular solution for the ODE $y'' - 2y' - 3y = 3x + 1$.

Solution. We have $f(x) = 3x + 1 = (3x + 1)e^{0x}$, so $\lambda = 0$ and the polynomial $3x + 1$ has degree 1. The characteristic equation of the associated homogeneous ODE is

$$r^2 - 2r - 3 = 0.$$

We know $\lambda = 0$ is not a root of this equation, so we can assume a particular solution takes the form

$$y_p = (Ax + B)e^{0x} = Ax + B.$$

Substitution into $y'' - 2y' - 3y = 3x + 1$ gives

$$0 - 2A - 3(Ax + B) = 3x + 1.$$

We equate the coefficients involving the same power of x, so we have

$$\begin{cases} -3A = 3, \\ -2A - 3B = 1, \end{cases}$$

with solution

$$A = -1 \quad \text{and} \quad B = \frac{1}{3}.$$

Thus, a particular solution for the ODE is $y_p = -x + \frac{1}{3}$, and a general solution is $y = C_1 e^{3x} + C_2 e^{-x} - x + \frac{1}{3}$.

Example 5.3.15. Find a general solution of

$$y'' - 3y' + 2y = xe^x.$$

Solution. The characteristic equation of the associated homogeneous equation is $r^2 - 3r + 2 = 0$ with roots $r = 1$ and 2. Hence, a complementary function is

$$y_c = C_1 e^x + C_2 e^{2x}.$$

The right-hand side xe^x is a polynomial of degree 1 multiplied by $e^{\lambda x}$ with $\lambda = 1$. Since $\lambda = 1$ is one of the roots of the characteristic equation, the particular solution is chosen to be a polynomial of degree 1, multiplied by x, and then by e^x, i.e.,

$$y_p = x(Ax + B)e^x.$$

Substitution into $y'' - 3y' + 2y = xe^x$, simplifying, and dividing by e^x leads to

$$-2Ax + (2A - B) = x.$$

Equating the coefficients gives

$$-2A = 1 \quad \text{and} \quad 2A - B = 0.$$

Solving the system of equations gives $A = -1/2$ and $B = -1$. Thus, a particular solution is

$$y_p = x\left(-\frac{1}{2}x - 1\right)e^x,$$

and a general solution is

$$y = y_c + y_p = C_1 e^x + C_2 e^{2x} - \left(\frac{1}{2}x^2 + x\right)e^x.$$

Example 5.3.16. Find a general solution of $y'' + 6y' + 9y = 5e^{-3x}$.

Solution. The characteristic equation of the associated homogeneous equation is $r^2 + 6r + 9 = 0$, with repeated root $r = -3$. Hence, a complementary function is

$$y_c = (C_1 + xC_2)e^{-3x}.$$

The right-hand side is $P(x)e^{\lambda x} = 5e^{-3x}$, with a polynomial $P(x) = 5$ of degree 0 and $\lambda = -3$. Since $\lambda = -3$ is a double root of the characteristic equation, a particular solution can be chosen to be an arbitrary polynomial of degree 0 (that is, a constant) multiplied by x^2 and then by e^{-3x}, i.e.,

$$y_p = Ax^2 e^{-3x}.$$

Pluging y_p into the original equation to get $A = \frac{5}{2}$, so a particular solution is

$$y_p = \frac{5}{2}x^2 e^{-3x}.$$

Hence, a general solution is

$$y = y_c + y_p = (C_1 + xC_2)e^{-3x} + \frac{5}{2}x^2 e^{-3x}.$$

Example 5.3.17. Solve $y'' - y = 4x \sin x$.

Solution. The characteristic equation $r^2 - 1 = 0$ has roots $r = \pm 1$. Hence, a complementary function is

$$y_c = C_1 e^x + C_2 e^{-x}.$$

To find a particular solution, we note that the right-hand side, $4x \sin x$, is the imaginary part of $4xe^{ix}$ since $e^{ix} = \cos x + i \sin x$. So we consider

$$\begin{cases} y_1'' - y_1 = 4x \cos x, \\ (iy_2)'' - iy_2 = 4x(i \sin x), \end{cases}$$

and we add them up, so we obtain

$$(y_1 + iy_2)'' - (y_1 + iy_2) = 4x(\cos x + i \sin x) = 4xe^{ix}.$$

So, if we can solve

$$y'' - 4y = 4xe^{ix}$$

for a particular solution y_p, then y_p will be $y_1(x) + iy_2(x)$. The real part of y_p, $y_1(x)$, must be a particular solution of $y'' - y = 4x \cos x$, and its imaginary part, $y_2(x)$, must be a particular solution of $y'' - y = 4x \sin x$.

Since $\lambda = i$ is not a root of the characteristic equation, we use a modified right hand side and choose a particular solution that is a polynomial of degree 1 multiplied by e^{ix}, i. e.,

$$y_p = (Ax + B)e^{ix}.$$

Substituting this into $y'' - y = 4xe^{ix}$, simplifying, and dividing by e^{ix} leads to

$$-2Ax - 2B + 2iA = 4x.$$

Thus, we obtain the equations

$$-2A = 4 \quad \text{and} \quad -2B + 2iA = 0,$$

with solutions $A = -2$ and $B = -2i$. Hence, a particular solution of $y'' - y = 4xe^{ix}$ is

$$y_p = (-2x - 2i)e^{ix}$$
$$= (-2x - 2i)(\cos x + i \sin x)$$
$$= -2(x \cos x - \sin x) - 2(x \sin x + \cos x)i.$$

The original right-hand side is the imaginary part of $4xe^{ix}$, so we take the imaginary part of y_p to get a particular solution of the original problem:

$$y_p = -2x \sin x - 2 \cos x.$$

Hence, a general solution is

$$y = y_c + y_p = C_1 e^x + C_2 e^{-x} - 2x \sin x - 2 \cos x.$$

Note. This example shows a special case of a method for finding a particular solution of equation (5.21) when the right-hand side is of the form $f(x) = e^{\lambda x} P(x) \cos mx$ or $f(x) = e^{\lambda x} P(x) \sin mx$. This example shows that $f(x)$ is replaced by the function $g(x) = e^{(\lambda+mi)x} P(x)$ (of which $f(x)$ is the real or imaginary part). The particular solution y_p of the new ODE, with $g(x)$ on the right-hand side, can be found using the methods developed before. The real or imaginary part of y_p is a particular solution of the original problem.

Some books give an alternative procedure, using only real-valued functions, where the trial solution (particular solution) is taken to be of the form

$$y_p(x) = e^{\lambda x} Q_1(x) \cos mx + e^{\lambda x} Q_2(x) \sin mx,$$

where $Q_1(x)$ and $Q_2(x)$ are polynomials with unknown coefficients and of the same degree as $P(x)$, but multiplied by x or x^2 if λ is a single root or repeated root of the corresponding auxiliary equation, respectively.

Example 5.3.18. Solve $y'' - y = 3e^{2x} + 4x \sin x$.

Solution. The associated auxiliary equation is $r^2 - 1 = 0$, and a complementary function is the same as in the previous example, so we take

$$y_c = C_1 e^x + C_2 e^{-x}.$$

In order to find a particular solution, we separately find particular solutions of the two equations, and then add them, so we have

$$y'' - y = 3e^{2x},$$
$$y'' - y = 4x \sin x.$$

The first has a particular solution (check this for yourself),

$$y_1 = e^{2x},$$

and the second has the particular solution found in the previous example,

$$y_2 = -2x \sin x - 2 \cos x.$$

We now add them to give a particular solution for the original equation in this example:

$$y_p = y_1 + y_2 = e^{2x} - 2x \sin x - 2 \cos x.$$

So a general solution is

$$y = y_c + y_p = C_1 e^x + C_2 e^{-x} + e^{2x} - 2x \sin x - 2 \cos x.$$

5.3.3 Variation of parameters

The method of undetermined coefficients is often useful to solve problems when $f(x) = P_m(x)e^{\lambda x}$. We now introduce the method of variation of parameters, which is another way to find a particular solution to the differential equation

$$ay'' + by' + cy = f(x). \tag{5.23}$$

Assume a general solution to $ay'' + by' + cy = 0$ is

$$y = C_1 y_1(x) + C_2 y_2(x). \tag{5.24}$$

Now we look for a particular solution to $ay'' + by' + cy = f(x)$ of the form

$$y_p = u_1(x)y_1(x) + u_2(x)y_2(x). \tag{5.25}$$

Then,

$$y_p' = u_1' y_1 + u_1 y_1' + u_2' y_2 + u_2 y_2'.$$

To solve for $u_1(x)$ and $u_2(x)$, we need two equations. We already have the condition that y_p is a particular solution, but we need an extra one. Let us impose

$$u_1' y_1 + u_2' y_2 = 0 \tag{5.26}$$

in order to simplify our calculation. Therefore,

$$y_p' = u_1 y_1' + u_2 y_2' \quad \text{and}$$
$$y_p'' = u_1' y_1' + u_1 y_1'' + u_2' y_2' + u_2 y_2''.$$

We substitute these into the differential equation

$$a(u_1' y_1' + u_1 y_1'' + u_2' y_2' + u_2 y_2'') + b(u_1 y_1' + u_2 y_2') + c(u_1 y_1 + u_2 y_2) = f(x)$$

to obtain

$$u_1(ay_1'' + by_1' + cy_1) + u_2(ay_2'' + by_2' + cy_2) + a(u_1' y_1' + u_2' y_2') = f(x).$$

This means

$$a(u_1' y_1' + u_2' y_2') = f(x). \tag{5.27}$$

In view of equation (5.26) and equation (5.27), by Cramer's rule we have

$$u_1'(x) = \frac{\begin{vmatrix} 0 & y_2 \\ f(x)/a & y_2' \end{vmatrix}}{\begin{vmatrix} y_1 & y_2 \\ y_1' & y_2' \end{vmatrix}} \quad \text{and} \quad u_2'(x) = \frac{\begin{vmatrix} y_1 & 0 \\ y_1' & f(x)/a \end{vmatrix}}{\begin{vmatrix} y_1 & y_2 \\ y_1' & y_2' \end{vmatrix}}.$$

Integration gives

$$u_1(x) = \int \frac{\begin{vmatrix} 0 & y_2 \\ f(x)/a & y_2' \end{vmatrix}}{\begin{vmatrix} y_1 & y_2 \\ y_1' & y_2' \end{vmatrix}} dx \quad \text{and} \quad u_2(x) = \int \frac{\begin{vmatrix} y_1 & 0 \\ y_1' & f(x)/a \end{vmatrix}}{\begin{vmatrix} y_1 & y_2 \\ y_1' & y_2' \end{vmatrix}} dx.$$

Thus, a particular solution is given by

$$y_p = y_1(x) \cdot \int \frac{\begin{vmatrix} 0 & y_2 \\ f(x)/a & y_2' \end{vmatrix}}{\begin{vmatrix} y_1 & y_2 \\ y_1' & y_2' \end{vmatrix}} dx + y_2(x) \cdot \int \frac{\begin{vmatrix} y_1 & 0 \\ y_1' & f(x)/a \end{vmatrix}}{\begin{vmatrix} y_1 & y_2 \\ y_1' & y_2' \end{vmatrix}} dx. \tag{5.28}$$

Example 5.3.19. Solve $y'' - y = 4xe^x$.

Solution. Since $r^2 - 1 = 0$, we have $r = \pm 1$, and the complementary function y_c is

$$y_c = C_1 e^{-x} + C_2 e^x.$$

By equation (5.28), a particular solution y_p is given by

$$y_p = e^{-x} \int \frac{\begin{vmatrix} 0 & e^x \\ 4xe^x & e^x \end{vmatrix}}{\begin{vmatrix} e^{-x} & e^x \\ -e^{-x} & e^x \end{vmatrix}} dx + e^x \int \frac{\begin{vmatrix} e^{-x} & 0 \\ -e^{-x} & 4xe^x \end{vmatrix}}{\begin{vmatrix} e^{-x} & e^x \\ -e^{-x} & e^x \end{vmatrix}} dx$$

$$= e^{-x} \int \frac{-4xe^{2x}}{2} dx + e^x \int \frac{4x}{2} dx$$

$$= 2e^{-x} \int -xe^{2x} dx + e^x \int 2x dx$$

$$= 2e^{-x} \left(-\frac{1}{2} xe^{2x} + \frac{1}{4} e^{2x}\right) + x^2 e^x$$

$$y_p = \frac{1}{2} e^x (2x^2 - 2x + 1).$$

So, a general solution is given by

$$y = C_1 e^{-x} + C_2 e^x + \frac{1}{2} e^x (2x^2 - 2x + 1).$$

5.4 Other ways of solving differential equations

Very often, it is hard or even impossible to solve a differential equation exactly. That is, there is in theory a function that is the solution of the differential equation, but we cannot obtain an explicit formula for this solution. This is true even for a simple-looking equation like

$$y'' - xy' + x^2 y = 0. \tag{5.29}$$

In this section, we introduce, very briefly, two ways to find exact or approximate solutions to differential equations: the *power series method* and *Euler's method*.

5.4.1 Power series method

When we cannot find an explicit expression for the solution of a differential equation, we try to get information about the solution in other ways. One way is to express the solution in the form of a power series,

$$y = f(x) = \sum_{n=0}^{\infty} c_n x^n = c_0 + c_1 x + c_2 x^2 + \cdots + c_n x^n + \cdots.$$

The method is to substitute this expression into the differential equation and use the equation to determine the values of the coefficients $c_0, c_1, c_2 \cdots$. This technique resembles the method of undetermined coefficients discussed previously. Once a Taylor series solution or some of the initial terms of that Taylor series have been found, this can be used to compute numerical approximations to the solution of the ODE.

We now illustrate the method on the equation $y' - y = x$. We already know how to solve this equation exactly by techniques introduced before, but it is a simple example, helping us to understand the power series method.

Example 5.4.1. Use a power series to solve the initial value problem $y' - y = x$ and $y(0) = 1$.

Solution. We assume there is a solution of the form

$$y = c_0 + c_1 x + c_2 x^2 + c_3 x^3 + \cdots + c_{n-1} x^{n-1} + c_n x^n + \cdots = \sum_{n=0}^{\infty} c_n x^n.$$

We can differentiate the power series term by term to get

$$y' = c_1 + 2c_2 x + 3c_3 x^2 + 4c_4 x^3 + \cdots + (n-1)c_{n-1} x^{n-2} + n c_n x^{n-1} + \cdots = \sum_{n=1}^{\infty} n c_n x^{n-1}.$$

So, $y' - y = x$ becomes

$$y' - y = (c_1 - c_0) + (2c_2 - c_1)x + (3c_3 - c_2)x^2 + \cdots + (nc_n - c_{n-1})x^{n-1} + \cdots = x.$$

Now, equating the coefficients gives

$$c_1 = c_0, \quad 2c_2 - c_1 = 1, \quad 3c_3 - c_2 = 0, \quad \ldots, \quad nc_n - c_{n-1} = 0, \quad \ldots.$$

The initial value $y(0) = 1$ gives $c_0 = 1$. Thus,

$$c_1 = 1, \quad c_2 = 1 = \frac{2}{2!}, \quad c_3 = \frac{2}{3!}, \quad \ldots \quad c_n = \frac{2}{n!}, \quad \ldots.$$

Therefore, the complete Taylor series for this solution of $y' - y = x$ is

$$y = 1 + x + \frac{2}{2!}x^2 + \frac{2}{3!}x^3 + \cdots + \frac{2}{n!}x^n + \cdots.$$

Since $e^x = 1 + x + \frac{x^2}{2!} + \frac{x^3}{3!} + \cdots$, this solution is

$$y = 2\left(1 + x + \frac{x^2}{2!} + \frac{x^3}{3!} + \cdots\right) - x - 1$$
$$= 2e^x - x - 1.$$

5.4.2 Numerical approximation: Euler's method

As previously mentioned, it is the exception rather than the rule when a first-order ODE of the general form

$$\frac{dy}{dx} = f(x, y)$$

can be solved exactly and explicitly by elementary methods like those discussed earlier. Even the simple equations

$$\frac{dy}{dx} = e^{-x^2} \quad \text{and} \quad \frac{dy}{dx} = \frac{\sin x}{x}$$

cannot be solved this way, since it can be proved that the antiderivatives of e^{-x^2} and $\frac{\sin x}{x}$ are not elementary functions. However, if a solution exists, then we can always find numerical approximations to the solution. The most basic of the approximation methods is *Euler's method*.

We consider the initial value problem of the form

$$\frac{dy}{dx} = f(x, y), \quad y(x_0) = y_0.$$

In Euler's method we first choose a small step size h, and we use this to define a sequence of x-values, starting with some initial value (x_0, y_0) and separated by h, giving

$$x_0, \quad x_1 = x_0 + h, \quad x_2 = x_0 + 2h, \quad x_3 = x_0 + 3h, \quad \ldots \quad x_n = x_0 + nh, \quad \ldots.$$

We compute a succession of approximate y-values, y_1 at x_1, y_2 at x_2, y_3 at x_3, and so on, using the iterative formula

$$y_{n+1} = y_n + h f(x_n, y_n) \quad \text{for } n = 0, 1, 2, 3, \ldots.$$

Euler's method works because, by Taylor's theorem, for the solution function $y = y(x)$ we have

$$y(x_n + h) = y(x_n) + y'(x_n)h + o(h^2).$$

Thus,

$$y(x_n + h) \approx y(x_n) + y'(x_n)h \quad \text{when } h \text{ is small,}$$
$$y_{n+1} \approx y_n + f(x_n, y_n)h.$$

5.4 Other ways of solving differential equations

Example 5.4.2. Use Euler's method with step size 0.1, and then step size 0.05, to construct a table of approximate values for the solution on the interval $0 \le x \le 1$ for the initial value problem

$$y' = x - y \quad \text{and} \quad y(0) = 1.$$

Solution. We start with $h = 0.1$, $x_0 = 0$, and $y_0 = 1$, so that for $n = 0, 1, 2, 3, \ldots, 10$, $x_n = 0, 0.1, 0.2, 0.3, \ldots 0.9, 1.0$. Hence, we compute using the formula

$$y_{n+1} = y_n + 0.1(x_n - y_n).$$

We obtain

$$y_1 = y_0 + 0.1(x_0 - y_0) = 1 + 0.1(0 - 1) = 0.9,$$
$$y_2 = y_1 + 0.1(x_1 - y_1) = 0.9 + 0.1(0.1 - 0.9) = 0.82,$$
$$\vdots$$

Proceeding with similar calculations, we find the values in the two tables (for $h = 0.1$ and $h = 0.05$). We have also included the corresponding values of the exact solution, $y = x + 2e^{-x} - 1$, and the deviation (error) of the approximate solution from the exact solution.

x	y_n ($h=0.1$)	$x+2e^{-x}-1$	Error	x	y_n ($h=0.05$)	$x+2e^{-x}-1$	Error
0	1	1.0	0	0	1	1.0	0
0.1	0.9	0.909 674 8	0.009 674 8	0.05	0.95	0.952 458 8	0.002 458 8
0.2	0.82	0.837 461 5	0.017 461 5	0.1	0.905	0.909 674 8	0.004 674 8
0.3	0.758	0.781 636 4	0.023 636 4	0.15	0.864 75	0.871 416 0	0.006 666
0.4	0.712 2	0.740 640 1	0.028 440 1	0.2	0.829 012 5	0.837 461 5	0.008 449
0.5	0.680 98	0.713 061 3	0.032 081 3	0.25	0.797 561 9	0.807 601 6	0.010 039 7
0.6	0.662 882	0.697 623 3	0.034 741 3	0.3	0.770 183 8	0.781 636 4	0.011 452 6
0.7	0.656 593 8	0.693 170 6	0.036 576 8	0.35	0.746 674 6	0.759 376 2	0.012 701 6
0.8	0.660 934 4	0.698 657 9	0.037 723 5	0.4	0.726 840 9	0.740 640 1	0.013 799 2
0.9	0.674 841 0	0.713 139 3	0.038 298 3	0.45	0.710 498 8	0.725 256 3	0.014 757 5
1.0	0.697 356 9	0.735 758 9	0.038 402	0.5	0.697 473 9	0.713 061 3	0.015 587 4
				0.55	0.687 600 2	0.703 899 6	0.016 299 4
				0.6	0.680 720 2	0.697 623 3	0.016 903 1
				0.65	0.676 684 2	0.694 091 6	0.017 407 4
				0.7	0.675 350 0	0.693 170 6	0.017 820 6
				0.75	0.676 582 4	0.694 733 1	0.018 150 7
				0.8	0.680 253 3	0.698 657 9	0.018 404 6
				0.85	0.686 240 7	0.704 829 9	0.018 589 2
				0.9	0.694 428 6	0.713 139 3	0.018 710 7
				0.95	0.704 707 2	0.723 482	0.018 774 8
				1.0	0.716 971 8	0.735 758 9	0.018 787 1

Graphs of the approximate solutions with $h = 0.1$ (diamonds), $h = 0.05$ (crosses), and the exact solution (solid line) are shown in Figure 5.5.

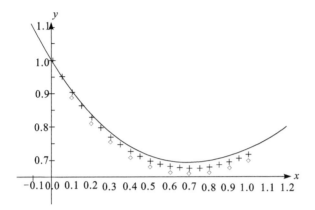

Figure 5.5: Euler's method, Example 5.4.2.

Note. Euler's method is subject to the numerical errors experienced by most iterative methods. The small errors caused by the approximate solution at each step are incorporated into the calculations for the next step, and so can gradually build up into large errors. This is illustrated in the figure above. This error build-up can be reduced by decreasing the size of the step h. However, as h gets smaller, the number of computations increases, and this can cause another kind of error during computer computations. This is because computers approximate numbers by rounding them to a certain precision, and this introduces minute errors (round-off errors). If an iterative method requires an extremely large number of computations, then the round-off errors can build up into a significant error.

Example 5.4.3. Apply Euler's method to approximate the solution of the initial value problem

$$\begin{cases} \frac{dy}{dx} = \sqrt{x^2 + y^2}, \\ y(0) = -1 \end{cases}$$

with step size $h = 0.1$ on the interval $[0, 1]$.

Solution. In this case the iterative formula is $y_{n+1} = y_n + 0.1\sqrt{x_n^2 + y_n^2}$, starting from $x_0 = 0$ and $y_0 = -1$, and for $n = 0, 1, 2, 3, \ldots$, the values of x_n are $0, 0.1, 0.2, 0.3, \ldots, 0.9, 1$. A table of the computed approximate solution values is shown.

n	x_n	y_n	$\sqrt{x_n^2 + y_n^2}$
0	0.0	−1.0000	1.0000
1	0.1	−0.9000	0.9055
2	0.2	−0.8094	0.8337
3	0.3	−0.7260	0.7855
4	0.4	−0.6474	0.7610
5	0.5	−0.5713	0.7592
6	0.6	−0.4954	0.7781
7	0.7	−0.4176	0.8151
8	0.8	−0.3361	0.8677
9	0.9	−0.2493	0.9339
10	1.0	−0.1559	1.0121

Note. In this example it is not possible to find an exact solution formula, so a numerical approach is the only way to investigate the solution.

5.5 Review

Main concepts discussed in this chapter are listed below.
1. Separable differential equations, $\frac{dy}{dx} = f(x,y) = g(x)h(y)$, have the solution

$$\int \frac{1}{h(y)} dy = \int g(x) dx.$$

2. Substitution method: if

$$\frac{dy}{dx} = F\left(\frac{y}{x}\right),$$

then $y = xv$ will transform the ODE into a separable one.
3. Exact differential equation:

$$f(x,y)dx + g(x,y)dy \text{ is exact}$$
$$\iff g_x = f_y$$
$$\iff \text{there is a function } \varphi \text{ such that } d\varphi = f(x,y)dx + g(x,y)dy.$$

Thus, a general solution is $\varphi(x,y) = C$.
4. A first-order linear ODE $y' + P(x)y = Q(x)$ has the general solution

$$y = e^{-\int P(x)dx}\left(\int e^{\int P(x)dx} Q(x) dx + C\right).$$

5. For a reducible differential equation $F(x,y,y',y'') = 0$:

$$F(x,y',y'') = 0 \quad y \text{ is missing, let } p(x) = y',$$
$$F(y,y',y'') = 0 \quad x \text{ is missing, let } p(y) = y'.$$

6. A homogeneous second-order linear differential equation with constant coefficients,

$$ay'' + by' + cy = 0,$$

has a general solution given by

$y = C_1 e^{r_1 x} + C_2 e^{r_2 x}$ if r_1 and r_2 are two distinct roots of $ar^2 + br + c = 0$,

$y = C_1 e^{rx} + C_2 x e^{rx}$ if r is a repeated root of $ar^2 + br + c = 0$,

$y = C_1 e^{\alpha x} \cos \beta x + C_2 e^{\alpha x} \sin \beta x$ if $\alpha \pm \beta x$ are two complex roots of $ar^2 + br + c = 0$.

7. A nonhomogeneous second-order linear differential equation

$$ay'' + by' + cy = f(x),$$

has a general solution given by

$$y = \underbrace{\text{a general solution of } ay'' + by' + cy = 0}_{\text{complementary function}}$$
$$+ \underbrace{\text{a particular solution}}_{\text{particular integral}}.$$

8. To find a particular solution y^* for $ay'' + by' + cy = f(x)e^{\lambda x}$, where $f(x)$ is a polynomial of degree m, assume $Q(x)$ is a polynomial of degree m, with unknown coefficients, we try

$y^* = Q(x)e^{\lambda x}$ if λ is not a root of $ar^2 + br + c = 0$,

$y^* = xQ(x)e^{\lambda x}$ if λ is a single root of $ar^2 + br + c = 0$,

$y^* = x^2 Q(x)e^{\lambda x}$ if λ is a double root of $ar^2 + br + c = 0$.

9. A particular solution of $ay'' + by' + cy = f(x)$ is given by

$$y^* = y_1(x) \int \frac{\begin{vmatrix} 0 & y_2 \\ f(x)/a & y_2' \end{vmatrix}}{\begin{vmatrix} y_1 & y_2 \\ y_1' & y_2' \end{vmatrix}} dx + y_2(x) \int \frac{\begin{vmatrix} y_1 & 0 \\ y_1' & f(x)/a \end{vmatrix}}{\begin{vmatrix} y_1 & y_2 \\ y_1' & y_2' \end{vmatrix}} dx,$$

where y_1 and y_2 are two independent solutions of $ay'' + by' + cy = 0$.

10. There are some other ways to solve an ODE, such as the power series method and Euler's method.

5.6 Exercises

5.6.1 Introduction to differential equations

1. Which of the following equations are differential equations? For those that are, state whether they are ODE or PDE. For those that are ODE, give their orders and degrees.
   (1) $y' = 2x + 6$,
   (2) $y = 2x + 3$,
   (3) $\frac{d^2y}{dx^2} = y + 2x$,
   (4) $x^2 - 3t = 0$,
   (5) $y' = x + y + y^2 \cos x$,
   (6) $y^x + 8(y')^2 + 6y^8 = e^{2t}$,
   (7) $y(y')^2 = 1$,
   (8) $x^2 dx + y dx = 0$,
   (9) $y^{(4)} + 2y' + 3x = 5$,
   (10) $\frac{\partial^2 u}{\partial x^2} + (\frac{\partial u}{\partial t})^2 = x^2 - t$.

2. Verify that $x = 2(\sin 2t - \sin 3t)$ is the solution of the initial value problem $\frac{d^2x}{dt^2} + 4x = 10 \sin 3t$, $x(0) = 0$, $x'(0) = -2$.

3. Graph the slope fields for the following differential equations using computer software:
   (1) $y' = \frac{x-y}{x+y}$,
   (2) $\frac{dy}{dx} = (x + y - 2)^2$,
   (3) $\frac{dy}{dx} = \sin x$,
   (4) $\frac{dy}{dx} = x(6 - x)$.

4. Find an equation of the curve that passes through the point $(1, 0)$ and whose slope at each point (x, y) is x^2.

5.6.2 First-order differential equations

1. Solve each of the following separable differential equations:
   (1) $(xy^2 + x)dx + (y - x^2y)dy = 0$,
   (2) $y' = \frac{x^2}{\cos y}$,
   (3) $\frac{du}{dt} = \frac{2t + \sec^2 t}{2u}$, $u(0) = -5$,
   (4) $y' = \frac{x}{y \ln y}$,
   (5) $xy(y - xy') = x + yy'$,
   (6) $\sec^2 x \tan y dx + \sec^2 y \tan x dy = 0$.

2. Find a general solution for the logistic equation
   $$\frac{dP}{dt} = kP(M - P).$$
   Solve the initial value problem
   $$\frac{dP}{dt} = 0.18P(20 - P), \quad P(0) = 4.$$
   Sketch the graph of this particular solution. When does $\frac{dP}{dt}$ change fastest?

3. Use a suitable substitution to solve each of the following the differential equations:
   (1) $xy' - y - \sqrt{y^2 - x^2} = 0$,
   (2) $xy' = y \ln \frac{y}{x}$,
   (3) $y' = e^{\frac{y}{x}} + \frac{y}{x}$, $y(1) = 0$,
   (4) $x^2 y' + y^2 = xyy'$,
   (5) $\frac{dy}{dx} = (x + y - 2)^3$,
   (6) $\frac{dy}{dx} = \frac{2y+4}{x+y-1}$.

4. Determine whether each of the following differential equations is linear:
   (1) $y' + e^x y = x^2 y^2$, (2) $y' = \tan y$,
   (3) $x(\frac{dx}{dt} + 2) = t^2$, (4) $3x^2 + 5y - 5y' = 0$,
   (5) $y' - \frac{3y}{x} = x$, (6) $y' = \ln x$.
5. Solve each of the following exact differential equations:
   (1) $(3x^2 + 6xy^2)dx + (6x^2y + y^2)dy = 0$, (2) $\frac{dy}{dx} = \frac{x-y^2}{2xy+2y^3}$.
6. Solve each of the following first-order differential equations:
   (1) $xy' + y = e^x, y(1) = e$, (2) $y' + y\cos x = e^{-\sin x}$,
   (3) $(x^2 + 1)y' + 2xy = 4x^2$, (4) $t\frac{dy}{dt} + 2y = t^3, t > 0, y(1) = 0$,
   (5) $y' + \frac{2}{x}y = \frac{y^3}{x^2}$, (6) $x\frac{dy}{dx} - 4y = x^2\sqrt{y}$.

5.6.3 Second-order differential equations

1. Solve each of the following reducible differential equations:
   (1) $y''' = xe^x + 2$, (2) $y'' = y' + x$,
   (3) $y'' = 1 + (y')^2$, (4) $xy'' + y' = 0$,
   (5) $y'' = 3\sqrt{y}, y(0) = 1, y'(0) = 2$, (6) $\frac{1}{(y')^2}y'' = \cot y$,
   (7) $y'' = y'(1 + y'^2)$.
2. To solve the second-order differential equation $y^3 y'' + 1 = 0, y(1) = 1, y'(1) = 0$, a student provided the following solution. Since
$$y'' = -\frac{1}{y^3},$$
we must have
$$y'' dy = -\frac{1}{y^3} dy \to \int y'' dy = \int -\frac{1}{y^3} dy.$$
Thus, he obtained $y' = -\frac{1}{y^2} + C$. Using the initial conditions, he had $C = 1$. So
$$y' = 1 - \frac{1}{y^2}.$$
This becomes a separable ODE now. Is this solution correct?
3. Solve each of the following differential equations:
   (1) $y'' - 3y' + 2y = 0$, (2) $y'' + a^2 y = 0$,
   (3) $4y'' + 4y' + y = 0, y(0) = 2, y'(0) = 0$, (4) $y'' - 6y' + 8y = 0$,
   (5) $\frac{d^2y}{dt^2} + 3\frac{dy}{dt} + 2 = 0$, (6) $\frac{d^2x}{d\theta^2} + 4\frac{dx}{d\theta} + 4x = 0$,
   (7) $\ddot{y} + \dot{y} + y = 0$.
4. Solve $\frac{d^2y}{dx^2} - 2\frac{dy}{dx} + 3y = 0$, given that $y = 0$ and $\frac{dy}{dx} = 6$ when $x = 0$.

5. Solve each of the following differential equations using the method of undetermined coefficients:
   (1) $y'' - 7y' + 6y = 4x$,
   (2) $y'' - 2y' - 3y = 6e^{2x}$,
   (3) $y'' + 4y = x\cos x$,
   (4) $y'' - y = 4xe^x$, $y(0) = 0$, $y'(0) = 1$,
   (5) $y'' - 2y' + 5y = e^x \sin 2x$,
   (6) $y'' + y = e^x + \cos x$,
   (7) $\ddot{y} + \dot{y} + y = 0$,
   (8) $\frac{d^2y}{dt^2} + 16y = 3\cos 4t$,
   (9) $2\frac{d^2x}{dt^2} - 3\frac{dx}{dt} - 5x = 10t^2 + 1$,
   (10) $\frac{d^2y}{dx^2} + 2\frac{dy}{dx} + y = e^{-x}$.

6. Solve $20\frac{d^2x}{dt^2} + 4\frac{dx}{dt} + x = 2t + 11$, given that $x = 1$ and $\frac{dx}{dt} = 2.8$ when $t = 0$. Describe the behavior of x when $t \to \infty$.

7. Solve the differential equation $y'' + y = \tan x$ using variation of parameters.

8. A spring with a mass of 2 kg is put on a table. One of the ends of the spring is fixed on a wall and the other end is attached to the mass. It is held stretched 0.2 m beyond its natural length by a force of 40 N. Now, suppose the mass is at its equilibrium point, a push gives the mass an initial velocity of 2 m/s. Find the position of the mass after t seconds.

9. The Kirchhoff voltage law says that

$$L\frac{d^2Q}{dt^2} + R\frac{dQ}{dt} + \frac{1}{C}Q = E(t),$$

where L is an inductor, R is a resistor, C is a capacitor, Q is the charge, and E is the electromotive force. The current I is always equal to $\frac{dQ}{dt}$. Find the charge and current at t in a circuit if the initial charge and current are both 0, and $L = 1$, $R = 40$, $C = 16 \times 10^{-4}$, and $E(t) = 10\sin(2t)$.

10. Attempt to find a general solution to the **Euler equation**

$$ax^2\frac{d^2y}{dx^2} + bx\frac{dy}{dx} + cy = 0,$$

where $a \neq 0$, b, and c are constants. Hint: Try $y = x^r$.

11. (**Solving a simple PDE**) In general, solving a PDE for an analytical solution is not easy. Numerical methods are widely used in obtaining approximate solutions. However, in some cases, we may be able to find an exact solution to a PDE. Consider heat conduction in a cube, which can be modeled by

$$\frac{\partial^2 u}{\partial x^2} + \frac{\partial^2 u}{\partial y^2} + \frac{\partial^2 u}{\partial z^2} = 0, \quad \text{for } 0 < x, y, z < a.$$

Note that $\frac{\partial^2 u}{\partial x^2} + \frac{\partial^2 u}{\partial y^2} + \frac{\partial^2 u}{\partial z^2}$ is often denoted by $\nabla^2 u$. Suppose $u = P(x)Q(z)$ and boundary conditions are

$u = 0$ on $x = 0$ and a,
$u = 0$ on $z = 0$,
$u = 1$ on $z = a$.

(a) Show that $P''Q + PQ'' = 0$ and that there is a constant λ such that $Q'' = \lambda Q$ and $P'' + \lambda P = 0$.
(b) Show that a general solution for $P(x)$ is

$$P(x) = C_1 \cos \sqrt{\lambda} x + C_2 \sin \sqrt{\lambda} x.$$

(c) Show that $\lambda = n^2 \pi^2 / a^2$.
(d) Show that a general solution for $Q(z)$ is

$$Q(z) = A e^{\sqrt{\lambda} z} + B e^{-\sqrt{\lambda} z}, \quad \text{where } A \text{ and } B \text{ are two arbitrary constants.}$$

(e) Recall that

$$\cosh x = \frac{e^x + e^{-x}}{2} \quad \text{and} \quad \sinh = \frac{e^x - e^{-x}}{2}.$$

Show that the general solution for $Q(z)$ shown in (d) can be rewritten as

$$Q(z) = C_3 \cosh \sqrt{\lambda} z + C_4 \sinh \sqrt{\lambda} z.$$

(f) Show that $C_3 = 0$ by using some of the boundary conditions.
(g) Show that by superimposing solutions,

$$u(x,z) = \sum_{n=1}^{\infty} a_n \sin \frac{n\pi x}{a} \sinh \frac{n\pi z}{a}.$$

(h) Using the condition $u = 1$ when $z = a$, show that

$$1 = \sum_{n=1}^{\infty} a_n \sin \frac{n\pi x}{a} \sinh(n\pi).$$

Thus, by using knowledge about Fourier series, prove that

$$a_n = \begin{cases} \frac{4}{n\pi \sinh(n\pi)} & \text{if } n \text{ is odd}, \\ 0 & \text{if } n \text{ is even}. \end{cases}$$

(i) Thus, show that the desired particular solution is

$$u(x,z) = \sum_{k=1}^{\infty} \frac{1}{(2k-1)\pi} \frac{\sinh \frac{(2k-1)\pi z}{a}}{\sinh((2k-1)\pi)} \sin \frac{(2k-1)\pi x}{a}.$$

Further reading

1. Gilbert Strang. Calculus. Wellesley: Wellesley-Cambridge Press, 1991.
2. Alex Himonas, Alan Howard. Calculus: Ideas and Applications. New Jersey: Wiley, 2002.
3. Michael Spivak. Calculus. 3rd edtion. London: Cambridge University Press, 2006.
4. Robert A. Adams, Christopher Essex. Calculus. 7th edition. Toronto: Pearson Eduction, 2007.
5. James Stewart. Calculus. 6th edition. California: Brooks Cole, 2017.
6. Donald Trim. Calculus for Engineers. 4th edition. Toronto: Pearson Education, 2008.
7. Ross L. Finney, Franlin D. Demana, Bert K. Waits, Daniel Kennedy. Calculus: Graphical, Numerical, Algebraic. 4th edition. New Jersey: Prentice Hal, 2012.
8. Ron Larson, Bruce H. Edwards. Calculus, 10th edition. California: Brooks Cole, 2013.
9. James Stewart. Calculus. 5th edition. Beijing: Higher Education Press, 2004.
10. Thomas's Calculus, 10th edition. Beijing: Higher Education Press, 2004.
11. Department of mathematics, Sichuan University. Higher Mathematics. 4th edition. Beijing: Higher Education Press, 2009.
12. Department of mathematics, Sichuan University. Higher Mathematics. 2nd edition. Chengdu: Sichuan University Press, 2013.
13. Department of applied mathematics, Tongji University. Higher Mathematics. 7th edition. Beijing: Higher Education Press, 2014.
14. Ma Jigang, Zou Yunzhi, P. W. Aitchison. Calculus II. Beijing: Higher Education Press, 2010.
15. William Briggs, Lyle Cochran, Bernard Gillett. Calculus: Early Transcendentals. 2nd edition. Malaysia: Pearson Education, 2015.
16. Elgin H. Johnston, Jerold C. Mathews. Calculus (Annotated Instructor's edition). USA: Pearson education, 2002.

Index

absolute maximum 123
absolute minimum 123
angle between two lines 21
angle between two planes 25
angle between two vectors 8

Bernoulli equation 286
boundary 66
boundary point 66
bounded region 66
bounded region test 127

candidate theorem 124
Cartesian equation of a plane 24
center of mass 187
chain rule 92
chain rule with more than one independent
    variable 94
chain rule with one independent variable 92
change of variables 161
change of variables in triple integrals 179
change the order of integration 156
circulation 217
circulation density 217
circulation integral 217
Clairaut theorem 83
closed region 66
complementary function 298
components of a vector 5
conservative field 211
constrained maximum 130
constrained minimum 130
continuous functions of two variables 75
coordinate planes 4
criteria for exactness 281
critical point 124
cross product 13
curl 219
curl of a 3D vector field 256
curvature 39
cylinder 44
cylindrical coordinates 175

degree 274
dependent variables 65
difference of vectors 3
differentiability 83

differentiable 84
differential approximation 90
direction 1
direction angle 8
direction cosine 8
direction field 277
direction numbers 20
directional derivative 113
divergence 233
divergence of a 3D vector field 248
divergence theorem 248, 250
domain 65
dot product 9
double integral 147
double integral in polar coordinates 157
double integral in rectangular coordinates 150

ellipsoid 46
elliptic cone 46
elliptic cylinder 45
elliptic paraboloid 46
equivalent vectors 1
Euler's method 310
exact differential equation 281
extrema of functions of several variables 122

first-order differential equation 273
first-order linear differential equation 283
flux 231
flux density 232
flux integral 242
Fubini theorem 152
functions of multiple variables 65
functions of two variables 65
fundamental theorem of line integrals 208

general solution 275
generalized Green's theorem 228
global maximum 123
global minimum 123
gradient vector 113
gravitational field 202
Green's theorem 216
Green's theorem: circulation-curl form 219
Green's theorem: flux-divergence form 231

Hessian matrix 122

homogeneous equation 279
homogeneous second-order linear differential equations 293
hyperbolic cylinder 45
hyperbolic paraboloid 46
hyperboloid of one sheet 46
hyperboloid of two sheets 46

implicit differentiation 101
independent variables 65
initial point 1
initial value problem 276
integrating factor 283
interior 66
interior point 66
intermediate variable 95
intersecting curves 50
iterated limits 74

Jabobian determinant 162

lagrange multiplier 130
length of curves 37
level curves 67
level surface 70
limits for functions of two variables 70
line integral of a vector field 201
line integral with respect to arc length 195, 196
linear approximation 90
linear differential equation 283
linear equation of a plane 24
linear independence of two functions 292
linearization 83
lines in space 18
local linearization 91
local maximum 123
local minimum 123

magnitude 1
method of undetermined coefficients 302
moment of inertia 187, 188

negative of a vector 2
nonhomogeneous second-order linear differential equations 299
normal line 23, 109
normal plane 34, 106

objective function 130

octant 4
ODE 274
open region 66
optimization problem 131
order 274
ordinary differential equation 273, 274
orientable surfaces 241

parabolic cylinder 44
parallelepiped 17
parallelogram law 2
parameterization by arc length 38
parameterized surfaces 181
parametric equations of lines 19
partial derivative 77
partial derivatives of higher order 82
partial differential equation 274
particular solution 275
path independence 208
PDE 274
perpendicular vectors 8
plane 23
position vector 5
positive orientation of a curve 38
positive oriented curve 216
potential function 211
power series method 309
principal unit normal vector 41
principle of superposition for homogeneous equations 292
projection 12
projection curves 50
projection region 56

quadratic approximation 98
quadric surfaces 46

range 65
reducible second-order equations 287
regions bounded by surfaces 56
relative maximum 123
relative minimum 123
right-hand rule 3
ruled surface 63

saddle point 125
scalar multiplication 2
scalar projection 12
scalar triple product 17

second derivative test 126
second-order differential equation 274, 287
second-order linear differential equation 291
second-order ODE 287
separable differential equation 277
simple curve 216
simply connected region 216
skew lines 21
spherical coordinates 175, 177
steepest ascent/descent 116
Stokes theorem 256
surface area 181
surface in space 42
surface integral of vector fields 241
surface integral with respect to surface area 237
symmetric equations of lines 19

tangent line 106
tangent plane 109
tangent vector 34
Taylor expansion 98
terminal point 1
TNB frame 39

total differential 89
tree diagram 93
triangle law 2
triple integral 165
triple integral in rectangular coordinates 165
type I region 153
type II region 153

unbounded region 66
unit binormal vector 42
unit tangent vector 34
unit vector 1

variation of parameters 294, 307
vector 1
vector addition 2
vector equation of a line 20
vector equations of planes 28
vector field 201
vector-valued functions 30
Viviani curve 54

zero vector 1